Operations Research

Advances in Applied Mathematics
SERIES EDITOR: Daniel Zwillinger

Operations Research

A Practical Introduction

Second Edition

Michael W. Carter

Camille C. Price

Ghaith Rabadi

CRC Press
Taylor & Francis Group
Boca Raton   London   New York

CRC Press is an imprint of the
Taylor & Francis Group, an **informa** business

CRC Press
Taylor & Francis Group
6000 Broken Sound Parkway NW, Suite 300
Boca Raton, FL 33487-2742

First issued in paperback 2022

© 2019 by Taylor & Francis Group, LLC
CRC Press is an imprint of Taylor & Francis Group, an Informa business

No claim to original U.S. Government works

ISBN 13: 978-1-03-247606-3 (pbk)
ISBN 13: 978-1-4987-8010-0 (hbk)

DOI: 10.1201/9781315153223

Publisher's Note
The publisher has gone to great lengths to ensure the quality of this reprint but points out that some imperfections in the original copies may be apparent.

Visit the Taylor & Francis Web site at
http://www.taylorandfrancis.com

and the CRC Press Web site at
http://www.crcpress.com

Contents

Preface

This book presents a practical introduction to the field of Operations Research and serves as a guide to the use of Operations Research techniques in scientific decision making, design, analysis, and management. Our aim has been to create a readable and useful text that provides not only an introduction to standard mathematical models and algorithms, but also an up-to-date examination of practical issues pertinent to the development and use of computational methods for solving problems. We offer a sound yet practical introduction to the mathematical models and the traditional as well as innovative solution methods underlying the modern software tools that are used for quantitative analysis and decision-making. Our presentations of problem formulations, solution methods, and software tools are accompanied by illustrative applications of Operations Research techniques.

The First Edition of this book has been thoroughly updated and expanded through the inclusion of new and timely topics, more modern perspectives on fundamental material, revised and updated descriptions of currently available software, and the addition of numerous new case studies that illustrate the application of Operations Research techniques for solving important problems. This Second Edition extends the purpose of the previous edition as a textbook for students and a professional reference for practitioners.

We have designed this book as a **text for an introductory course** in Operations Research. We target specifically the needs of students who are taking only one course on the subject of Operations Research, and accordingly we have chosen to include just those topics that provide the best possible one-semester exposure to the broad discipline of Operations Research. An introductory course in Operations Research may be a required, elective, or auxiliary course for many degree programs. In various institutions, the course may be taught in Industrial or Mechanical Engineering, Computer Science, Engineering Management, Management Science, Applied Mathematics, or Operations Research departments, at either the intermediate or advanced undergraduate or graduate levels.

This book may also serve as a **professional reference book** for corporate managers and technical consultants. We welcome readers from a variety of subject disciplines who recognize the potential value of incorporating the tools of Operations Research into their primary body of knowledge. Because the mathematical models and processes of Operations Research are used so pervasively in all areas of engineering, science, management, economics and finance, and computer science, we are confident that students and professionals from many different fields of study will be at a substantial advantage by having these analytical tools at hand. We hope that, in the course of studying the material in this book, readers will be struck not only by fascination with the mathematical principles which we will present, but also by the many and varied applications of the methods and techniques. With the preparation provided by material in this book, readers should be in a position to identify problems in their own special areas of expertise which can be solved with the methods of Operations Research. In addition, this book may encourage some readers to pursue more advanced studies in Operations Research; our presentation provides an adequate foundation for continued study at higher levels.

Some engineering and management professionals received their formal academic training before personal computing devices and powerful workstations became so readily available and before the subsequent rapid increase in the number of sophisticated yet accessible new software products. Such experienced practitioners, educated in traditional mathematics, operations research or quantitative management, will find that many parts of this book will provide them with the opportunity to sharpen and refresh their skills with an **up-to-date perspective on current methodologies** in the field of Operations Research.

Important mathematical principles are included in this book where necessary, in order to facilitate and promote a firm grasp of underlying principles. At the same time, we have tried to minimize abstract material in favor of an applied presentation. Because our readers may have quite diverse backgrounds and interests, we anticipate a considerable mixture of motivations, expectations, and mathematical preparation within our audience. Since this book addresses optimization and quantitative analysis techniques, users should have some knowledge of calculus and a familiarity with certain topics in linear algebra, probability, and statistics. More advanced calculus is useful in the chapters on integer programming and nonlinear optimization. Many of our students will take only *one* course in the techniques of Operations Research, and we believe that the greatest benefit for those individuals is obtained through a very broad survey of the many techniques and tools available for quantitative decision making. Such breadth of coverage, together with the mixture of mathematical backgrounds in our audience of readers, necessitates that we temper the level of mathematical rigor and sophistication in our presentation of the material.

Special Features

The field of Operations Research has experienced a dramatic shift in the availability of software, from software support primarily for large mainframe computing systems to the current proliferation of convenient software for a variety of desktop computers and workstations. With such an abundance of software products, practitioners of Operations Research techniques need to be aware of the capabilities and limitations of the wide variety of software available to support today's style of analysis and decision-making. Associated with each chapter in this book is a section devoted to **Software** in which we offer a brief description of some of the most popular software currently available specifically for solving the various types of problems presented in that chapter. (The **Software** guide contained in Chapter 1 elaborates more fully on the purpose and use of the guides to software in subsequent chapters.) Because software packages generally focus on a particular type of problem rather than on a specific application area, we will organize our discussions of software implementations according to the chapter topics which are indicative of the problem type. Most of the cited software packages and products are applicable to a wide array of application areas.

The information contained in these Software descriptions is not intended to represent an endorsement of any particular software product, nor to instruct readers in the detailed use of any specific software package. We merely mention a representative few of the broad range of available software packages and libraries, in order to create an awareness of the issues and questions that might arise during the development or selection of software for solving real problems.

Computing capabilities are almost ubiquitous, and the software available for student use is often the same *industrial strength* software that practitioners use for solving large practical problems. Educational discounts in pricing may reflect minor limitations in the sizes of problems that can be solved with the software, but the software used in an educational environment is likely to be very typical of software designed and distributed for commercial application.

Instructors who wish to supplement the introductory course in Operations Research with computing exercises and projects should have no difficulty in finding reasonably-priced software with appropriate educational site licenses, or even free and open software. Although computer usage has become a popular aspect of many introductory courses in Operations Research, our intention in developing this book has been to provide support for learning the foundations necessary for building appropriate models, and to encourage an adequate understanding of solution methods so that students can become self-reliant and judicious users of the many software products that have been and will be developed for practical use.

Each of the chapters in this book is enriched by several **Illustrative Applications**, drawn from the industrial, computing, engineering, and business disciplines. These miniature case studies are intended to give the reader some insight into how the problem solving tools of Operations Research have been used successfully to help solve real problems in public and private scientific, economic, and industrial settings. Details are omitted in some cases, but references are provided for all of the illustrative applications, which may serve as the inspiration for term projects or further studies that expand on the brief sketches given in this book. Our Illustrative Applications include examples from the petroleum industry, wildlife habitat management, forestry, space exploration, humanitarian relief, manufacturing, agriculture production, mining, waste management, military operations, shipping and transportation planning, computing systems, finance, and health care.

Near the end of each chapter, is a brief summary of the important topics presented in the chapter. To further assist students in their review and assimilation of chapter material, each chapter in the book contains a list of **Key Terms**. Definitions or explanations of these key terms are found in the chapter discussion, and typically the key term appears highlighted in bold type. Mastery of the content of the chapter material requires a recognition and understanding of these important terms, and the key terms should be used as a checklist during review of the subject matter contained in each chapter.

A selection of **Exercises** appears in each chapter. Many of these problems and questions provide a straight-forward review of chapter material, and allow the student to practice and apply what has been learned from the text. In addition, some of the exercises prompt the discovery of mathematical and computational phenomena that may not be explicitly mentioned in the chapter material, but which offer important practical insights. Exercises are an essential and integral part of learning, and the exercises included in this book have been chosen to give students a thorough appreciation for and understanding of the text material.

References and Suggested Readings are included at the end of each chapter. These reference lists contain titles of general and specialized books, scholarly papers, and other articles, which may be used to follow up on interesting, difficult, or more advanced topics related to material presented in the chapter. In case the reader would like to consult still other authorities, or perhaps see alternative explanations from different sources, we maintain a website for this book at www.operationsresearch.us. The website also includes additional support material for both instructors and students.

An **Appendix** at the end of the book contains a review of mathematical notation and definitions, and a brief overview of matrix algebra. Readers having marginal mathematical preparation for the material in this book may find that the appendix provides an adequate review of the mathematics essential for comprehension of introductory Operations Research. Additional references are listed in the Appendix for those who need a more complete review or study of mathematics.

Book Overview

This book contains material that can be covered in a single semester. A course based on this book would cover a range of topics that collectively provide the basis for a scientific approach to decision making and systems analysis. Over half of the book is directed toward the various subclasses of mathematical programming models and methods, while the remainder is devoted to probabilistic areas such as Markov processes, queueing systems, simulation, decision analysis, heuristics, and metaheuristics.

We recommend that, if time permits, the topics be studied in the order in which they appear in the book. In particular, Chapter 2 on Linear Programming, Chapter 4 on Integer Programming and Chapter 5 on Nonlinear Optimization might reasonably be treated as a sequence. Similarly, Chapter 6 on Markov Processes, Chapter 7 on Queueing Models, and Chapter 8 on Simulation form a natural sequence, since the discussions on simulation build on the two preceding chapters. However, readers with more specific interests will find that, after reading the first chapter, it is possible to read almost any of the chapters without having thoroughly studied all the preceding ones.

Chapter 1 describes the nature of Operations Research, the history of the field, and how the techniques of Operations Research are used. Since the analysis and optimization of systems requires that mathematical models of real systems be built, we discuss some of the principles of system modeling, a topic that will be re-visited frequently in the book. Solving problems involves the use of computational processes, and we take this opportunity to introduce algorithms and their efficiency, and the inherent complexity of some of the problems that are solvable with the tools of Operations Research.

In **Chapter 2**, we study what is undoubtedly the most popular topic in Operations Research, the creation and solution of linear programming problems. Many practical problems can indeed be modeled as linear systems: optimizing a linear function subject to linear constraints on the variables. Fortunately, a great deal of work has resulted in practical and effective methods for solving these types of problems. We first look at the formulation of problems in the linear programming form, then study the simplex, and other, solution methods and identify several computational phenomena that can take place when the methods are applied to problems.

Network analysis is the subject of **Chapter 3**. A wide variety of problems can be modeled as graph or network problems, and many algorithms have been developed for finding paths, routes and flow patterns through networks of all sorts. Some network problems have obvious tangible applications in the areas of transportation and distribution. Other views of networks inspire solutions to more abstract problems such as the matching or assignment of the entities in a system, or the planning, scheduling, and management of the phases of projects.

In the next two chapters of the book, we study problems that are in some respects just harder to solve than the problems seen earlier. Some of the problems are conceptually more difficult, while some require more sophisticated mathematical solution techniques. On the other hand, some types of problems are quite simple to describe but the solution methods seem to be prohibitively time-consuming to carry out. **Chapter 4** introduces the subject of Integer Programming, in which the problem formulations may look remarkably similar to the linear and network formulations seen in Chapters 2 and 3, but with the exception that the decision variables are now constrained to have integer values. This additional requirement almost always implies that these problems require solution methods that are in a different league, computationally, from the methods previously considered in this book. Many interesting and practical problems are modeled as integer programming problems, and in this chapter we introduce the best known ways to find exact solutions to such problems.

In **Chapter 5**, we study an even larger and more unwieldy class of problems. Nonlinear optimization actually includes all mathematical programming problems whose objective or constraints cannot be expressed as linear functions of the decision variables. Because there are so many forms of these problems, no one optimization method works for all problems, but several representative and useful solution methods are presented.

Stochastic processes are studied in the next several chapters. In **Chapter 6**, we study processes having probabilistic characteristics and behaviors, known as Markov processes. Many practical dynamic systems can be described by simple probabilities of moving from one state to another. For example, in a clinical setting, probabilities may be used to define how patients respond to various treatments. Or in nature, certain weather phenomena may occur with known probabilities during certain times of the year or under certain conditions. Systems exhibiting Markov properties can be analyzed in order to determine what the system's most likely state is and how long it takes for a dynamic system to resolve into this state. Some stochastic processes however never settle into any predictable set of states. The analytical tools presented in this chapter are not tools that are directly used to *optimize* a system, but rather to analyze a system and identify a system's most likely properties. An understanding of the most probable behavior of a system may then be used to modify and improve the system's performance.

Many systems can be described in terms of customers waiting to be served in some way: human customers waiting to be served by a cashier, computational processes waiting to be executed by a processor in a computer, or manufactured products waiting to be worked on by a machine in an assembly-line process. **Chapter 7** deals with the performance of systems that involve waiting lines, or queues. In this chapter we study queueing models and the properties of queueing systems that can be computed on the basis of parameters that describe the arrival rates of customers into the system and the service rates of the servers.

For some special cases, these computations can be made easily, but for more complicated systems, analysts often resort to the use of simulation techniques. **Chapter 8** presents simulation as a modeling process in which we use the computer to simulate the activities in a real system, in order to discover the behavioral properties of the system.

Although practically all of the techniques of Operations Research can become involved in decision-making processes, **Chapter 9** takes a closer look at some of the theories and psychological issues that are specifically related to decision making. Game theory, decision trees, and utility theory are among the more formal topics in this chapter. We then discuss some of the human factors influencing decision making, the effects of human misconceptions of probabilities, the irrational behaviors of human decision makers, and how these difficulties can be dealt with to improve the decision making process in practice.

In the last chapter, **Chapter 10**, we give an overview of some of the recently developed approaches to problem solving that practitioners have resorted to because of the inadequacy or ineffectiveness of the more formal traditional methods. Inasmuch as perfect methodologies for some *known-to-be-difficult* problems have so far eluded analysts (and in fact may never be forthcoming!), the heuristic and metaheuristic methods presented here are often used to obtain solutions that may be sub-optimal but often acceptable in practice.

This book contains a comprehensive collection of topics that we believe provide an accurate and useful introductory perspective on the discipline and practice of Operations Research. We, the authors, have prepared this book on the basis of our various experiences in teaching, research, technical consulting, and systems analysis. Significant credit goes to our own professors whose excellent instruction once introduced us to the field of Operations Research, and whose knowledgeable enthusiasm initially sparked our interest in the subject. Research and consulting opportunities have sustained and broadened our awareness and appreciation of the importance of these topics.

The immediate motivation for developing this book arose from our many years of teaching courses in various areas of operations research, mathematics, computer science, business analysis, and systems engineering.

In the preparation of this edition of the book, we particularly appreciate and gratefully acknowledge the contributions of Mariam Kotachi, Max Siangchokyoo, and Chris Knight for their assistance with formatting the references and equations, and the help of Paul Ticu, June Au Yeung and Kavin Fong for their help with the problems and solutions for the first edition. Many of our students have been introduced to Operations Research through courses in which early drafts of this book were used as text material. We appreciate these students, notably Avinash Atholi and Russell Hyland among others, for their interest in the subject and their careful reading of the chapters. Their constructive and insightful responses and suggestions have contributed substantially to improvements in the presentation of the material in this book. We continue to welcome feedback from our readers, and invite comments that will assist us in keeping this book correct, up-to-date, educational, and of practical value.

The artwork on the front cover of this book captures the philosophy and illustrates the context in which we as Operations Researchers attempt to formulate and solve problems. Our models and methodologies (represented in the cover art by a poetic assembly of graphs and figures) are often not firmly anchored to an idealized grid, but rather rest upon a ground full of ups and downs, uncertainties, constant change, and incomplete knowledge (suggested in the cover art by photographic excerpts of the Grand Canyon). The elements in the illustration are drawn from Figures 2.6, 3.7, and 10.1 in this book; the 3D graph is a model for Exercise 5.3, and was plotted using GeoGebra. In the cover image, the diagrams appear to arise from the predictable grid foundation, but are actually perilously close to the cliffs and canyons.

In order to take best advantage of our circumstances, we make fundamental assumptions that we know may not always be completely justifiable. But nevertheless on the basis of this seemingly frail foundation, we have built sophisticated and reliable tools for solving important practical problems. The field of Operations Research consists of a broad variety of analytical tools and methods which can provide essential assistance in making informed and responsible decisions and reaching worthy goals.

About the Authors

Michael W. Carter is a professor in the Department of Mechanical and Industrial Engineering at the University of Toronto, Toronto, Ontario (since 1981) and founding director of the Centre for Healthcare Engineering (in 2009). He received his PhD in Combinatorics and Optimization from the University of Waterloo, Waterloo, Ontario. He also spent seven years at Waterloo as a full-time Systems Analyst in the Data Processing Department. He is a member of the Canadian Operational Research Society (CORS), the Institute for Operations Research and the Management Sciences (INFORMS), the Health Applications Society (of INFORMS), the Institute of Industrial and Systems Engineering (IISE) and the Society for Health Systems (SHS). He is the Canadian representative for ORAHS (EURO: Operations Research Applied to Health Services).

Since 1989, his research focus has been in the area of health care resource modeling and capacity planning. As of January 2018, Dr. Carter had supervised 23 PhD students and 90 Masters and directed more than 250 undergraduate engineering students in over 100 projects with industry partners. He has over 100 former students who now work in the healthcare industry. He is cross appointed to the Institute of Health Policy, Management and Evaluation (IHPME) and the School of Public Policy & Governance at the University of Toronto.

Dr. Carter teaches undergraduate courses in Healthcare Systems and Engineering Economics. Graduate courses include Healthcare Engineering, Healthcare Research and an Introduction to Operations Research for students in a part-time Master of Health Administration (MHSc) in IHPME.

He was the winner of the Annual Practice Prize from the Canadian Operational Research Society (CORS) four times (1988, 1992, 1996, and 2009). In 2000, he received the CORS Award of Merit for lifetime contributions to Canadian Operational Research. He also received an *Excellence in Teaching* Award from the University of Toronto Student Administrative Council. He is on the editorial board for the journals *Health Care Management Science*, *Operations Research for Health Care*, *Health Systems*, and *IISE Transactions on Healthcare Systems*. He is an adjunct scientist with the Institute for Clinical Evaluative Sciences in Toronto (www.ices.on.ca) and a member of the Faculty Advisory Council for the University of Toronto Chapter of the Institute for Healthcare Improvement (IHI). He is a member of the Professional Engineers of Ontario. In 2012, he was inducted as a Fellow of the Canadian Academy of Engineering and in 2013, he was inducted as a Fellow of INFORMS, the international society for Operations Research and Management Science.

Camille C. Price has been a professor of Computer Science at Stephen F. Austin State University, Nacogdoches, Texas, and she now continues her academic association as emeritus professor. She has also held faculty appointments at the University of Texas at Dallas, Richardson, Texas; Southern Methodist University, Dallas, Texas; Colby College, Waterville, Maine; and Williams College, Williamstown, Massachusetts; and was a Visiting Scholar in the Center for Cybernetic Studies at the University of Texas at Austin, Austin, Texas.

She holds BA and MA degrees in Mathematics from the University of Texas at Austin, and the PhD degree from Texas A&M University, College Station, Texas, with graduate specializations in Computing Science and Operations Research. She held a research fellowship at the Jet Propulsion Laboratory of California Institute of Technology, Pasadena,

California, and subsequently was engaged as a technical consultant for research projects at the JPL. Professional memberships include the Institute for Operations Research and the Management Sciences (INFORMS) and the INFORMS Computing Society, life membership in the Institute of Electrical and Electronics Engineers and the IEEE Computer Society, the Association for Computing Machinery, and the Sigma Xi Scientific Research Society.

Dr. Price has been the principal investigator on a variety of research projects funded by the National Science Foundation and the State of Texas. She has twice received NASA Awards in recognition of technical innovation in task scheduling and resource allocation in specialized computer networks. She reviews research proposals for the National Science Foundation and the Canadian Natural Sciences and Engineering Research Council. She has served as an advisory consultant for program accreditation assessments and curriculum reviews at universities in Texas, Oklahoma, Georgia, and Jordan; and as a member of the research advisory board for the Texas Department of Transportation. As a consultant for IBM Corporation, she has taught courses in advanced operating systems to IBM technical employees in Tokyo, Rome, Texas, and Florida. She has been an editorial consultant and Series Editor in Operations Research for CRC Press, and is currently the Series Editor of the Springer International Series in Operations Research and Management Science.

Her primary responsibilities as a faculty member have involved teaching undergraduate and graduate courses in computer science and operations research, serving as graduate advisor for computer science and directing graduate student research projects. She is the recipient of Teaching Excellence Awards from her college and department; and her research interests and activities have resulted in numerous papers published in scientific journals and presented at conferences.

Dr. Price's research projects have addressed various topics in Operations Research. Her work on heuristic algorithms for mathematical programming problems has been applied to scheduling and allocation of tasks and resources in distributed computing systems, novel computer architectures, load balancing in multiprocessor computer systems, flow control, routing, fault-tolerance in parallel computing systems, and design and analysis of parallel methods for combinatorial optimization.

Ghaith Rabadi is a professor of Engineering Management & Systems Engineering (EMSE) at Old Dominion University (ODU), Norfolk, Virginia. He received his PhD and MS in Industrial Engineering from the University of Central Florida (UCF), Orlando, Florida, in 1999 and 1996 respectively, and his BSc in Industrial Engineering from the University of Jordan, Amman, Jordan, in 1992. Prior to joining ODU in 2002, he worked at UCF as Post Doc where he led NASA funded projects on developing discrete-event simulations of the Space Shuttle ground processes. He was then a visiting assistant professor at the department of Industrial Engineering & Management Systems at UCF. He then worked as a research director at Productivity Apex, a modeling and simulation firm based in Orlando, Florida.

In summer 2003, he received the NASA Faculty Fellowship where he worked on operation modeling and simulation of future space launch vehicles at NASA Langley Research Center in Hampton, Virginia. For their work with NASA, he and his colleagues were awarded the NASA Software Invention Award and the NASA Board Action Invention Award. In 2008, he received the Fulbright Specialist Program Award to work with the faculty at the German-Jordanian University in Amman, Jordan.

He was a visiting professor for one year at the Department of Mechanical and Industrial Engineering at Qatar University, Doha, Qatar, in 2013–2014 academic year. He taught graduate and undergraduate courses in Operations Research, Engineering Economics, and Simulation, and collaborated with the faculty on research pertaining to port operation simulation and optimization.

In 2016, he received ODU's Doctoral Mentoring Award for advising 14 PhD students to graduation over the past 14 years, and for continuing to work closely and publish with his students. Most recently, he with a team of professors and PhD students received NATO's Global Innovation Challenge Award for their work on humanitarian logistics optimization.

Dr. Rabadi's research has been funded by NASA, NATO Allied Transformation Command, Department of Homeland Security, Army Corps of Engineers, Department of the Army, Virginia Port Authority, Northrop Grumman Shipbuilding, MITRE Corporation, Boeing, STIHL, CACI, Sentara Hospitals and Qatar Foundation.

His research and teaching interests include Planning & Scheduling, Operations Research, Simulation Modeling and Analysis, Supply Chain Management & Logistics, and Data Analytics. He has published a book, and over 100 peer reviewed journal and conference articles and book chapters. He is a co-founder and is currently the chief editor for the *International Journal of Planning and Scheduling*. More information is available at www. ghaithrabadi.com.

1

Introduction to Operations Research

1.1 The Origins and Applications of Operations Research

Operations Research can be defined as the use of quantitative methods to assist analysts and decision-makers in designing, analyzing, and improving the performance or operation of systems. The systems being studied may be any kind of financial systems, scientific or engineering systems, or industrial systems; but regardless of the context, practically all such systems lend themselves to scrutiny within the systematic framework of the scientific method.

The field of Operations Research incorporates analytical tools from many different disciplines, which can be applied in a rational way to help decision-makers solve problems and control the operations of systems and organizations in the most practical or advantageous way. The tools of Operations Research can be used to optimize the performance of systems that are already well-understood, or to investigate the performance of systems that are ill-defined or poorly understood, perhaps to identify which aspects of the system are controllable (and to what extent) and which are not. In any case, mathematical, computational, and analytical tools and devices are employed merely to provide information and insight; and ultimately, it is the human decision-makers who will utilize and implement what has been learned through the analysis process to achieve the most favorable performance of the system.

The ideas and methodologies of Operations Research have been taking shape throughout the history of science and mathematics, but most notably since the Industrial Revolution. In various ways, all of human knowledge seems to play a role in determining the goals and limitations underlying the decisions people make. Physical laws (such as gravity and the properties of material substances), human motivations (such as greed, compassion, and philanthropy), economic concepts (supply and demand, resource scarcity, division of labor, skill levels, and wage differentials), the apparent fragility of the environment (erosion, species decline), and political issues (territorial aggression, democratic ideals) all eventually are evident, at least indirectly, in the many types of systems that are studied using the techniques of Operations Research. Some of these are the natural, physical, and mathematical laws that are inherent and that have been discovered through observation, while others have emerged as a result of the development of our society and civilization. Within the context of these grand themes, decision-makers are called upon to make specific decisions—whether to launch a missile, introduce a new commercial product, build a factory, drill a well, or plant a crop.

Operations Research (also called Management Science) became an identifiable discipline during the days leading up to World War II. In the 1930s, the British military buildup centered around the development of weapons, devices, and other support equipment. The buildup was, however, of an unprecedented magnitude, and it became clear that there was also an urgent need to devise systems to ensure the most advantageous deployment and management of material and labor.

Some of the earliest investigations led to the development and use of radar for detecting and tracking aircraft. This project required the cooperative efforts of the British military and scientific communities. In 1938, the scientific experts named their component of this project *operational research*. The term *operations analysis* was soon used in the U.S. military to refer to the work done by teams of analysts from various traditional disciplines who cooperated during the war.

Wartime military operations and supporting activities included contributions from many scientific fields. Chemists were at work developing processes for producing high octane fuels; physicists were developing systems for the detection of submarines and aircraft; and statisticians were making contributions in the area of utility theory, game theory, and models for various strategic and tactical problems. To coordinate the effectiveness of these diverse scientific endeavors, mathematicians and planners developed quantitative management techniques for allocating scarce resources (raw materials, parts, time, and labor) among all the critical activities in order to achieve military and industrial goals. Informative overviews of the history of Operations Research in military operations are to be found in White (1985) and McArthur (1990).

The new analytical research on how best to conduct military operations had been remarkably successful, and after the conclusion of World War II, the skill and talent of the scientists that had been focused on military applications were immediately available for redirection to industrial, financial, and government applications. At nearly the same time, the advent of high speed electronic computers made feasible the complex and time consuming calculations required for many operations research techniques. Thus, the methodologies developed earlier for other purposes now became practical and profitable in business and industry.

In the early 1950s, interest in the subject was so widespread, both in academia and in industry, that professional societies sprang up to foster and promote the development and exchange of new ideas. The first was the Operational Research Society in Britain. In the U.S., the Operations Research Society of America (ORSA) and The Institute of Management Science (TIMS) were formed and operated more or less as separate societies until the 1990s. These two organizations, however, had a large and overlapping membership and served somewhat similar purposes, and have now merged into a single organization known as INFORMS (Institute for Operations Research and the Management Sciences). National societies in many other countries are active and are related through IFORS (the International Federation of Operational Research Societies). Within INFORMS, there are numerous special interest groups, and some specialized groups of researchers and practitioners have created separate societies to promote professional and scholarly endeavors in such areas as simulation, transportation, computation, optimization, decision sciences, and artificial intelligence. Furthermore, many mathematicians, computer scientists and engineers have interests that overlap those of operations researchers. Thus, the field of Operations Research is large and diverse. Some of the many activities and areas of research sponsored by INFORMS can be found at the website http://www.informs.org or in the journals associated with that organization. As will be apparent from the many illustrative applications presented throughout this book, the quantitative analysis techniques that found their first application nearly a hundred years ago are now used in many ways to influence our quality of life today.

1.2 System Modeling Principles

Central to the practice of Operations Research is the process of building mathematical models. A **model** is a simplified, idealized representation of a real object, a real process, or a real system. The models used here are called **mathematical models** because the building blocks of the models are mathematical structures such as equations, inequalities, matrices, functions, and operators. In developing a model, these mathematical structures are used to capture and describe the most salient features of the entity that is being modeled. For example, a financial balance sheet may model the financial position of a corporation; mathematical formulas may serve as models of market activity or trends; and a probability distribution can be used to describe the frequency with which certain asynchronous events occur in a multiprocessor computer. Mathematical models may look very different, depending on the structure of the system or problem being modeled and the application area. In studying the various topics in this book, we will see that models do indeed take on various forms. Each chapter provides the opportunity to build different kinds of models. This chapter merely makes a few general observations pertinent to all modeling.

The first step in building a model often lies in discovering an area that is in need of study or improvement. Having established a need and a target for investigation, the analyst must determine which aspects of the system are controllable and which are not, and identify the goals or purpose of the system, and the constraints or limitations that govern the operation of the system. These limitations may result from physical, financial, political, or human factors. The next step is to create a model that implicitly or explicitly embodies alternative courses of action, and to collect data that characterize the particular system being modeled.

The process of *solving* the model or the problem depends entirely on the type of problem. Solving the problem may involve applying a mathematical process to obtain a *best* answer. This approach is sometimes called **mathematical optimization,** *or* **mathematical programming**. In other cases, the solution process may necessitate the use of other specialized quantitative tools to determine, estimate, or project the behavior of the system being modeled. Realizing that the data may have been only approximate, and that the model may have been an imperfect representation of the real system, a successful analyst ultimately has the obligation to assess the practical applicability and flexibility of the solution suggested by the foregoing analysis. Merely finding an *optimal* solution to a model may be just the beginning of a manager's job; a good manager must constantly reevaluate current practices, seek better ways to operate a system or organization, and notice trends in problem data that may not explicitly appear as part of a mathematical solution, such as excess production capacity, under-utilized labor, or a decreasing product demand over time. The entire modeling process is likely to require the skill and knowledge of a variety of individuals who are able to work effectively in teams and communicate clearly and convincingly among themselves, and then to explain and *sell* their recommendations to management.

Considerable skill is required in determining just how much detail to incorporate into a mathematical model. A very accurate representation of a system can be obtained with a large and sophisticated mathematical model. But if too many details are included, the model may be so complex and unwieldy that it becomes impossible to analyze or solve the system being modeled. Therefore, we do not even try to make a model as realistic as possible. On the other hand, a very simplistic model may not carry enough detail to provide an accurate representation of the real object or system; in that case, any analysis that is performed on the model may not apply to the reality.

It is tempting to confuse detail (or precision) with accuracy. They are not the same, although many people are under the impression that the more detailed or complex a model, the more accurately it reflects reality. Not all details are correct, and not all details are relevant. The availability of powerful computing hardware and user-friendly software for building computer models almost seem to encourage runaway complexity and detail, as there seems to be no limit to what can be included almost effortlessly in a model. Nevertheless, it is possible to build models that are both realistic and simple, and doing so may spare an analyst from losing sight of the purpose of building the model in the first place.

The best model is one that strikes a practical compromise in representing a system as realistically as possible, while still being understandable and computationally tractable. It is, therefore, not surprising that developing a mathematical model is itself an iterative process, and a model can assume numerous forms during its development before an acceptable model emerges. An analyst might in fact need to see some numerical results of a solution to a problem in order to begin to recognize that the underlying model is incomplete or inaccurate.

The purpose of building models of systems is to develop an understanding of the real system, to predict its behavior, to learn the limiting capabilities of a system, and eventually to make decisions about the design, development, fabrication, modification, or operation of the real system being modeled. A thorough understanding of a model may make it unnecessary to build and experiment with the real system, and thus may avoid expense or alleviate exposure to dangerous situations.

Operations Research deals with decision-making. **Decision-making** is a human process that is often aided by intuition as well as facts. Intuition may serve well in personal decisions, but decisions made in political, governmental, commercial, and institutional settings that will affect large numbers of people require something more than intuition. A more systematic methodology is needed. Mathematical models that can be analyzed by well-understood methods and algorithms inspire more confidence and are easier to justify to the people affected by the decisions that are made.

Experience in modeling reveals that, although quantitative models are based on mathematical truths and logically valid processes and such models may command the respect of management, solutions to mathematical problems are typically interpreted and implemented under a variety of compromising influences. Management is guided by political, legal, and ethical concerns, human intuition, common sense, and numerous personality traits. Problems and systems can be represented by mathematical models, and these formulations can be solved by various means. However, final decisions and actions are taken by humans who have the obligation to consider the well-being of an organization and the people in it. Ideally, if these factors are going to influence the decisions that are made, then these human concerns, as well as technological and financial goals and constraints, should be incorporated in an honest way into the models that are created and analyzed. In this way, we can gain the greatest value from our efforts in applying quantitative methods.

As a final word of advice and caution, it is suggested that before expending any substantial effort in solving or analyzing a problem or system, analysts and managers should try to confront and answer a few preliminary questions:

Does the problem need to be solved?

Will it be possible to determine what the *real* problem is?

If a model were developed and a solution proposed, would anybody care?

Would anybody try to implement the solution?

How much of the analyst's time and expense is it worth to try to solve this problem?

Is there enough time and are there adequate resources available to make any significant progress toward solving this problem?

Will the solution create other serious problems for which there is no apparent remedy?

These are difficult questions, often overlooked by an eager and motivated analyst, but they are issues that an analyst should try to confront frankly and candidly before becoming irreversibly involved in a large problem-solving project.

1.3 Algorithm Efficiency and Problem Complexity

An **algorithm** is a sequence of operations that can be carried out in a finite amount of time. An algorithm prescribes a process that may be repetitive in some sense (perhaps iterative or recursive), but that will eventually terminate. Practical examples of algorithms include recipes for cooking, the instructions in an owner's manual for connecting a new sound system component, and computer programs that do not contain infinite loops. Algorithms are the processes that software developers put into action when they create computer programs for solving all kinds of problems.

In the 1930s, a mathematician by the name of Alan Turing developed a general computational model (which now bears his name) that is powerful enough to represent all possible numeric and symbolic computational procedures. Turing also demonstrated the existence of problems for which no algorithms exist that successfully handle all possible instances of the problem. Such problems are called **unsolvable** or **undecidable problems**. (It had been previously assumed that an algorithm could be developed for any problem if the problem-solver were merely clever enough.) Some of the earliest problems to be classified as unsolvable were of only theoretical interest. However, more recently, other more practical **unsolvable problems** have been identified.

When such problems do arise in actual practice, we might just try to deal with special or limited cases, rather than with the general problem. Special cases of unsolvable problems, perhaps involving highly restricted inputs, may not be unsolvable, and therefore it may be entirely possible to design algorithms for these cases. Alternatively, we might find it fairly simple to use human *ingenuity* (a very poorly defined talent that cannot be easily automated) to deal with individual problem instances on a case-by-case basis.

While unsolvable (or undecidable) problems do exist, most analysts would prefer to concentrate on the many important **solvable problems** that face us; that is, problems for which algorithms *can* be developed. With this in mind, the next question to arise might be: are all solvable problems of similar difficulty, or are there some that are truly more difficult than others? What is meant by a *difficult* problem? And just what is known about algorithms, and the complexity (or computational behavior) of algorithms? This is a topic of study that has undergone enormous progress during the past several decades, and the advances that have been made in this field have provided valuable concepts, notations, and tools that allow for discussion and analysis of an algorithm's performance.

Several factors influence the amount of time it takes for a computer program to execute to solve a problem: the programming language used, the programmer's skill, the hardware used in executing the program, and the task load on the computer system during execution.

But none of these factors is a direct consequence of the underlying algorithm that has been implemented in software. Given a particular algorithm, its performance is strongly dependent on the *size* of the problem being solved. For example, we would expect a sorting algorithm to take longer to sort a list of 10,000 names than to sort a list of 100 names. Similarly, we recognize that solving a system of equations takes longer when there are more equations and more variables. For this reason, the performance of an algorithm is often described as a function of a variable denoting the **problem size**, which denotes the size of the data set that is input to the algorithm.

During the early years of the discipline of Operations Research, relatively little was understood about the formal properties of algorithms and the inherent complexity of problems. However, the 1970s and 1980s witnessed remarkable developments in this area. Two interesting classes of problems have been defined. One class of problems (called class P) contains those problems that can be solved by an algorithm within an amount of computation time proportional to some polynomial function of problem size; that is, the problems are solvable by **polynomial-time algorithms**. The other class (called class NP for nondeterministic polynomial time) contains problems that may require the computation time to be proportional to some exponential (or larger) function of problem size; these algorithms are called **exponential-time algorithms**. For a more precise description of these problem classes, based on the notions of deterministic and nondeterministic Turing machines, refer to any elementary textbook on algorithms or theory of computation, such as Cormen et al. (2009), Baase and Gelder (2000), Manber (1989), and Hein (1995).

Within the class NP, there is another special class of important problems called **NP-complete**, which are characterized as being the most difficult problems in NP. This class includes many very practical problems and so has received considerable attention from analysts. Another class of NP problems, known as **NP-hard**, are at least as hard as the hardest NP problem. Some of these NP problems, and their practical applications, are described in Chapters 3, 4, and 10.

The problems in class P are generally considered to be *easy* problems—not necessarily in the conceptual sense but in the sense that efficient algorithms for these problems exist that execute in reasonably small amounts of computation time. NP-complete and NP-hard problems, in contrast, appear to require computation time that grows as an exponential function of problem size. This implies that unacceptably large amounts of computation time could be required for solving problems of any practical size, and therefore such problems have been termed *intractable* problems. Solutions for such problems are not necessarily difficult to conceptualize or even to implement in computer code, but the execution time may be completely unaffordable—both physically and financially.

It is known that $P \subseteq NP$, but it is an open question whether $P = NP$. In other words, are the NP-complete problems truly more costly to solve than the problems in P, or have analysts just not yet been clever enough to discover efficient algorithms for these apparently difficult problems? Discovery of an efficient (polynomial-time) algorithm for any NP-complete problem would be sufficient to establish that $P = NP$ and, therefore, that all the NP-complete problems can be solved efficiently. In the absence of any such discovery, analysts are faced daily with the need to solve practical problems that are computationally intractable. Chapter 10 reveals how some of these problems are dealt with in effective and affordable ways. An informative overview of this subject is available in Garey and Johnson (1979).

Most of the problem models presented in this book are not intractable, and the solution methods for these problems are based on polynomial-time algorithms. These methods find optimal solutions in an amount of time proportional to a polynomial function of the problem size. Depending on the nature of the data (e.g., the distribution of data values or the arrangement of the data values in the data set), the execution time for a given algorithm may vary. Sorting a list of 10,000 names that are already in order *may* take less time than to sort 10,000 names that are scrambled—if the algorithm is sensitive to the initial ordering and can take advantage of it. Similarly, finding the *best* solution to a system of equations *may* be rather easy if a *reasonably good* solution is already known.

Thus, we will see that, under different circumstances, the same algorithm may require an execution time that is a *different* function of problem size. If so, which of these different functions should analysts use to characterize the performance of the algorithm? There are several possibilities: the most favorable (fastest) case, the average case, or the most unfavorable (slowest) case.

To help phrase an answer to this question, special notations have been developed that facilitate describing the computation time required to execute an algorithm to completion. For this particular purpose, we do not want to try to capture specific information about how an algorithm is implemented (programmed), or on what type of computer it is to be executed; rather, we should focus on the algorithm itself and, in particular, the **step count**, or the number of steps inherent in carrying out the algorithm. For some purposes, one might want to characterize the **best case performance** of an algorithm (the fewest number of steps that it could ever need under any circumstances). Best case might be the choice of an optimist, but using this as an indicator of algorithm performance could be misleading; and in any case, this is rarely indicative of what analysts need to know in order to assess the dependable performance of the algorithm. For example, multiplying two $n \times n$ matrices generally takes time proportional to n^3; but of course, if one of the matrices is the identity matrix, this could be discovered in only n^2 steps (inspecting each element of the matrix) and the rest of the process could be omitted. Using the function n^2 to describe the step count, or run-time, of a matrix multiplication routine does give an accurate measure of this best case, but it is not generally indicative of the time required for matrix multiplication.

An algorithm's **average case performance** may seem to be the most practical characterization because it indicates the typical, or expected, step count. It would certainly be useful to know the *most likely* amount of time required to execute an algorithm. However, because such an analysis must be based on statistical assumptions about the nature, order, or distribution of the data on which the algorithm operates, the validity of such assumptions may be on shaky ground for any particular set of data. Indeed, the expected performance may never actually be observed for any given set of input data. In addition, the statistical analysis that must be carried out in order to characterize an algorithm's average behavior is often quite a mathematically difficult analysis.

The characterization of an algorithm that is both straightforward and often of greatest practical value is the **worst case performance**, that is, the greatest number of steps that may be necessary for guaranteed completion of the execution of the algorithm. For this purpose, we introduce **big-Oh notation**, which is written as $O(f(n))$ and pronounced "big Oh of f of n," where n denotes problem size and $f(n)$ is some function of problem size. The meaning of the notation is as follows. An algorithm is said to be $O(f(n))$ if there exist constants c and n_0 such that for all $n > n_0$, the execution time of the algorithm is $\leq c \cdot f(n)$. The function $f(n)$ is the algorithm's worst case step count, measured as a function of the problem size. The constant c is called a constant of proportionality and is intended to account for the various extraneous factors that influence execution time, such as hardware speed, programming style, and computer system load during execution of the algorithm. The problem size threshold n_0 accounts for the fact that for very small problem sizes, the algorithm may not reveal its characteristic worst case performance. Paraphrased, the definition given above may be stated as follows: To say that an algorithm is $O(f(n))$, or "of order $f(n)$," means that for large enough problem sizes, the execution time is proportional to at most $f(n)$.

Thus, a matrix multiplication algorithm is $O(n^3)$ because the process may take n^3 steps, although the algorithm *could* be programmed to look for special input forms that may in certain cases permit completion of the task in fewer than n^3 steps. Some algorithms may operate in such a way that their worst case performance is also the best case; the performance of such algorithms does not vary depending on the nature of the data, but, of course, does vary with problem size.

There are even some algorithms whose performance is independent of problem size, and therefore not really dependent on any function of problem size n (e.g., retrieving the first item in a list takes the same amount of time regardless of the length of the list). If we need to describe the worst-case performance of such an algorithm, we could use the notation $O(1)$, where $f(n)$ is just the constant function 1. Where appropriate throughout this book, the big-Oh notation is used to describe the worst case performance of the algorithms that are presented.

Many of the methods studied in this book are based on algorithms whose complexity functions range from n, n^2, n^3, up to 2^n, $n!$, and n^n. To give an idea of the relative growth rates of these functions as n increases, Table 1.1 shows indications of function values. Instead of raw numeric values, we can impose a more practical interpretation and assume that the function values $f(n)$ denote the step count of some algorithm, and that each *step* can be executed in 1 second on a computer. The entries in the table can then be viewed as estimates of actual amounts of the computation times required to apply algorithms of different complexities to increasingly larger problems of size n. The great differences that are evident between the polynomial functions and the exponential functions are quite dramatic, and the execution times for the exponential algorithms are indeed staggering.

In practical applications, problem sizes may well range into the hundreds of thousands, and we will encounter a number of important practical problems for which the only known algorithms have worst case complexity that is exponential. It is obvious from the table that such exponential-time algorithms are completely useless for solving problems of any reasonably large size. Given this dilemma, what are the options? It is pretty clear that, for these types of problems, faster hardware does not offer an immediate solution; CPU chips whose processing speeds are doubled, or even increased by several orders of magnitude, will not make a dent in these formidable execution times. Until some theoretical breakthroughs come to the rescue that suggest new algorithms for solving such problems, we

TABLE 1.1

Computation Times

$f(n)$	$n = 10$	$n = 20$	$n = 50$	$n = 100$
n	10 s	20 s	50 s	100 s
n^2	100 s	400 s $\approx$ 7 min	2,500 s $\approx$ 42 min	10,000 s $\approx$ 2.8 h
n^3	1,000 s $\approx$ 17 min	8,000 s $\approx$ 2 h	125,000 s $\approx$ 35 h	1,000,000 s $\approx$ 12 d
2^n	1,024 s $\approx$ 17 min	1,048,576 s $\approx$ 12 d	1.126×10^{15} s $\approx$ 350,000 centuries	1.268×10^{30} s $\approx 10^{21}$ centuries
$n!$	3,628,800 s $\approx$ 1 month	2.433×10^{18} s $\approx 10^9$ centuries	3.041×10^{64} s $\approx 10^{55}$ centuries	
n^n	10^{10} s $\approx$ 300 yr	1.049×10^{26} s $\approx 10^{17}$ centuries	8.882×10^{84} s $\approx 10^{75}$ centuries	

may have to settle for using methods that do not solve the problems perfectly, but which yield acceptable solutions in an affordable amount of computation time. This may seem to be a disappointing direction to follow, but the discussion in Section 1.4 might provide convincing arguments in defense of suboptimal solutions.

1.4 Optimality and Practicality

Everyone with a mathematical education has been trained to search for exact solutions to problems. If we are solving a quadratic equation, there is a formula which, if applied correctly, yields *exact* results, namely values that *satisfy* the equation. If a list of names needs to be sorted, we employ an algorithm that gets the list *perfectly* ordered. And if we need to find the maximum point in a continuous, differentiable function, we may be able to use the methods of calculus to find that *optimal* point. And certainly in the case of giving proofs of mathematical theorems, a respect for truth and perfection has been developed, and a nearly correct but incomplete or slightly flawed proof is of little or no value at all. Against this backdrop, the idea of solving a problem and not getting the *right* answer is indeed disappointing and disturbing. Yet there are justifiable reasons for accepting computational results that are imperfect or suboptimal.

First, it has already been pointed out that the models created by an analyst are not perfect representations of a system being analyzed. So, even if we could obtain exact solutions to the model, such solutions would not necessarily constitute exact solutions or perfect managerial advice to be applied within the real system. Hence, costly efforts to achieve perfect solutions to a mathematical model may not be warranted.

Contributing to the imperfection in our problem-solving endeavors is the use of automatic computing devices to assist in the calculations. The exact representation of real numbers requires the use of an arbitrarily large number of binary digits. However, the finite number of bits, sometimes known as word length, typically used for storing numerical values in computer memories implies that real numeric data values cannot always be represented exactly in computers. As an example, a correct representation of the value pi requires infinitely many digits, but we often settle for a truncated approximation using seven or eight significant digits (such as 3.141592) and tolerate the resulting inaccuracy in the results. This is known as **round-off error**, and after repeated calculations involving many inexact values, the **accumulated round-off error** can so distort the final results that they bear little resemblance to the pure theoretically correct answers that were anticipated. Hardware standards, such as the IEEE Floating-Point Standard, and results from the well-developed field of numerical analysis have provided analysts with tools to define, measure, and place bounds on the effects of accumulated computational errors, but being able to predict these errors does not necessarily suggest any method for avoiding or correcting erroneous results.

It is known that the data values associated with some types of problems, such as matrix problems and solving systems of equations, are inherently *ill-conditioned*, and certain computational procedures, such as matrix operations or iterative searches designed to converge to a solution, are inherently *unstable*. In some cases, although the algorithm underlying a solution process might be proven to yield optimal results, ill-conditioned problem data and numerical instability can practically preclude obtaining solutions of any reasonable quality. For further discussions on the successful use of numerical techniques with computers, refer to any reputable book on numerical analysis, such as by Cheney and Kincaid (2013), Sauer (2011), and Wilkinson (1963).

Finally, the innate difficulty of some problems might suggest that accepting suboptimal solutions is the only practical approach. Problems whose algorithms take an exponential amount of computation time to guarantee a perfect, or optimal, solution leave us little alternative but to look for faster ways of obtaining solutions, even at the price of getting solutions of lesser quality. Suppose we are faced with the choice of expending an exponential amount of time (perhaps translating into centuries of computation time) to obtain an *optimal* result, or expending polynomial-time computational effort to obtain a solution that is *adequate*. In some cases, there may be a guarantee that the polynomial-time solution will be within some specified percentage of the optimal solution. In other cases, there may be no such guarantee, but perhaps experience has shown that in common practice the results are considered to be good enough for the context in which the solution is to be applied. Realizing also that the *optimal* result may be the solution to the wrong model, that the *optimal* result may be infused with round-off error, and that the data used as parameters might have been flawed and could have changed over time, a realistic analyst would probably feel completely justified in applying the polynomial-time algorithm to obtain a practical solution quickly, and feel no remorse whatsoever over having foregone the chance to obtain a slightly better solution. Given our very imperfect grasp on the concept and reality of *perfection*, the price of optimality—in this case and in many others—is entirely impractical.

Settling for solutions of merely *good enough* quality may at first seem to be an inexcusable lowering of one's standards and expectations. Yet in a complex and in some ways subjective world, compromise should not necessarily be seen as evidence of mediocrity. In the real world of imperfect models, precarious data, unavoidable numerical inaccuracies, and time constraints, insistence upon so-called optimal solutions may border on the compulsive. A rational analyst with a comprehensive view of the problem-solving process would prefer to spend a reasonable amount of time in search of good, practical solutions, and then proceed to put the results into practice to achieve the original goal of improving the performance or operation of the system being studied. Chapter 10 introduces some of the inspiration and influences behind solution methods that incorporate pragmatic approaches to solving difficult problems.

1.5 Software for Operations Research

Each chapter in this book contains a section on software tools, in which there is a brief description of some of the most popular software currently available for solving the types of problems studied in the chapter. The principles and methods presented in each chapter are intended to provide the foundations necessary for building and understanding appropriate models. The authors' aim is to encourage an adequate understanding of the mathematical principles and methods for solving problems so that students can become informed users of the software that is available to them.

Because there is no single software system that is capable of solving all optimization and system analysis problems, the user must be knowledgeable enough about the various classes of problems to make a selection of appropriate software packages. Thus, being able to build a mathematical model of a problem *and* being able to identify that model as a linear program, integer program, or network problem, for example, not only helps to clarify the model, but also puts the analyst well on the way to selecting the right software for solving the problem.

The most visible *users* of commercial software may be the people who actually *run* application systems that contain optimization modules. However, playing even more essential roles in the process are the analysts who formulate the mathematical models and who adapt and refine the standard algorithms, and the developers of the software packages who incorporate optimization modules (sometimes called *solvers*), together with application systems and user interfaces. In our discussions, we will address various practical issues that are important to all software users.

The references to software products in this and subsequent chapters are by no means exhaustive and are not intended to comprise a comprehensive catalog of available software. Instead, we hope to give readers a feel for the types of products that are on the market and that may deserve their consideration when selecting implementations for practical applications.

Note also that our references to software tools are not intended to represent endorsement of any specific software products. Rather, we merely mention examples from the broad range of software available for the various application areas and offer short descriptions of selected software packages and libraries, in order to create an awareness of the general capabilities of typical software, as well as some of the questions, difficulties, or limitations that might arise during the development or use of software for solving real problems.

New products are being introduced rapidly, and it would be impossible to maintain a perfectly up-to-date list of software tools. Advertisements and published product reviews are helpful and, in particular, the software reviews that appear frequently in issues of *OR/MS Today* are an extremely valuable source of information.

We have avoided making any comparisons of products on the basis of performance or cost. Performance depends on the underlying hardware as well as on the frequent updates and modifications that occur during the evolutionary development of the software. Software prices vary rapidly, depending on competition in the market, whether the purchaser or user is in academia or industry, and whether copies are sold for installations in individual workstations, client/server, or cloud-based versions intended for multiple users. More expensive commercial versions of some software may handle larger problem models and solutions, while the less expensive personal versions or student versions may be limited in the size of problems that can be solved.

In light of the above considerations, a few of the pertinent characteristics and features that will likely play a role in the reader's consideration of software products are highlighted. Each chapter's discussion covers software related to the topics covered in that chapter. In this first chapter, no specific solution methods are introduced; however, there is discussion of some of the general principles of building mathematical models. Thus, some software systems that facilitate the construction of models (i.e., modeling languages and environments) and the preparation of model parameters and characteristics are identified. These representations of models can then be introduced as input to various other software solution generators, or **solvers.**

One way to create a problem model to be solved with a specialized solver is to use a general-purpose programming language (such as C, C++, Python, or Java) and write a program to format input parameters appropriately and to generate output reports in the desired form. The advantages of this approach are that such languages are typically available and usable on any hardware, and there is no need to purchase and learn a new language or *package.*

An analyst who creates models in this way can then choose to solve the problem using available software such as is found in the **IMSL Mathematical Subroutine Library**. A comprehensive collection of approximately 1300 mathematical and statistical functions and user-callable subroutines is capable of solving most of the types of problems that will be studied later in this book. The IMSL libraries are ideal for programmers skilled in C, C#, Java, and Fortran, and are available for use on Windows, Unix, Linux and MAC computers. The IMSL software system has been used by developers worldwide for four decades, and is still considered by many to offer valuable autonomy to the user and thereby accelerate development of applications in many contexts (Demirci 1996).

The initial simplicity and low-cost investment associated with this approach, however, may be paid for in the long term, as code written and customized for one modeling project may not be directly transferrable and reusable on subsequent projects. Nevertheless, there can be some value in maintaining direct in-house control over the development and construction of software solutions.

For some types of problems, the row and column (tabular) orientation of problem parameters offered by many spreadsheet programs is easy to create and read; and although the analyst loses some flexibility, many problems lend themselves nicely to the spreadsheet framework. Moreover, many solvers can read and write directly to spreadsheet files.

A much more powerful means for creating models is through the use of **algebraic modeling languages**. These languages permit the user to define the structure of a model and declare the data to be incorporated into the structure. An algebraic modeling language accepts as input the analyst's algebraic view of the model, and creates a representation of the model in a form that the solver algorithm can use. It also allows the analyst to design the desired output reports to be produced after the solver has completed its work. Modeling languages can be bundled with a solver or optimization module, or can allow the user to customize an application system by selecting the best optimization component for the job. Among the most commonly used modeling languages are the following.

AMPL, a modeling language for mathematical programming, is an integrated software package for describing and solving a variety of types of problems. Developed initially by AT&T Bell Labs, it is a complex and powerful language that enables model developers to effectively utilize the system's sophisticated underlying capabilities. AMPL is a command and model interpreter that is available in Windows, Linux, MacOS, and several Unix-based workstations, and interfaces with over 30 powerful optimization engines including MINOS, CPLEX, OSL, GUROBI, and many of the most widely used large-scale solvers. AMPL features an integrated scripting language, provides access to spreadsheet and database files, and has application programming interfaces for embedding within larger systems. A review of AMPL and its use can be found in Fourer et al. (1993) and at www.ampl.com.

MPL is a mathematical programming language that is considered one of the earliest integrated model development systems that supports input and output through interfaces with databases and spreadsheets. MPL is most commonly used with Windows and interfaces with and supports almost all commercial solvers.

LINGO is a thoroughly integrated modeling language and solver that interfaces with the entire LINDO system family of linear and nonlinear problem-solvers. (LINDO products are mentioned in several subsequent chapters, as this line of software offers application tools for a wide variety of types of problems, as further described at www.lindo.com.) This powerful modeling language features a convenient environment for expressing problems, facilitates using information directly from text files, spreadsheets, and databases, provides access to a comprehensive set of built-in solvers that handles a wide range of model types, and generates output reports as well as graphical displays during and upon completion of the solution process. LINGO runs on Windows, Mac and Linux systems.

AIMMS has emerged from its original role as a basic modeling language into a comprehensive, innovative technology company offering sophisticated modeling and solution platforms that support both strategic and operational optimization, decisions, planning and scheduling. A full description of AIMMS is available at www.aimms.com.

SAS/OR OPTMODEL is an optimization modeling language that uses a flexible algebraic syntax for model formulation for different types of mathematical programming problems including linear, mixed integer and nonlinear programming.

GAMS, a general algebraic modeling system, was one of the earliest developed modeling languages, and is now among the most well known and widely used modeling systems for large scale optimization. GAMS links to libraries and programming languages, databases and spreadsheet files, and runs on Windows, Macintosh, Linux, and IBM platforms. GAMS is best known for its sophisticated solvers for the full range of optimization problems and for its graphical interface generator. More information on this system may be found at www.gams.com.

Software for Operations Research is also available through the Internet. As any knowledgeable computer user must know, products (be they information, software, or more tangible items) offered on the Internet may not always be subject to the same standards of quality and control that are imposed on other avenues of commerce. The source, authenticity, quality, and reliability of software or any other information posted on the Internet may be difficult to confirm. Despite these concerns, the Internet has nevertheless become one of the most exciting sources of information available today. With so many kinds of services available online, it makes sense that computational and analytical services and tools should be found there, too. For example, in 1994, a group of researchers at Argonne National Laboratory and Northwestern University launched a project known as the **Network-Enabled Optimization System** (NEOS), which now includes a large number of solvers that accepts models in various formats, solves them on remote servers, and presents the results to the user for free. The NEOS server is hosted by the Wisconsin Institute for Discovery at the University of Wisconsin in Madison, and provides access to more than 60 state-of-the-art solvers in more than a dozen optimization categories. Solvers hosted by the University of Wisconsin run on distributed high-performance machines; remote solvers run on machines at Argonne National Laboratory, Arizona State University, the University of Klagenfurt in Austria, and the University of Minho in Portugal. The NEOS project has been effective in providing information, communication, and high quality software as a valuable service to the operations research community.

Of great interest also is the **CO**mputational **IN**frastructure for **O**perations **R**esearch, known as **COIN-OR**, which is a project dedicated to providing open-source software for the Operations Research community (Lougee-Heimer 2008). It encourages and supports the development of high-quality software suitable for use by a broad range of practitioners, educators, and students working in industry, academia and government. This collection of robust and portable software includes computational tools powerful enough for large collaborative project development, yet accessible to less experienced users as well. Much of the software is structured into building blocks which may be modified to suit the needs of an individual user and combined to create customized application packages. Software components have been used compatibly with proprietary languages and software products. COIN-OR software modules are available for constrained optimization, linear and nonlinear, continuous and discrete problems. Source distributions are provided in standard open source configuration, and precompiled binary distributions are available for Windows and Linux on Intel and AMD platforms, and for Mac OS X on Intel and Power PC platforms.

COIN-OR began in the year 2000 as an initiative of IBM Research, and was incorporated four years later as an independent nonprofit foundation responsible for directing the activities of the organization. Professional technical leaders from universities and research laboratories have continued to work diligently since the founding of COIN-OR to standardize the infrastructure and maintain a stable and reliable repository of software. INFORMS computing and optimization societies regularly sponsor workshops and conference clusters to acquaint prospective users with the wide variety of freely available software that serves the computational needs of operations researchers. Further information about this ambitious and valuable project may be found at www.coin-or.org.

The open source movement has demonstrated over the years that high quality software systems can actually be produced by contributors who volunteer their time and experience to make their products available for other people, hoping that in return people will contribute back. This has been an interesting approach that showed tremendous success and even for-profit companies started to participate in this model as it turned out that it pays off on the long run. For example, Google offers open source codes and binaries for Operations Research tools (solvers, interfaces, algorithms) in different computer languages and for different operating systems. More is available on Google's website.

1.6 Illustrative Applications

1.6.1 Analytical Innovation in the Food and Agribusiness Industries (Byrum 2015)

Food and agribusiness currently represent a $5 trillion industry that amounts to 10% of consumer spending globally. Food production broadly demands about 40% of employment worldwide. And yet, despite the enormity and apparent success of the industry, there is still hunger in many parts of the world.

The global population, currently at over seven billion, is expected to increase by around two billion over the next few decades, and the demand for food crops needed for consumption by humans and their animals is predicted to double. These staggering requirements for nutrition must be met in a context of changing climate and environmental conditions, without further uncontrolled greenhouse gas emissions and destruction of arable land and other natural resources, and with an amount of water that will likely be only about two-thirds of what is actually needed for crop irrigation. The challenges of meeting the increasing demand for food production seem daunting, however we can look toward a radically more ambitious application of Operations Research techniques that can improve efficiency and productivity within the food and agribusiness industries.

Agriculture already is a very information intensive enterprise. Data are gathered regularly on soil conditions, weather, market demands, and prices. Livestock feeds are routinely measured for weight, moisture, and nutritional content. On another level, farmers must deal with data that describe their own specific operational processes and associated risk management as well as with technical, regulatory, and policy issues. Information technology advances such as mobile and remote sensing devices, and satellite image data analysis, all contribute to the mix of inputs that must be processed by powerful analytical capabilities. This vast amount of accumulated information will require increasing amounts of database storage, networking, communication, and more powerful and specialized computational and optimization capabilities.

Food producers are technologically sophisticated. Advances in science, technology, and Operations Research all play a role in addressing problems of economic efficiency, social responsibility, and gainful productivity in agribusiness. Leading international innovators in plant genetics have created customized operations research tools in developing new specialized breeds of seeds that produce higher yields which approximately tripled their annual increases in yield over what had been achieved before the use of these more elegant and powerful analytic techniques. Improvements in their seed products led to genetic gains valued at nearly $300 million over a recent four-year period.

New advances in Operations Research, including theoretical and abstract concepts, can be expected to contribute new analytical tools that can be skillfully applied to real problems. Formal methods will have to be adapted by knowledgeable analysts and applied to the actual problems faced by farmers, ranchers and related food production practitioners to produce practical and tangible results.

Revolutionary changes in agriculture are going to be critical to our ability to provide food for the increasing world population. Researchers and practitioners in agricultural production will benefit from their acquired knowledge and experience with traditional and innovative methodologies in operations research, but they nevertheless will face difficult challenges as they apply these tools to create practical solutions that will be effective and workable in a context of new technologies, changing human needs, environmental transitions, and evolving political factors.

1.6.2 Humanitarian Relief in Natural Disasters (Battini et al. 2014)

Humanitarian relief operations play an increasingly important role in a world stressed by population growth, urban residential density, natural resource use and depletion, global warming, and economic and political factors. Urgent humanitarian needs occur in places where food, water and medical supplies are constantly in demand, requiring routine and sustainable distribution of supplies to save lives and mitigate human suffering. In such situations, analysts regularly study available data to assess the needs, identify sources for supplies, evaluate transportation options, and plan for timely and predictable delivery of appropriate supplies to the most critically vulnerable and to those most urgently in need.

Even greater logistical challenges are presented when natural disasters occur (Wex et al. 2014). Earthquakes, floods, hurricanes, tsunamis, and fires, for example, often cause sudden and immediate injuries and loss of life, destruction of basic shelter perhaps requiring evacuation and relocation of victims, and interruption of normal availability of food and supplies. And in just such circumstances, relief operations may be seriously hampered: analysts may have only limited access to reliable information with which to identify the locations where rescue crews are needed, the extent of injury and destruction, the status of resources and supplies, and the usability of various modes of transport. Communications, water, and basic elements of infrastructure may have fallen prey to the disaster, and local decision making may have become impossible. Managing and executing the logistics of an efficient humanitarian supply chain in response to emergency needs arising from natural catastrophic destruction is an enormous and complex challenge.

In 2010, Haiti experienced a devastating earthquake that measured 7.2 on the Richter scale. Casualties were high with approximately a quarter of a million people killed and an even greater number of injured survivors. More than half of government and school buildings in Port-au-Prince and in the south and west districts of the country were destroyed or damaged. Financial loss related to the quake exceeded Haiti's entire 2009 gross domestic product. Overall, nearly 3.5 million people were affected by this catastrophic event.

Relief efforts typically begin by dispatching emergency rescue units into the areas of destruction, with the aim of reducing casualties and identifying longer term needs. Indeed, relief teams arrived in Haiti from various agencies such as the United Nations, International Red Cross and Red Crescent, the World Food Programme, and UNICEF. Their immediate focus was on delivering temporary shelter such as: blankets, tents, tarpaulins, and mosquito nets; food kits and water cans; and sanitation/hygiene kits.

The Haitian transport infrastructure was reported to have been very weak even before the earthquake hit, therefore delivering supplies through uncharted damaged areas to the earthquake victims was a difficult challenge. A plan to distribute relief supplies had to be devised, but as is often the case in the humanitarian field, data was incomplete or non-existent. A preliminary step in providing humanitarian aid is to find a means of collecting data, defining the type, extent, and locations in need of help, assessing the status of communication and transportation systems, and identifying sources capable of providing food and supplies and knowledgeable emergency staff personnel.

Pre-existing road network data were helpful in identifying all available routes and the current condition of roads. And from an inventory of available fleet vehicles (trucks and helicopters), it was possible to determine the cost of operation of each type of vehicle, which ones were undamaged, where they were currently located, and estimates of the time required for each type of vehicle to follow each available route.

Through cooperation among the agencies, food kits and hygiene kits were packaged in containers of the same size and shape for ease of transport, storage and delivery. Although food supplies were provided by different agencies than were the hygiene supplies, the uniformly shaped kits could be efficiently stacked and mixed together arbitrarily on the different types of delivery vehicles as needed.

Research analysts had already developed an elaborate network routing model to describe the logistics of general distribution processes, and this previous work was successfully amended to address the Haitian disaster requirements. The purpose ultimately is to find the best and most efficient possible way to deliver supplies to meet the needs of disaster victims; and this was accomplished by varying the type and number of vehicles allocated to achieve the lowest cost distribution plan. Further modifications to the model allowed for consideration of changes in the availability of supplies at their source (based on when and how much assistance could be mustered by the international agencies) and changes in the expected number of people assisted for each delivery to a given location.

Based on the acquired data, the demand for supplies, and the operational constraints, a mathematical model was developed, and was optimized for the Haitian earthquake scenario. The complex problem described in this way was then expressed in a special form using the GAMS modeling system, and was solved with CPLEX software executed on an Intel-based PC running Windows 7. The computational results were then interpreted by analysts, and a distribution plan was created to guide the efficient and effective delivery of food and supplies.

Analytical approaches have been applied to some extent in the past to create workable distribution systems. However, modern research to significantly improve the effectiveness and efficiency of relief operations is relatively recent. Although every natural disaster presents its own characteristic details, it has been shown here that the modeling and computational tools developed in basic Operations Research can be adapted to the specific needs of distributing available supplies for humanitarian relief in the wake of a natural disaster.

1.6.3 Mining and Social Conflicts (Charles 2016)

Peru has become one of the best performing economies in Latin America during recent years. Peru's model of economic growth has been driven by its mining industry and the associated potential for remarkable productivity. This country contains approximately 22% of the world's silver, 13% of copper reserves, and smaller but globally significant percentages of zinc, lead, tin, and gold reserves. Productivity is high, with Peru being the world's third largest producer of copper, silver, and zinc.

Although investment commitments in mining operations increased and reached over $40 billion during the period 2011–2016 to support a portfolio of mining projects, there have nevertheless been delays or lapses in implementing many of the projected mining activities. The delays have frequently been related to uncertainties involving social issues and conflicts. Local community concerns seem to be centered around environmental issues such as contamination of land and water, and the failure to improve everyday services such as health and education for the local populations. Local communities had anticipated greater benefits and services to accrue from the lucrative mining industry, but were disappointed by the lack of actual and apparent improvements in their daily living. Peru's wealth of natural mineral resources did not seem to have transformed and enriched the social and environmental structure nor brought to Peruvian communities the general prosperity that had been hoped for.

Poor communication and a perceived mismatch between mining priorities and social concerns led to conflicts that have resulted in the inability of some mines to continue operations. Through the years, many attempts were made to resolve conflicts, including forcing consultation between indigenous communities and the mining industry concerning infrastructure impact prior to mine development. It became evident that the various parties held vastly divergent perspectives on underlying problems. As an example, some Peruvian communities view the land as sacred, so any disturbance or relocation due to mining activities is considered a sacrilege and yet seems to be an inevitable aspect of any possible economic development.

Interactions among conflicting parties were mired in a complex mixture of misunderstandings, ambiguities, uncertainties, and insensitivities, so that attempts at meaningful communication and cooperation were often unsuccessful. Expectations and perceptions were so unclear or at cross-purposes that goals and objectives could not be well defined. The traditional mathematical modeling tools and established practices of Operations Research were useful only for studying specific and narrow avenues for progress in rather small contexts, but proved ineffective in addressing and overcoming most of the larger and more difficult issues.

With initiatives from the CENTRUM Católica Graduate Business School, some reasonable approaches were defined and followed to try to deal with the unstructured aspects of the dilemma. It had been generally assumed that mining companies that were perceived as having a more socially and environmentally responsible position were less likely to be involved in social conflicts, but analysts initially found little hard data to support or clarify this perception. Mining firms file corporate social responsibility (CSR) reports annually or periodically over many years, but the content of these reports had not been analyzed to determine the companies' actual commitment and discipline in adhering to the stated strategies. The job of reviewing the huge volume of accumulated reports was overwhelming, but the challenge was addressed by CENTRUM in collaboration with Cornell University. This team of researchers cooperatively devised machine learning approaches to extract data from the CSRs for analyzing and profiling the mining companies' practical commitment to sustainability. Preliminary results of this analysis proved to be an extremely important first big step toward matching actual practice with the ideals of sustainability.

In an effort to better understand and address socio-cultural issues, these analysts identified the following constituencies whose positions needed to be heard:

- Local communities and their needs for water, land, and respect for their cultural values
- Mining and associated industries and companies
- Government and state organizations
- Environmentalists with credible environmental constraints assisted and advised by technical innovation centers which included experts in Operations Research

Perhaps for the first time, researchers were able to take actions to help define and state the needs, expectations, goals, and tolerances of each of these constituencies. A platform was created for stating and discussing each group's ideals, and for comparing ideals versus currently existing conditions. By formally allowing and facilitating interaction among the various parties involved, it became possible to encourage cooperative analysis of feasible and desirable changes that could be made in the mining industry.

Conflicts based on uncertainty and misunderstandings were now being replaced by meaningful discussions aimed toward structuring and realistically conceptualizing the problems and goals expressed by both the mining companies and the local communities. With better understanding all around, and with well founded expectations for continued further progress, it is hoped that future collaboration will lead to formulating new models for solving the technical problems in operations, economics, social order, and sustainability for the development of Peru's natural resources.

1.7 Summary

Operations Research consists of the use of quantitative methods for analysis, optimization, and decision-making. The ideas and methods of Operations Research began to take shape during World War II, and thereafter have been put to good use in a wide variety of industrial, financial, government, nonprofit, and scientific endeavors.

Central to the theory and practice of Operations Research is the use of mathematical models to represent real systems or processes. A skillfully constructed model embodies enough of the details of the real entity being modeled so that it captures the essential characteristics of the entity, yet is simple enough so that the model can be studied using standard analytical techniques. In addition, successful modeling depends on a human analyst's knowledge, experience, intuition, and good judgment.

Algorithms are computational processes that can be applied to the structures within mathematical models. The performance of algorithms is often measured by the amount of computer time required to apply the algorithm. Depending on the type of problem being solved, algorithms may execute very rapidly (efficiently), or their execution may take so long that the algorithm is essentially worthless for actual problems. This book makes a special point of indicating, where possible, just what level of performance can be expected of each of the computational methods presented in this and subsequent chapters.

Many algorithms are designed to solve their targeted problems perfectly; but with imperfect or incomplete models and uncertain data, and the limited numerical accuracy of computer hardware, it should be recognized that it may be more sensible and easily justifiable to develop problem solutions that are less than optimal, but adequate for a given application. It may be necessary to compromise the *quality* of solutions in order to obtain solutions within a reasonable amount of computation time.

Key Terms

accumulated round-off error
algebraic modeling languages
algorithm
average case performance
best case performance
big-Oh notation
decision making
exponential-time algorithms
mathematical model
mathematical optimization
mathematical programming
model
NP-complete
NP-hard
polynomial-time algorithms
problem size
round-off error
solvable problems
solvers
step count
undecidable problems
unsolvable problems
worst case performance

References and Suggested Readings

Adam, E. E., and R. J. Ebert. 1992. *Production and Operations Management: Concepts, Models, and Behavior.* Englewood Cliffs, NJ: Prentice Hall.

Assad, A. A., and S. I. Gass. 2011. *Profiles in Operations Research: Pioneers and Innovators,* Vol. 147. New York: Springer Science & Business Media.

Baase, S., and A. van Gelder. 2000. *Computer Algorithms: Introduction to Design and Analysis,* 3rd ed. Reading, MA: Addison-Wesley.

Balci, O. 2014. *Computer Science and Operations Research: New Developments in Their Interfaces.* New York: Elsevier.

Barr, R. S., R. V. Helgason, and J. L. Kennington. 1997. *Interfaces in Computer Science and Operations Research.* Boston, MA: Kluwer Academic.

Battini, D., U. Peretti, A. Persona, and F. Sgarbossa. 2014. Application of humanitarian last mile distribution model. *Journal of Humanitarian Logistics and Supply Chain Management* 4 (1): 131–148.

Bhargava, H. K., and R. Krishnan. 1993. Computer-aided model construction. *Decision Support Systems* 9 (1): 91–111.

Buffa, E. S. 1981. *Elements of Production/Operations Management.* New York: John Wiley & Sons.

Byrum, J. 2015. Agriculture: Fertile ground for analytics and innovation. *OR/MS Today* 42 (6): 28–31.

Charles, V. 2016. Mining and mitigating social conflicts in Peru. *OR/MS Today* 43 (2): 34–38.

Cheney, E. W., and D. Kincaid. 2013. *Numerical Mathematics and Computing,* 7th ed. Boston, MA: Thompson Brooks Cole.

Chong, E. K. P., and S. H. Zak. 2013. *An Introduction to Optimization,* 4th ed. New York: Wiley.

Clauss, F. J. 1997. The trouble with optimal. *OR/MS Today* 24 (1): 32–35.

Cochran, J. J., L. A. Jr. Cox, P. Keskinocak, J. P. Kharoufeh, and J. Cole Smith. 2011. *Wiley Encyclopedia of Operations Research and Management Science, 8 Volume Set.* New York: Wiley.

Connell, J. L., and L. Shafer. 1987. *The Professional User's Guide to Acquiring Software.* New York: Van Nostrand Reinhold.

Cook, T. M., and R. A. Russell. 1989. *Introduction to Management Science.* Englewood Cliffs, NJ: Prentice-Hall.

Cormen, T. H., C. E. Leiserson, R. Rivest, and C. Stein. 2009. *Introduction to Algorithms,* 3rd ed. Cambridge, MA: MIT Press.

Czyzyk, J., J. H. Owen, and S. J. Wright. 1997. Optimization on the Internet. *OR/MS Today* 24: 48–51.

Dannenbring, D., and M. Starr. 1981. *Management Science: An Introduction.* New York: McGraw-Hill.

Demirci, M. 1996. IMSL C numerical libraries, version 2.0. *Computer* 29: 100–102.

Ecker, J. G., and M. Kupferschmid. 1988. *Introduction to Operations Research.* New York: John Wiley & Sons.

Fabrycky, W. J., P. M. Ghare, and P. E. Torgersen. 1984. *Applied Operations Research and Management Science.* Englewood Cliffs, NJ: Prentice-Hall.

Fourer, R. 1996. Software for optimization: A buyer's guide. *INFORMS Computer Science Technical Section Newsletter* 17 (1 and 2): 14–17.

Fourer, R. 1998. Software for optimization: A survey of recent trends in mathematical programming systems. *OR/MS Today* 25: 40–43.

Fourer, R., D. M. Gay, and B. W. Kernighan. 1993. *AMPL: A Modeling Language for Mathematical Programming.* South San Francisco, CA: The Scientific Press.

Garey, M. R., and D. S. Johnson. 1979. *Computers and Intractability: A Guide to the Theory of NP Completeness.* San Francisco, CA: W. H. Freeman Press.

Gass, S. 1987. Managing the modeling process: A personal reflection. *European Journal of Operational Research* 31: 1–8.

Gass, S., and A. Assad. 2004. *An Annotated Timeline of Operations Research: An Informal History.* New York: Springer.

Gass, S., H. Greenberg, K. Hoffman, and R. W. Langley (Eds.). 1986. *Impacts of Microcomputers on Operations Research*. New York: North-Holland.

Geoffrion, A. M. 1987. An introduction to structured modeling. *Management Science* 33: 547–588.

Gould, F. J., G. D. Eppen, and C. P. Schmidt. 1991. *Introductory Management Science*, 3rd ed. Englewood Cliffs, NJ: Prentice-Hall.

Greenberg, H., and F. H. Murphy. 1992. A comparison of mathematical programming modeling systems. *Annals of Operations Research* 38: 177–238.

Greenberg, H., and F. H. Murphy. 1995. Views of mathematical programming models and their instances. *Decision Support Systems* 13 (1): 3–34.

Gupta, S. K., and J. M. Cozzolino. 1974. *Fundamentals of Operations Research for Management: An Introduction to Quantitative Methods*. San Francisco, CA: Holden-Day.

Hall, O. P., Jr. 1993. *Computer Models for Operations Management*, 2nd ed. Reading, MA: Addison-Wesley.

Hein, J. L. 1995. *Discrete Structures, Logic, and Computability*. Boston, MA: Jones and Bartlett.

Hillier, F. S., and G. J. Lieberman. 2010. *Introduction to Operations Research*, 9th ed. Boston, MA: McGraw-Hill.

Horner, P. 2002. History in the making. *OR/MS Today* 29 (5): 30–39.

Howard, R. A. 2001. The ethical OR/MS professional. *Interfaces* 31 (6): 69–82.

Lenstra, J. K., A. H. G. Rinnooy Kan, and A. Schrijver. 1991. *History of Mathematical Programming: A Collection of Personal Reminiscences*. New York: Elsevier/North Holland.

Lougee-Heimer, R. 2008. COIN-OR in 2008. *OR/MS Today* 35: 46.

Manber, U. 1989. *Introduction to Algorithms: A Creative Approach*. Reading, MA: Addison-Wesley.

Matula, D. W. 1986. Arithmetic for microcomputers—Some recent trends. In S. I. Gass, H. J. Greenberg, L. L. Hoffman, and R. W. Langley (Eds.), *Impacts of Microcomputers on Operations Research*. New York: Elsevier.

McArthur, C. W. 1990. *Operations Analysis in the U.S. Army Eighth Air Force in World War II*. Providence, RI: American Mathematical Society.

McCloskey, J. F. 1987. The beginnings of operations research: 1934–1941. *Operations Research* 35 (1): 143–151.

More, J. J., and S. J. Wright. 1993. *Optimization Software Guide*. Philadelphia, PA: SIAM Publications.

Morse, P. M. 1986. The beginnings of operations research in the United States. *Operations Research* 34 (1): 10–17.

Murphy, F. H. 2005. ASP, The art and science of practice: Elements of the practice of operations research: A framework. *Interfaces* 35 (2): 154–163.

Murty, K. G. (Ed.). 2015. *Case Studies in Operations Research: Applications of Optimal Decision Making*. New York: Springer.

Orchard-Hays, W. 1978. History of mathematical programming systems. In H. J. Greenberg (Ed.), *Design and Implementation of Optimization Software*. Alphen aan den Rijn, the Netherlands: Sijthoff and Noordhoff.

Pidd, M. 1999. Just modeling through: A rough guide to modeling. *Interfaces* 29 (2): 118–132.

Ragsdale, C. T. 1998. *Spreadsheet Modeling and Decision Analysis: A Practical Introduction to Management Science*, 2nd ed. Cincinnati, OH: Southwestern College Publishing.

Ravindran, A. (Ed.). 2008. *Operations Research Applications*. Boca Raton, FL: CRC Press.

Ravindran, A., D. T. Phillips, and J. J. Solberg. 1987. *Operations Research: Principles and Practice*. New York: John Wiley & Sons.

Salvendy, G. (Ed.). 1982. *Handbook of Industrial Engineering*. New York: John Wiley & Sons.

Sauer, T. 2011. *Numerical Analysis*, 2nd ed. Boston, MA: Addison Wesley Longman.

Sharda, R., and G. Rampal. 1995. Algebraic modeling languages on PCs. *OR/MS Today* 22 (3): 58–63.

Taha, H. A. 2011. *Operations Research: An Introduction*, 9th ed. Upper Saddle River, NJ: Pearson.

Wagner, H. M. 1975. *Principles of Operations Research with Applications to Managerial Decisions.* Englewood Cliffs, NJ: Prentice-Hall.

Wex, F., G. Schryen, S. Feuerriegal, and D. Neumann. 2014. Emergency response in natural disaster management: Allocation and scheduling of rescue units. *European Journal of Operational Research* 235: 697–708.

White, D. J. 1985. *Operational Research.* New York: John Wiley & Sons.

Wilkinson, J. H. 1963. *Rounding Errors in Algebraic Processes.* Englewood Cliffs, NJ: Prentice-Hall.

Willemain, T. R. 1994. Insights on modeling from a dozen experts. *Operations Research* 42 (2): 213–222.

Williams, H. P. 1999. *Model Building in Mathematical Programming,* 4th ed. New York: Wiley.

Winston, W. L. 2004. *Operations Research: Applications and Algorithms,* 4th ed. Boston, MA: Brooks/ Cole.

2

Linear Programming

2.1 The Linear Programming Model

Linear programming is a special class of mathematical programming models in which the **objective function** and the **constraints** can be expressed as *linear* functions of the decision variables. As with the more general mathematical programming models, the decision variables represent quantities that are, in some sense, controllable inputs to the system being modeled. An **objective function** represents some principal objective criterion or goal that measures the effectiveness of the system (such as maximizing profits or productivity, or minimizing cost or consumption). There is always some practical limitation on the availability of resources (time, materials, machines, energy, or manpower) for the system, and such **constraints** are expressed as linear inequalities or equations involving the decision variables. *Solving* a linear programming problem means determining actual values of the decision variables that optimize the objective function, subject to the limitations imposed by the constraints.

The use of linear programming models for system optimization arises quite naturally in a wide variety of applications. Some models may not be strictly linear, but can be made linear by applying appropriate mathematical transformations. Still other applications are admittedly not at all linear, but can be effectively approximated by linear models. The ease with which linear programming problems can usually be solved makes this an attractive means of dealing with otherwise intractable nonlinear problems.

In the following section, we will see examples of the wide variety of applications that can be modeled with linear programming. In each case, the first task will be to identify the controllable decision variables x_i, where $i = 1, \ldots, n$. Then the objective criterion will be established: to either maximize or minimize some function of the form

$$z = c_1 x_1 + c_2 x_2 + \cdots + c_n x_n = \sum_{i=1}^{n} c_i x_i$$

where c_i represents problem dependent constants. Finally, resource limitations and bounds on decision variables will be written as equations or inequalities relating a linear function of the decision variables to some problem dependent constant; for example,

$$a_1 x_1 + a_2 x_2 + \cdots + a_n x_n \leq b$$

Although the primary purpose of this chapter will be to present methods of solving linear programming problems, the first critical step in successful problem-solving lies in the correct **formulation** of an application problem into the linear programming framework.

2.2 The Art and Skill of Problem Formulation

A combination of practical insight and technical skill is required in order to recognize *which* problems can be appropriately modeled in a linear programming format, and then to formulate those problems accurately. Because of the wide variety of problems that can be made to fall into the linear programming mold, it is difficult to give guidelines that are universally applicable to the process of problem formulation. Rather, problem formulation is an art that must be cultivated through practice and experience. Several examples are given to point the way, and to illustrate the creativity that is sometimes helpful in framing problems as linear programs. The exercises at the end of the chapter should then provide some of the practice necessary to develop the skill of formulating linear programming models.

Example 2.2.1

A manufacturer of computer system components assembles two models of wireless routers, model A and model B. The amounts of materials and labor required for each assembly, and the total amounts available, are shown in the following table. The profits that can be realized from the sale of each router are $22 and $28 for models A and B, respectively, and we assume there is a market for as many routers as can be manufactured.

	Resources Required per Unit		Resources Available
	A	**B**	
Materials	8	10	3400
Labor	2	3	960

The manufacturer would like to determine how many of each model to assemble in order to maximize profits.

Formulation 2.2.1

Because the solution to this problem involves establishing the number of routers to be assembled, we define the decision variables as follows:

Let x_A = number of model A routers to be assembled

and

x_B = number of model B routers to be assembled

In order to maximize profits, we establish the objective criterion as

$$\text{maximize } z = 22x_A + 28x_B$$

Two types of resource limitations are in effect. The materials constraint is expressed by the inequality

$$8x_A + 10x_B \leq 3400$$

and the labor constraint by

$$2x_A + 3x_B \leq 960$$

Finally, as it would be meaningless to have a negative number of terminals manufactured, we also include the constraints $x_A \geq 0$ and $x_B \geq 0$.

Example 2.2.2

A space agency planning team wishes to set up a schedule for launching satellites over a period of three years. Experimental payloads are of two types (say, T1 and T2), and each launch carries only one experiment. Externally negotiated agency policies dictate that at most 88 of payload type T1 and 126 of type T2 can be supported. For each launch, type T1 payloads will operate successfully with probability 0.85 and type T2 payloads with probability 0.75. In order for the program to be viable, there must be a total of at least 60 successful deployments. The agency is paid $1.5 million for each successful T1 payload, and $1.2 million for each successful T2 payload. The costs to the agency to prepare and launch the two types of payloads are $1.05 million for each T1 and $0.7 million for each T2. One week of time must be devoted to the preparation of each T2 launch payload and two weeks are required for T1 launch payloads. The agency, while providing a public service, wishes to maximize its expected net income from the satellite program.

Formulation 2.2.2

Let x_1 = number of satellites launched carrying a type T1 payload, and x_2 = number of satellites launched carrying a type T2 payload. Income is realized only when launches are successful, but costs are incurred for all launches. Therefore, the expected net income is

$$(1.5)(0.85)x_1 + (1.2)(0.75)x_2 - (1.05)x_1 - (0.7)x_2 \text{million dollars}$$

The objective is then to maximize $z = 0.225x_1 + 0.2x_2$. Problem constraints in this case are of various types. Agency policies impose the two simple constraints

$$x_1 \leq 88 \text{ and } x_2 \leq 126$$

The successful deployment quota yields the constraint

$$0.85x_1 + 0.75x_2 \geq 60$$

If we assume that 52 weeks per year (for three years) can be applied to the satellite program, then the launch preparation time constraint is

$$2x_1 + 1x_2 \leq 156$$

As in the previous example, we include the non-negativity constraints $x_1 \geq 0$ and $x_2 \geq 0$.

Example 2.2.3

A company wishes to minimize its combined costs of production and inventory over a four-week time period. An item produced in a given week is available for consumption during that week, or it may be kept in inventory for use in later weeks. Initial inventory at the beginning of week 1 is 250 units. The minimum allowed inventory carried from one week to the next is 50 units. Unit production cost is $15, and the cost of storing a unit from one week to the next is $3. The following table shows production capacities and the demands that must be met during each week.

Period	Production Capacity	Demand
1	800	900
2	700	600
3	600	800
4	800	600

A minimum production of 500 items per week must be maintained. Inventory costs are not applied to items remaining at the end of the fourth production period, nor is the minimum inventory restriction applied after this final period.

Formulation 2.2.3

Let x_i be the number of units produced during the i-th week, for $i = 1, \ldots, 4$. The formulation is somewhat more manageable if we let A_i denote the number of items remaining at the end of each week (accounting for those held over from previous weeks, those produced during the current week, and those consumed during the current week). Note that the A_i values are not decision variables, but merely serve to simplify our written formulation. Thus,

$$A_1 = 250 + x_1 - 900$$

$$A_2 = A_1 + x_2 - 600$$

$$A_3 = A_2 + x_3 - 800$$

$$A_4 = A_3 + x_4 - 600$$

The objective is to minimize

$$z = \$15 \cdot (x_1 + x_2 + x_3 + x_4) + \$3 \cdot (A_1 + A_2 + A_3)$$

Minimum inventory constraints are expressed as $A_i \geq 50$ for $i = 1, 2,$ and 3, and $A_4 \geq 0$. Production capacities and minima during each period are enforced with the constraints

$$500 \leq x_1 \leq 700$$

$$500 \leq x_2 \leq 700$$

$$500 \leq x_3 \leq 600$$

$$500 \leq x_4 \leq 800$$

Finally, $x_i \geq 0$ for $i = 1, \ldots, 4$.

Example 2.2.4

A mixture of freeze-dried vegetables is to be composed of beans, corn, broccoli, cabbage, and potatoes. The mixture is to contain (by weight) at most 40% beans and at most 32% potatoes. The mixture should contain at least 5 grams iron, 36 grams phosphorus, and 28 grams calcium. The nutrients in each vegetable and the costs are shown in the following table.

Vegetable	Milligrams Nutrient per Pound of Vegetable			Cost per Pound (cents)
	Iron	Phosphorus	Calcium	
Beans	0.5	10	200	20
Corn	0.5	20	280	18
Broccoli	1.2	40	800	32
Cabbage	0.3	30	420	28
Potatoes	0.4	50	360	16

The amount of each vegetable to include should be determined so that the cost of the mixture is minimized.

Formulation 2.2.4

Let x_1, x_2, x_3, x_4, and x_5 be the number of pounds of beans, corn, broccoli, cabbage, and potatoes, respectively. To minimize the cost of the mixture, we wish to minimize $z = 20x_1 + 18x_2 + 32x_3 + 28x_4 + 16x_5$. The percentage of beans in the mixture is $x_1/(x_1 + x_2 + x_3 + x_4 + x_5)$, and must be less than 40%. Therefore,

$$x_1 \leq 0.4\left(x_1 + x_2 + x_3 + x_4 + x_5\right)$$

and similarly the potato restriction can be written as

$$x_5 \leq 0.32(x_1 + x_2 + x_3 + x_4 + x_5)$$

To achieve the required level of nutrients, we have three constraints (for iron, phosphorus, and calcium, respectively):

$$0.5x_1 + 0.5x_2 + 1.2x_3 + 0.3x_4 + 0.4x_5 \geq 5000$$

$$10x_1 + 20x_2 + 40x_3 + 30x_4 + 50x_5 \geq 36{,}000$$

$$200x_1 + 280x_2 + 800x_3 + 420x_4 + 360x_5 \geq 28{,}000$$

Negative amounts are not possible, so $x_i \geq 0$ for $i = 1, \ldots, 5$.

Example 2.2.5

A saw mill makes two products for log home kits: fir logs and spruce logs which can be sold at profits of $4 and $5, respectively. Spruce logs require two units of processing time on the bark peeler and six units of time on a slab saw. Fir logs require three units of time on the peeler and five units on the slab saw. Each then requires two units of time on the planer. Because of maintenance requirements and labor restrictions, the bark peeler is available 10 hours per day, the slab saw 12 hours per day, and the planer 14 hours per day. Bark and sawdust are by-products of these operations. All the bark can be put through a chipper and sold in unlimited quantities to a nearby horticulture supplier. Dried fir sawdust can be directed to a similar market, at a net profit of $0.38 per processed log. Limited amounts of the spruce sawdust can be made into marketable pressed wood products, but the rest must be destroyed. For each spruce log produced, enough sawdust (five pounds) is generated to make three pressed wood products, which after manufacturing can be sold at a unit profit of $0.22. However, the market can absorb only 60 of the pressed wood products per day and the remaining spruce sawdust must be destroyed at a cost of $0.15 per pound. The saw mill wishes to make the largest possible profit, considering the cost of destroying the unusable sawdust.

Formulation 2.2.5

The formulation of this problem cannot follow exactly the pattern established in previous examples because the profits to be maximized are not a linear function of the number of logs of each type produced. Spruce log production creates a by-product that is useful and profitable only up to a point, and thereafter any excess must be destroyed at a cost that diminishes total profits. Thus, profits are not a strictly increasing function of production levels. We can still let

$$x_1 = \text{number of fir logs produced}$$

$$x_2 = \text{number of spruce logs produced}$$

Because sawdust contributes nonlinearly to profits, we treat it in two parts and let

$$x_3 = \text{number of pounds of spruce sawdust used}$$

$$x_4 = \text{number of pounds of spruce sawdust destroyed}$$

Direct profit from the sale of logs is $4x_1 + 5x_2$. All the bark can be sold at a profit in unlimited quantities, therefore, although this affects the amount of profit, it does not affect our decision on how many logs of each type to produce. Fir sawdust brings in $0.38 for each processed log, or $0.38x_1$. For each $x_3/5$ spruce logs produced, there is enough sawdust to make three products at a profit of $0.22 each, if there is a market. Unmarketable spruce sawdust costs $0.15x_4$ to destroy. The objective is, therefore, to maximize

$$z = 4x_1 + 5x_2 + 0.38x_1 + \frac{3}{5}(0.22)x_3 - 0.15x_4$$

Relating the number of logs produced to pounds of sawdust by-product, we obtain the constraint

$$5x_2 = (x_3 + x_4)$$

Limitations on demand for the pressed wood product are expressed by

$$\frac{3}{5}x_3 \le 60$$

Constraints on availability of machinery are straightforward. For the bark peeler,

$$3x_1 + 2x_2 \le 10$$

On the slab saw,

$$5x_1 + 6x_2 \le 12$$

And on the planer,

$$2x_1 + 2x_2 \le 14$$

Because all production levels are non-negative, we also require $x_1 \ge 0$, $x_2 \ge 0$, $x_3 \ge 0$, and $x_4 \ge 0$.

Example 2.2.6

A dual processor computing facility is to be dedicated to administrative and scientific application jobs for at least 10 hours each day. Administrative jobs require two seconds of execution time on processor 1 and six seconds on processor 2, while scientific jobs require five seconds on processor one and three seconds on processor 2. A scheduler must choose how many of each type of job (administrative and scientific) to execute, in such a way as to minimize the amount of time that the system is occupied with these jobs. The system is considered to be occupied even if one processor is idle. (Assume that the *sequencing* of the jobs on each processor is not an issue here, just the selection of how many of each type of job.)

Formulation 2.2.6

Let x_1 and x_2 denote, respectively, the number of administrative and scientific jobs selected for execution on the dual processor system. Because policies require that each processor be available for a least 10 hours, we must write the two constraints as:

$$2x_1 + 5x_2 \ge 10 \cdot (3600) \quad \text{(Processor 1)}$$

$$6x_1 + 3x_2 \ge 10 \cdot (3600) \quad \text{(Processor 2)}$$

and

$$x_1 \ge 0 \text{ and } x_2 \ge 0$$

The system is considered occupied as long as either processor is busy. Therefore, to minimize the completion time for the set of jobs, we must

$$\text{minimize} \left\{ \text{maximum} \left(2x_1 + 5x_2, 6x_1 + 3x_2 \right) \right\}$$

This nonlinear objective can be made linear if we introduce a new variable x_3, where

$$x_3 = \max \left\{ 2x_1 + 5x_2, 6x_1 + 3x_2 \right\} \ge 0$$

Now if we require

$$x_3 \geq 2x_1 + 5x_2 \text{ and } x_3 \geq 6x_1 + 3x_2$$

and make our objective to minimize x_3, we have the desired linear formulation.

2.2.1 Integer and Nonlinear Models

There are many problems that appear to fall into the framework of linear programming problem formulations. In some problems, the decision variable values are meaningful only if they are *integer* values. (For example, it is not possible to launch a fractional number of satellites or to transport a fractional number of people.) However, general approaches to the solution of linear programming problems in no way guarantee integer solutions. The analyst must therefore be familiar enough with the actual application to determine whether it will be acceptable to round off a continuous (non-integer) **optimal solution** to an integer solution that may be suboptimal. In many applications, such practices yield solutions that are quite adequate. When rounding does not yield acceptable results, it may be necessary to resort to methods that are computationally more difficult than general linear programming solution methods, but which always yield integer solutions. Specialized methods for these cases will be introduced in Chapter 4 on Integer Programming.

More subtle nonlinearities exist inherently in almost all real applications. It is again left to the discretion of the analyst to determine whether the linear model can provide a sufficiently accurate approximation to the real situation. Because of the relative ease with which linear models can be solved, in some cases it may be worth making certain simplifying (albeit compromising) assumptions in order to formulate a real problem into a linear programming model.

2.3 Graphical Solution of Linear Programming Problems

2.3.1 General Definitions

Finding an optimal solution to a linear programming problem means assigning values to the decision variables in such a way as to achieve a specified goal and conform to certain constraints. For a problem with n decision variables, any **solution** can be specified by a point $(x_1, x_2, ..., x_n)$. The **feasible space** (or **feasible region**) for the problem is the set of all such points that satisfy the problem constraints. The feasible space is therefore the set of all **feasible solutions**. An **optimal feasible solution** is a point in the feasible space that is as effective as any other point in achieving the specified goal.

The solution of linear programming problems with only two decision variables can be illustrated graphically. In the following examples, we will see cases involving the maximization and minimization of functions. We will also see situations in which no feasible solution exists, some which have **multiple optimal solutions**, and others with no optimal solution.

Linear programming problems with more than two decision variables require more sophisticated methods of solution, and cannot be easily illustrated graphically. However, our graphical study of small problems will be helpful in providing insight into the more general solution method that will be presented later.

2.3.2 Graphical Solutions

Let us first consider a maximization problem:

maximize $\quad z = 3x_1 + x_2$

subject to $\quad$ (1) $x_2 \le 5$

$\quad\quad\quad\quad\quad$ (2) $x_1 + x_2 \le 10$

$\quad\quad\quad\quad\quad$ (3) $-x_1 + x_2 \ge -2$

$\quad\quad\quad\quad\quad x_1, x_2 \ge 0$

Each inequality constraint defines a half-plane in two dimensions, and the intersection of these half-planes comprises the feasible space for this case, as shown by the shaded area in Figure 2.1.

The points labeled A, B, C, D, and E are called **extreme points** of the feasible region. It is a property of linear programming problems that, if a unique optimal solution exists, it occurs at one of the extreme points of the feasible space.

For this small problem, it is not impractical simply to evaluate the objective function at each of these points, and select the maximum:

$$z_A = z(0,0) = 3 \times 0 + 0 = 0$$

$$z_B = z(0,5) = 3 \times 0 + 5 = 5$$

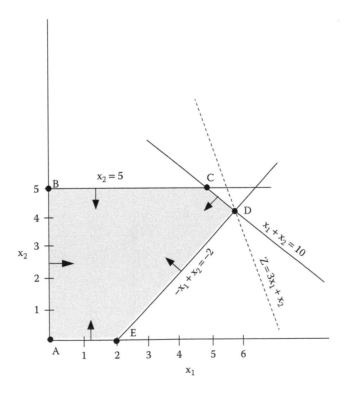

FIGURE 2.1
Graphical solution.

$$z_C = z(5,5) = 3 \times 5 + 5 = 20$$

$$z_D = z(6,4) = 3 \times 6 + 4 = 22$$

$$z_E = z(2,0) = 3 \times 2 + 0 = 6$$

The optimal solution lies at extreme point D where $x_1 = 6$ and $x_2 = 4$, and the optimal value of the objective function is denoted by $z^* = 22$.

Without evaluating z at every extreme point, we may more simply observe that the line specified by the objective function $3x_1 + x_2$ has a slope of -3. At optimality, this line is tangent to the feasible space at one of the extreme points. In Figure 2.1, the dashed line represents the objective function at the optimal point D.

Next, we use the same graphical technique to solve a minimization problem:

minimize $z = x_1 + x_2$

subject to $3x_1 + x_2 \geq 6$ (1)

$x_2 \geq 3$ (2)

$x_1 \leq 4$ (3)

$x_1, x_2 \geq 0$

The shaded area in Figure 2.2 denotes the feasible region, which in this case is unbounded.

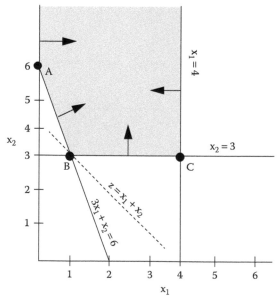

FIGURE 2.2
Unbounded feasible region.

The minimal solution must occur at one of the extreme points A, B, or C. The objective function $x_1 + x_2$, with a slope of -1, is tangent to the feasible region at extreme point B. Therefore, the optimal solution occurs at $x_1 = 1$ and $x_2 = 3$, and the optimal objective function value at that point is $z^* = 4$.

2.3.3 Multiple Optimal Solutions

Each of the problems that we have solved graphically had a unique optimal solution. The following example shows that it is possible for a linear programming problem to have multiple solutions that are all equally effective in achieving an objective. Consider the problem

$$\text{maximize} \qquad z = x_1 + 2x_2$$

$$\text{subject to} \qquad -x_1 + x_2 \le 2 \qquad (1)$$

$$x_1 + 2x_2 \le 8 \qquad (2)$$

$$x_1 \le 6 \qquad (3)$$

$$x_1, x_2 \ge 0$$

The feasible region is shown in Figure 2.3.

The line representing the objective function $x_1 + 2x_2$ can be made tangent to the feasible region at the origin, but clearly z is maximized by placing the line where the values of x_1 and x_2 are larger. Notice that the objective function line in this case is tangent to the feasible region not at a single extreme point, but rather along one of the boundaries of the feasible region.

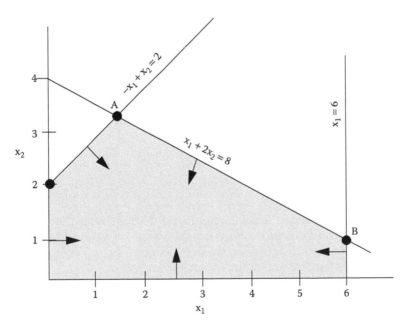

FIGURE 2.3
Multiple optimal solutions.

The values

$$z_A = z\left(\frac{4}{3}, \frac{10}{3}\right) = \frac{4}{3} + 2\left(\frac{10}{3}\right) = 8$$

and

$$z_B = z(6,1) = 6 + 2 \times (1) = 8$$

correspond to optimal solutions at points A and B; moreover, *all* points on the line between extreme points A and B are also optimal. Therefore, $z^* = 8$ and the optimal solutions can be expressed as a set

$$\left\{(x_1, x_2) \middle| \frac{4}{3} \le x_1 \le 6 \text{ and } 1 \le x_2 \le \frac{10}{3} \text{ and } x_1 + 2x_2 = 8\right\}$$

Such a situation may occur whenever the slope of the objective function line is the same as that of one of the constraints.

2.3.4 No Optimal Solution

When the feasible region is unbounded, a maximization problem may have no optimal solution, since the values of the decision variables may be increased arbitrarily. This can be illustrated by the problem:

$$\text{maximize} \quad z = 3x_1 + x_2$$

$$\text{subject to} \quad x_1 + x_2 \ge 4 \qquad\qquad (1)$$

$$-x_1 + x_2 \le 4 \qquad\qquad (2)$$

$$-x_1 + 2x_2 \ge -4 \qquad\qquad (3)$$

$$x_1, x_2 \ge 0$$

Figure 2.4 shows the unbounded feasible region and demonstrates that the objective function can be made arbitrarily large by allowing the values of x_1 and x_2 to grow within the unbounded feasible region. In this case, there is no point (x_1, x_2) that is optimal because there are always other feasible points for which z is larger.

Notice that it is not the unbounded feasible region alone that precludes an optimal solution. The *minimization* of the function subject to the constraints shown in Figure 2.4 would be solved at extreme point A.

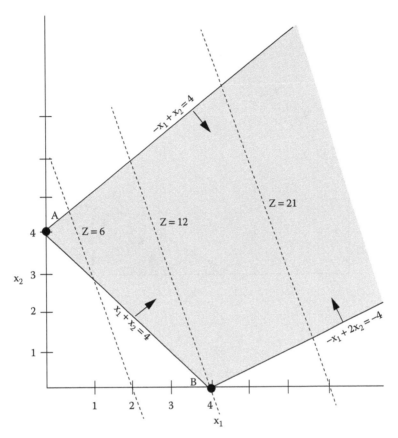

FIGURE 2.4
No optimal solution.

In practice, **unbounded solutions** typically arise because some real constraint, representing a practical resource limitation, has been omitted from the linear programming formulation. Because we do not realistically expect to be able to achieve unlimited profits or productivity, an indication of apparently unbounded solutions as seen in the previous example should be interpreted as evidence that the problem needs to be reconsidered more carefully, reformulated and re-solved.

2.3.5 No Feasible Solution

A linear programming problem has no feasible solution if the set of points corresponding to the feasible region is empty. For example, the constraints

$$-x_1 + x_2 \geq 4 \text{ and } -x_1 + 2x_1 \leq -4$$

where $x_1, x_2 \geq 0$, represent conditions that cannot simultaneously be satisfied by *any* point. Figure 2.5 shows the four half-planes whose intersection is empty.

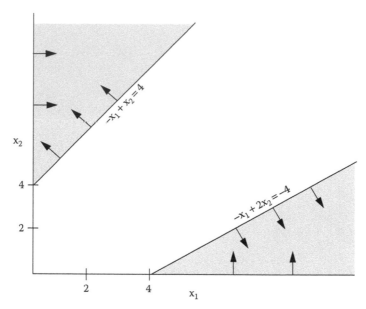

FIGURE 2.5
No feasible solution.

In small problems, infeasibilities such as this may be discovered visually during an attempted **graphical solution**. In larger problems, it may not be obvious, by inspecting a particular set of constraints, that no solution is possible. Fortunately, the general solution method to be described in the following sections is not only capable of solving typical maximization or minimization problems, but it also provides mechanisms for recognizing problems that have multiple optimal solutions, no optimal solution, or no feasible solution.

2.3.6 General Solution Method

We have seen in our graphical solutions that, if an optimal solution exists, it occurs at an extreme point of the feasible region. This fundamental property of linear programming problems is the foundation for a general solution method called the **Simplex method**. Because only the finitely many extreme points need be examined (rather than *all* the points in the feasible region), an optimal solution may be found systematically by considering the objective function values at the extreme points. In fact, in actual practice, only a small subset of the extreme points need be examined. The following sections will demonstrate how the Simplex method is able to locate optimal solutions with such efficiency.

2.4 Preparation for the Simplex Method

2.4.1 Standard Form of a Linear Programming Problem

In preparation for the use of the Simplex method, it is necessary to express the linear programming problem in **standard form**. For a linear program with n variables and m constraints, we will use the following standard form:

$$\begin{array}{ll}
\text{maximize} & z = c_1x_1 + c_2x_2 + \ldots + c_nx_n \\
\text{subject to} & a_{11}x_1 + a_{12}x_2 + \ldots + a_{1n}x_n = b_1 \\
& a_{21}x_1 + a_{22}x_2 + \ldots + a_{2n}x_n = b_2 \\
& \quad\vdots \qquad\qquad \vdots \qquad \vdots \\
& a_{m1}x_1 + a_{m2}x_2 + \ldots + a_{mn}x_n = b_m
\end{array}$$

where the variables $x_1, \ldots, x_n$ are non-negative, and the constants $b_1, \ldots, b_m$ on the right hand sides of the constraints are also non-negative. We can use matrix notation to represent the cost (or profit) vector $c = (c_1, c_2, \ldots, c_n)$ and the decision variable vector

$$x = \begin{bmatrix} x_1 \\ x_2 \\ \cdot \\ \cdot \\ \cdot \\ x_n \end{bmatrix}$$

The coefficient matrix is:

$$A = \begin{bmatrix} a_{11} \ldots a_{1n} \\ \cdot\cdot \\ \cdot\cdot \\ \cdot\cdot \\ a_{m1} \ldots a_{mn} \end{bmatrix}$$

and the requirement vector is:

$$b = \begin{bmatrix} b_1 \\ b_2 \\ \cdot \\ \cdot \\ \cdot \\ b_m \end{bmatrix}$$

Then the optimization problem can be expressed succinctly as:

$$\begin{array}{ll}
\text{maximize} & z = cx \\
\text{subject to} & Ax = b \\
& x \geq 0 \\
& b \geq 0
\end{array}$$

Although this standard form will be required by the Simplex method, it is not necessarily the form that arises naturally when we first formulate linear programming models. Several modifications may be necessary in order to transform an original linear programming formulation (as in Section 2.2) into the standard form.

To convert a minimization problem to a maximization problem, we can simply multiply the objective function by -1, and then maximize this function. (Recall that there are no sign restrictions on the c_i.) For example, the problem of minimizing $z = 3x_1 - 5x_2$ is equivalent to maximizing $z = -3x_1 + 5x_2$. Negative right hand sides of the constraints can be made positive by multiplying the constraint by -1 (reversing the sense of the inequality).

Equality constraints require no modification. Inequality constraints can be converted to equalities through the introduction of additional variables that make up the difference in the left and right sides of the inequalities. Less than or equal to ($\leq$) inequalities require the introduction of variables that we will call **slack variables**. For example, a constraint such as $3x_1 + 4x_2 \leq 7$ becomes the equality $3x_1 + 4x_2 + s_1 = 7$ when we introduce the slack variable s_1, where $s_1 \geq 0$. Greater than or equal to ($\geq$) constraints are modified by introducing **surplus variables**. For example, the constraint $14x_1 + 3x_2 \geq 12$ becomes the equality $14x_1 + 3x_2 - s_2 = 12$, where s_2 is the non-negative surplus variable. Although our notation (s_1 and s_2) may suggest otherwise, the slack and surplus variables are going to be treated exactly like any other decision variable throughout the solution process. In fact, their final values in the solution of the linear programming problem may be just as interesting to a systems manager or analyst as are the values of the original decision variables.

Finally, all variables are required to be non-negative in the standard form. In the event that the actual meaning associated with a decision variable is such that the variable should be unrestricted in sign, then that variable may be replaced by the *difference* of two new non-negative variables. For example, if x_1 is to be an unrestricted variable, then every occurrence of x_1 in the objective function or in any constraint will be replaced by $x_1' - x_1''$, where $x_1', x_1'', \geq 0$. Then in any solution, the sign of the value of x_1 is dependent on the relative values of x_1' and x_1''.

The reason for placing problems in standard form is that our general solution method will be seen to operate by finding and examining solutions to the system of linear equations $Ax = b$ (i.e., by finding values of the decision variables that are consistent with the problem constraints), with the aim of selecting a solution that is optimal with respect to the objective function.

2.4.2 Solutions of Linear Systems

We now have a system of linear equations, $Ax = b$, consisting of m equations and n unknowns. The n unknowns include the original decision variables *and* any other variables that may have been introduced in order to achieve standard form.

It may be useful at this point to review the material in the Appendix on solving systems of linear equations. If a system of independent equations has any solution, then $m \leq n$. If $m = n$ (and if rank $(A) = m$ and A is nonsingular), then there is the unique solution $x = A^{-1}b$. In this case, there is only one set of values for the x_i that is not in violation of problem constraints. Optimization of an objective function is not an issue here because there is only one feasible solution.

When m < n, there are infinitely many solutions to the system of equations. In this case, we have (n − m) **degrees of freedom** in solving the system. This means that we can arbitrarily assign *any* values to *any* (n − m) of the n variables, and then solve the m equations in terms of the remaining m unknowns.

A **basic solution** to the system of m equations and n unknowns is obtained by setting (n − m) of the variables to *zero*, and solving for the remaining m variables. The m variables that are not set equal to zero are called **basic variables**, and the (n − m) variables set to zero are **non-basic variables**. The number of basic solutions is just the number of ways we can choose n − m variables (or m variables) from the set of n variables, and this number is given by:

$$\binom{n}{n-m} = \binom{n}{m} = \frac{n!}{m!(n-m)!}$$

Not all of the basic solutions satisfy all problem constraints and non-negativity constraints. Those that do not meet these requirements are **infeasible solutions**. The ones that do meet the restrictions are called **basic feasible solutions**. An **optimal** basic feasible solution is a basic feasible solution that optimizes the objective function. The basic feasible solutions correspond precisely to the **extreme points** of the feasible region (as defined in our earlier discussion of graphical solutions). Because any optimal feasible solution is guaranteed to occur at an extreme point (and consequently is a basic feasible solution), the search for an optimal basic feasible solution could be carried out by an examination of the at most $\binom{n}{m}$ basic feasible solutions and a determination of which one yields the best objective function value.

The Simplex method performs such a search, but in a very efficient way. We define two extreme points of the feasible region (or two basic feasible solutions) as being **adjacent** if all but one of their basic variables are the same. Thus, a transition from one basic feasible solution to an adjacent basic feasible solution can be thought of as exchanging the roles of one basic variable and one non-basic variable. The Simplex method performs a sequence of such transitions and thereby examines a succession of **adjacent extreme points**. A transition to an adjacent extreme point will be made *only* if by doing so the objective function is improved (or stays the same). It is a property of *linear* programming problems that this type of search will lead us to the discovery of an optimal solution (if one exists). The Simplex method is not only successful in this sense, but it is remarkably efficient because it succeeds after examining only a fraction of the basic feasible solutions.

2.5 The Simplex Method

The Simplex method is a general solution method for solving linear programming problems. It was developed in 1947 by George B. Dantzig and, with some modifications for efficiency, has become *the* standard method for solving very large linear programming problems on computers. Most *real* problems are so large that a manual solution via the

Simplex method is impractical, and these problems must be solved with Simplex programs implemented on a computer. Small problems, however, are quite useful in demonstrating how the Simplex method operates; therefore, we will use such problems to illustrate the various features of the method.

The Simplex method is an iterative algorithm that begins with an initial feasible solution, repeatedly moves to a better solution, and stops when an optimal solution has been found and, therefore, no improvement can be made.

To describe the mechanics of the algorithm, we must specify how an initial feasible solution is obtained, how a transition is made to a better basic feasible solution, and how to recognize an optimal solution. From any basic feasible solution, we have the assurance that, if a better solution exists at all, then there is an *adjacent* solution that is better than the current one. This is the principle on which the Simplex method is based; thus, an optimal solution is accessible from any starting basic feasible solution.

We will use the following simple problem as an illustration as we describe the Simplex method:

$$\text{maximize} \qquad z = 8x_1 + 5x_2$$

$$\text{subject to} \qquad x_1 \leq 150$$

$$x_2 \leq 250$$

$$2x_1 + x_2 \leq 500$$

$$x_1,\ x_2 \geq 0$$

The standard form for this problem is:

$$\text{maximize} \qquad z = 8x_1 + 5x_2 + 0s_1 + 0s_2 + 0s$$

$$\text{subject to} \qquad x_1 + s_1 = 150$$

$$x_2 + s_2 = 250$$

$$2x_1 + x_2 + s_3 = 500$$

(Zero coefficients are given to the slack variables in the objective function because slack variables do not contribute to z.) The constraints constitute a system of m = 3 equations in n = 5 unknowns. In order to obtain an initial basic feasible solution, we need to select n − m = 5 − 3 = 2 variables as non-basic variables. We can readily see in this case that by choosing the two variables x_1 and x_2 as the non-basic variables, and setting their values to zero, then no significant computation is required in order to solve for the three basic variables: $s_1 = 150$, $s_2 = 250$, and $s_3 = 500$. The value of the objective function at this solution is 0.

In fact, a starting solution is just this easy to obtain whenever we have m variables, each of which has a coefficient of one in one equation and zero coefficients in all other equations (a unit vector of coefficients), and *each equation* has such a variable with a coefficient of one in it. Thus, whenever a slack variable has been added to each constraint, we may choose all the slack variables as the m basic variables, set the remaining (n − m) variables to zero, and the starting values of the basic variables are simply given by the constants b on the right hand sides of the constraints. (For cases in which slack variables are not present and, therefore, do not provide a starting basic feasible solution, further techniques will be discussed in Section 2.6.)

Once we have a solution, a transition to an adjacent solution is made by a pivot operation. A **pivot operation** is a sequence of elementary row operations (see the Appendix) applied to the current system of equations, with the effect of creating an equivalent system in which one new (previously non-basic) variable now has a coefficient of one in one equation and zeros in all other equations.

During the process of applying **pivot operations** to a linear programming problem, it is convenient to use a tabular representation of the system of equations. This representation is referred to as a **Simplex tableau.**

In order to conveniently keep track of the value of the objective function as it is affected by the pivot operations, we treat the objective function as one of the equations in the system of equations, and we include it in the tableau. In our example, the objective function equation is written as:

$$1z - 8x_1 - 5x_2 - 0s_1 - 0s_2 - 0s_3 = 0$$

The tableau for the initial solution is as follows:

Basis	z	x_1	x_2	s_1	s_2	s_3	Solution
Z	1	−8	−5	0	0	0	0
s_1	0	1	0	1	0	0	150
s_2	0	0	1	0	1	0	250
s_3	0	2	$\boxed{1}$	0	0	1	500

The first column lists the current basic variables. The second column shows that z is (and will always be) a basic variable; and because these elements will never change, they really do not need to be explicitly maintained in the tableau. The next five columns are the constraint coefficients of each variable. And the last column is the solution vector; that is, the values of the basic variables. Using this representation of a current solution, we can now describe the purpose and function of each iteration of the Simplex method for a maximization problem.

Observe that the objective function row represents an equation that must be satisfied for any feasible solution. Since we want to maximize z, some other (non-basic) term must *decrease* in order to offset the increase in z. But all of the non-basic variables are already at their lowest value, zero. Therefore, we want to increase some non-basic variable that has a *negative* coefficient. As a simple rule, we will choose the variable with the most negative coefficient, because making this variable basic will give the largest (per unit) increase in z. (Refer to Steps 1 and 2 in the following.)

The chosen variable is called the **entering variable,** that is, the one that will enter the basis. If this variable increases, we must adjust *all* of the equations. Specifically, increasing the non-basic variable must be compensated for by using only the one basic variable in each row (having a coefficient of one). If the non-basic coefficient is negative, the corresponding basic variable increases. There is no limit to how much we can increase this. Clearly, if all coefficients are negative (or zero), then we can increase the non-basic variable, and hence the value of z, indefinitely. In this case, we say that the problem is unbounded, and there is no maximum solution.

If one or more of the coefficients are positive, then increasing the entering variable must be offset by a corresponding decrease in the basic variable. Specifically, if $a_{ik} > 0$, for basic variable x_i the non-basic column of x_k, then the new value of x_i, after x_k is increased, will be

$$x_i = b_i - a_{ik}x_k$$

But $x_i \geq 0$; therefore, we can increase x_k only to that point where

$$x_k = \frac{b_i}{a_{ik}}$$

Define $\theta_i = b_i/a_{ik}$ for all equations i for which $a_{ik} > 0$. Because we want to maximize the increase in x_k, we increase precisely to the point at which *some* basic variable first becomes zero (the minimum value of θ_i). That variable now leaves the basis, and is called the **leaving variable**. (Refer to Steps 3 and 4 in the following.)

The Simplex method can be summarized succinctly as follows:

Step 1: Examine the elements in the top row (the objective function row). If all elements are ≥ 0, then the current solution is optimal; stop. Otherwise go to Step 2.

Step 2: Select as the non-basic variable to enter the basis that variable corresponding to the most negative coefficient in the top row. This identifies **the pivot column**.

Step 3: Examine the coefficients in the pivot column. If all elements are ≤ 0, then this problem has an unbounded solution (no optimal solution); stop. Otherwise go to Step 4.

Step 4: Calculate the ratios

$$\theta_i = b_i/a_{ik} \text{ for all } i = 1, \ldots, m \text{ for which } a_{ik} > 0$$

where a_{ik} is the i-th element in the pivot column k. Then select

$$\theta = \min \{\theta_i\}$$

This identifies the **pivot row** and defines the variable that will leave the basis. The **pivot element** is the element in the pivot row and pivot column.

Step 5: To obtain the next tableau (which will represent the new basic feasible solution), divide each element in the pivot row by the pivot element. Use this row now to perform row operations on the other rows in order to obtain zeros in the rest of the pivot column, including the z row. This constitutes a pivot operation, performed on the pivot element, for the purpose of creating a unit vector in the pivot column, with a coefficient of one for the variable chosen to enter the basis.

When we apply these steps to the initial tableau in our example problem, we select x_1 (with the most negative coefficient on the z row) as the entering variable:

Basis	z	x_1	x_2	s_1	s_2	s_3	Solution
z	1	−8	−5	0	0	0	0
s_1	0	1	0	1	0	0	150
s_2	0	0	1	0	1	0	250
s_3	0	2	1	0	0	1	500

We compute

$$\theta_1 = \frac{150}{1} = 150$$

$$\theta_3 = \frac{500}{2} = 250$$

and select the minimum $\theta = \theta_1$. Therefore, the leaving variable is the one corresponding to the first basic variable s_1. A pivot operation on the pivot element then produces the next tableau which shows the new basic feasible solution

$$x_1 = 150$$

$$s_2 = 250$$

$$s_3 = 200$$

$$x_2 = 0$$

$$s_1 = 0$$

$$z = 1200$$

Basis	z	x_1	x_2	s_1	s_2	s_3	Solution
z	1	0	−5	8	0	0	1200
x_1	0	1	0	1	0	0	150
s_2	0	0	1	0	1	0	250
s_3	0	0	1	−2	0	1	200

In the next iteration, x_2 is chosen as the entering variable. Based on the ratios $\theta_2 = 250/1$ and $\theta_3 = 200/1$, we select $\theta = \theta_3$, and, therefore, the third basic variable s_3 leaves the basis. The pivot element is shown in the previous tableau. A pivot operation produces the new tableau:

Basis	z	x_1	x_2	s_1	s_2	s_3	Solution
z	1	0	0	−2	0	5	2200
x_1	0	1	0	1	0	0	150
s_2	0	0	0	2	1	−1	50
x_2	0	0	1	−2	0	1	200

The solution represented by this tableau is

$$x_1 = 150$$

$$s_2 = 50$$

$$x_2 = 200$$

$$s_1 = 0$$

$$s_3 = 0$$

and

z is now 2200

From this tableau, we can now select s_1 as the entering variable. We compute $\theta_1 = 150/1$ and $\theta_2 = 50/2$, choose $\theta = \theta_2$, and, therefore, designate s_2 as the leaving variable. The resulting tableau after a pivot operation is:

Basis	z	x_1	x_2	s_1	s_2	s_3	Solution
z	1	0	0	0	1	4	2250
x_1	0	1	0	0	-1/2	1/2	125
s_1	0	0	0	1	1/2	-1/2	25
x_2	0	0	1	0	1	0	250

Because all of the objective function row coefficients are non-negative, the current solution is optimal. The decision variables are:

$$x_1 = 125$$

$$x_2 = 250$$

and the optimal objective function value, denoted as z^*, is:

$$z^* = 8x_1 + 5x_2 = 8(125) + 5(250) = 2250$$

The values of the slack variables at optimality also provide useful information. The slack variable s_1 for the first constraint has a value of 25, indicating that there is a difference of 25 in the right and left sides of the constraint; thus, $x_1 = 125$ is 25 less than 150. (This can typically be interpreted to mean that some resource corresponding to constraint 1 is not fully consumed at optimality; such a constraint is sometimes referred to as a **non-binding constraint**.) Since s_2 and s_3 are non-basic and, therefore, have a value of zero, we can see that the second and third constraints are met as equalities. (These resources are used to capacity at optimality, and these constraints are sometimes called **binding constraints**.)

If we examine a graphical representation of the feasible region of this linear programming problem in Figure 2.6, we can observe the progression from extreme point A (initial solution) to extreme point B, then C, and finally the optimal solution at point D. Extreme points F and G are infeasible, and point E is a basic feasible solution but is not examined by the Simplex method.

In summary, let us briefly review the steps of the Simplex algorithm and the rationale behind each step. Negative coefficients, corresponding to non-basic variables, in the objective function row indicate that the objective function can be increased by making those associated variables basic (non-zero). If in Step 1 we find no negative element, then no change of basis can improve the current solution. Optimality has been achieved and the algorithm terminates.

Otherwise, in Step 2, we select the non-basic variable to enter the basis that has the greatest potential to improve the objective function. The elements in the objective function row indicate the *per unit* improvement in the objective function that can be achieved by increasing the non-basic variables. Because these values are merely *indicators* of potential and do not reveal the actual total improvement in z, ties are broken

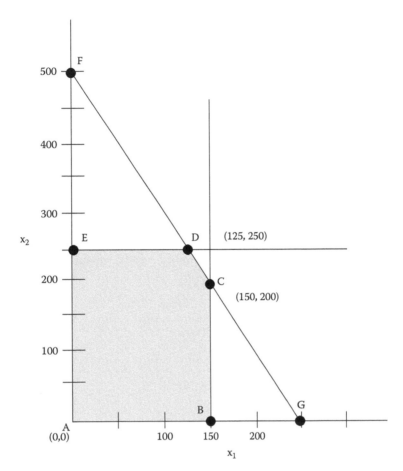

FIGURE 2.6
Simplex steps.

arbitrarily. In actual practice, choosing the *most* negative coefficient has been found to use about 20% more iterations than some more sophisticated criteria, such as are suggested by (Bixby 1994).

The basic variable to be replaced in the basis is chosen, in Step 4, to be the basic variable that reaches zero first as the entering variable is increased from zero. We restrict our examination of pivot column elements to positive values only (Step 3) because a pivot operation on a negative element would result in an unlimited increase in the basic variable. If the pivot column elements are all negative or zero, then the solution is unbounded and the algorithm terminates here. Otherwise, a pivot operation is performed as described in Step 5.

The Simplex tableau not only provides a convenient means of maintaining the system of equations during the iterations of the algorithm, but also contains a wealth of information about the linear programming problem that is being solved. In the following section, we will see various computational phenomena (indicating special problem cases) that may arise during application of the Simplex method, as well as information that may be obtained from an optimal tableau.

2.6 Initial Solutions for General Constraints

2.6.1 Artificial Variables

In the original presentation of the Simplex algorithm in Section 2.5, our sample problem was one in which all constraints were of the less-than-or-equal ($\leq$) type. In that case, we observed that by adding slack variables (in order to achieve equality constraints), we fortuitously also obtained an initial feasible set of basic variables. The coefficients of the slack variables provided the required unit vectors, embedded in the matrix of coefficients of the linear system of equations. In this section, we will see how to obtain an initial basic feasible solution for problems with more general forms of constraints, and to then use the Simplex method to solve such problems.

First of all, recall that all right hand sides b_i of constraints must be non-negative. Any constraint with a negative constant on the right hand side can be multiplied by -1 in order to satisfy this requirement. For example, an equality constraint such as:

$$-3x_1 + 4x_2 = -6$$

can be replaced by the constraint

$$3x_1 - 4x_2 = 6$$

An inequality such as:

$$5x_1 - 8x_2 \leq -10$$

can be replaced by

$$-5x_1 + 8x_2 \geq 10$$

At this point, it should be clear that typical linear programming problems in standard form contain equality constraints involving only the original decision variables as well as constraints that include slack variables and surplus variables. Slack variables can conveniently be used as basic variables; however, basic variables corresponding to equality constraints and greater than or equal ($\geq$) constraints are not always immediately available. Although it may be possible, by trial and error, to obtain a feasible starting basis for some problems, we prefer to use an approach that is straightforward and simple, and that can be used predictably in all cases.

We will deal with this situation by introducing additional variables, called **artificial variables**, solely for the purpose of obtaining an initial basis. These variables have no real meaning in the problem being solved, and will not be a part of the final solution. They merely provide a mechanism that will allow us to create a starting basic solution configuration, and then to apply the Simplex algorithm to the problem. (Note that it may not be necessary to add an artificial variable to every constraint; a constraint with a slack variable does not need an artificial variable.)

As an illustration, consider the following linear programming problem:

$$
\begin{array}{lll}
\text{maximize} & z = x_1 + 3x_2 & \\
\text{subject to} & 2x_1 - x_2 \leq -1 & (1) \\
& x_1 + x_2 = 3 & (2) \\
& x_1, x_2 \geq 0 &
\end{array}
$$

We multiply the first constraint by -1, to obtain $-2x_1 + x_2 \geq 1$, and then create an equality constraint by adding a (non-negative) surplus variable s_1 with a coefficient of -1. Now, the set of constraints

$$-2x_1 + x_2 - s_1 = 1$$

$$x_1 + x_2 = 3$$

is in standard form, but since there is no obvious starting solution (as there would have been if we had added *slack* variables in each constraint), we will introduce two artificial variables, R_1 and R_2, for this purpose. The constraint set becomes

$$-2x_1 + x_2 - s_1 + R_1 = 1$$

$$x_1 + x_2 + R_2 = 3$$

where x_1, x_2, s_1, R_1, $R_2 \geq 0$. We now have initial basic variables R_1 and R_2 for this enlarged problem; however, we must realize that the original equality constraint set is satisfied only if both R_1 and R_2 have values of zero. Therefore, the artificial variables must play only a temporary role in the solution.

There are two primary approaches that we can use to ensure that the artificial variables are not in the final solution. One method, commonly called the **Big-M method**, achieves this end by creating a modified objective function with huge negative coefficients $-M$ on the artificial variables. In our example, the modified objective function would be

$$z_M = x_1 + 3x_2 + 0s_1 - MR_1 - MR_2$$

When the Simplex method is applied to *maximize* this function, the heavy negative weights on the artificial variables will tend to drive R_1 and R_2 out of the basis, and the final solution will typically involve only the decision variables x_i and the slack or surplus variables.

For two reasons, the Big-M method is not considered to be a practical approach.

1. If the Simplex method terminates with an optimal solution (or with an indication that the linear program is unbounded), and at least one of the artificial variables is basic (positive) in the solution, then the original problem has no feasible solution. Moreover, in order to discover that no solution exists, we have had to solve an entire large (enlarged because of the additional artificial variables) linear programming problem.

2. A more serious difficulty with this method arises from a computational standpoint. The value of M must be chosen to be overwhelmingly large relative to all other problem parameters, in order to be sure that artificial variables do not remain in the basis of a feasible problem. However, as was pointed out in Chapter 1, computer arithmetic involving quantities of vastly different magnitudes leads to round-off error in which the smaller quantities (such as our original objective coefficients) are dwarfed by the artificial coefficients and are completely lost.

Thus, despite its intuitive appeal, the Big-M method is very poorly suited for computer implementation, and nowadays is rarely seen in commercial software.

The more practical alternative to solving linear programming problems having artificial variables is found in the **two-phase** Simplex method.

2.6.2 The Two Phase Method

Suppose we have a linear programming problem in standard form with artificial variables in the initial basic solution. Before expending the computational effort to solve the whole enlarged problem, it would be useful to know whether a feasible solution to the original problem exists. That is, we would like to know whether there is a solution, within the enlarged feasible region, in which the artificial variables are zero.

In order to make this determination, we first use the Simplex method to solve the problem of *minimizing* the sum of the artificial variables. If this sum can be minimized to *zero*, then there exists a solution not involving the artificial variables, and thus the original problem is feasible. Furthermore, in this case, we can use the final solution obtained from this computation as a starting solution for the original problem, and dispense with the artificial variables. On the other hand, if the optimized sum of the artificial variables is greater than zero, then at least one of the artificial variables remains basic, and we, therefore, know that the original problem constraint set cannot be satisfied. The two phases of this method can be summarized as follows.

Phase 1: Create a new objective function consisting of the sum of the artificial variables. Use the Simplex method to minimize this function, subject to the problem constraints. If this artificial objective function can be reduced to zero, then each of the (non-negative) artificial variables must be zero. In this case, all the original problem constraints are satisfied and we proceed to Phase 2. Otherwise, we know without further computation that the original problem is infeasible.

Phase 2: Use the basic feasible solution resulting from Phase 1 (ignoring the artificial variables which are no longer a part of any solution) as a starting solution for the original problem with the original objective function. Apply the ordinary Simplex method to obtain an optimal solution.

We will use the sample problem from Section 2.6.1 to illustrate the **two phase method**. In Phase 1, we seek to

$$\text{minimize } z_R = R_1 + R_2$$

which is equivalent to maximizing $z_R = -R_1 - R_2$. (Note that we minimize this sum regardless of whether the original problem is a minimization or a maximization problem.) Therefore, the top row of the tableau represents the equation

$$z_R + R_1 + R_2 = 0$$

With artificial variables in the constraints, the initial tableau for this phase is:

	x_1	x_2	s_1	R_1	R_2	Solution
z_R	0	0	0	1	1	0
R_1	−2	1	−1	1	0	1
R_2	1	1	0	0	1	3

Perform row operations to obtain a starting basis (i.e., with zero coefficient for R_1 and R_2 in the top row), and the tableau becomes:

	x_1	x_2	s_1	R_1	R_2	Solution
z_R	1	-2	1	0	0	-4
R_1	-2	1	-1	1	0	1
R_2	1	1	0	0	1	3

We then apply two iterations of the Simplex method to obtain the following two tableaus:

	x_1	x_2	s_1	R_1	R_2	Solution
z_R	-3	0	-1	2	0	-2
x_2	-2	1	-1	1	0	1
R_2	3	0	1	-1	1	2

	x_1	x_2	s_1	R_1	R_2	Solution
z_R	0	0	0	1	1	0
x_2	0	1	-1/3	1/3	2/3	7/3
x_1	1	0	1/3	-1/3	1/3	2/3

This is the optimal solution for the Phase 1 problem, and since R_1 and R_2 are zero and non-basic, this solution gives us a basic feasible starting solution for the original problem.

In Phase 2, artificial variables need not be considered and can be removed from the tableau. The top row of the starting tableau is replaced with the coefficients for the original (maximization) objective function:

	x_1	x_2	s_1	Solution
z	-1	-3	0	0
x_2	0	1	-1/3	7/3
x_1	1	0	1/3	2/3

Perform row operations to obtain an appropriate objective function row for a starting basis, and the Phase 2 tableau becomes:

	x_1	x_2	s_1	Solution
z	0	0	-2/3	23/3
x_2	0	1	-1/3	7/3
x_1	1	0	1/3	2/3

Now we apply the ordinary Simplex method, and in this case one iteration produces the optimal solution shown in the final tableau:

	x_1	x_2	s_1	Solution
z	2	0	0	9
x_2	1	1	0	3
s_1	3	0	1	2

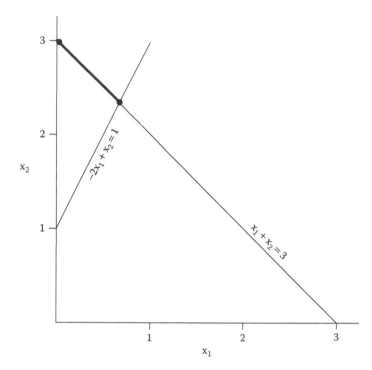

FIGURE 2.7
Infeasible origin.

It may be useful to look at a graphical solution of the problem we have just solved. Notice in Figure 2.7 that the feasible region consists only of points on the line $x_1 + x_2 = 3$, between the extreme points $(0, 3)$ and $(2/3, 7/3)$. The origin is not a feasible starting point, as was the case in several of our previous examples. Instead, we initially use an augmented feasible region (not visible in the graphical sketch) and a solution in which R_1 and R_2 are positive. During Phase 1, R_1 and R_2 become zero while the real variables x_1 and x_2 become positive. Phase 1 yielded the initial feasible solution $(2/3, 7/3)$ which *can* be shown in the two dimensional drawing; and Phase 2 found the optimal solution at $(0, 3)$.

2.7 Information in the Tableau

Several of the special cases introduced in Section 2.3 may reveal themselves in the Simplex tableau during the iteration phases of the Simplex algorithm. In particular, based on information that appears within the tableau, we can deduce certain characteristics of the linear programming problem being solved. These include linear programming problems with **multiple optimal solutions**, those with **unbounded solutions**, and problems having a property known as **degeneracy**. We will also find information in the tableau that provides insights concerning the roles played by the various resources in the system being modeled as a linear program.

2.7.1 Multiple Optimal Solutions

Recall from our example in Section 2.3.3 that when the line corresponding to the objective function is parallel to one of the straight lines bounding the feasible region, then the objective function can be optimized at all points on that edge of the feasible region. Thus, instead of a unique optimal solution, we have infinitely many optimal solutions from which to choose, thereby permitting management to select on the basis of secondary factors that do not appear in the model.

This situation can be recognized in the Simplex tableau during Step 2 of the Simplex algorithm. If a zero appears in the objective function row corresponding to a non-basic variable, then that non-basic variable can enter the basis without changing the value of the objective function. In other words, there are two distinct **adjacent extreme points** that yield the same value of z.

When we apply the Simplex algorithm to the problem illustrated in Figure 2.3, the initial solution is $x_1 = x_2 = 0$. In the first iteration, x_2 enters the basis and s_1 leaves, and this solution $x_1 = 0$, $x_2 = 2$ yields $z = 4$. Next, x_1 enters the basis and s_2 leaves, and we obtain the solution designated as point A in the figure where $x_1 = 4/3$, $x_2 = 10/3$, and $z = 8$. (Observe that s_3 is a basic variable and, therefore, constraint 3 is not *binding* at this point.) Now, the third Simplex tableau is as follows.

	z	x_1	x_2	s_1	s_2	s_3	Solution
z	1	0	0	0	1	0	8
x_2	0	0	1	1/3	1/3	0	10/3
x_1	0	1	0	−2/3	1/3	0	4/3
s_3	0	0	0	2/3	−1/3	1	14/3

This solution is optimal since all elements on the top row are non-negative. The zero in the top row corresponding to the non-basic variable s_1 signals that this problem has multiple optimal solutions. And, in fact, if we apply another pivot operation (by bringing s_1 into the basis and selecting s_3 to leave the basis), we obtain the fourth tableau

	z	x_1	x_2	s_1	s_2	s_3	Solution
z	1	0	0	0	1	0	8
x_2	0	0	1	0	1/2	−1/2	1
x_1	0	1	0	0	0	1	6
s_1	0	0	0	1	−1/2	3/2	7

This solution corresponds to point B in Figure 2.3 where $x_1 = 6$, $x_2 = 1$, and $z = 8$; and where s_1 is basic and consequently constraint 1 is not binding at this point.

2.7.2 Unbounded Solution (No Optimal Solution)

When the feasible region of a linear programming problem is unbounded, then it is also possible that the objective function value can be increased without bound. Evidence of both of these situations can be found in the Simplex tableau during Step 3 of the Simplex algorithm.

If in any tableau the constraint coefficients corresponding to a non-basic variable are all either negative or zero, then that non-basic variable can be increased arbitrarily without violating any constraint. Thus, the feasible region is unbounded in the direction of that variable.

Furthermore, if that variable is eligible to enter the basis (i.e., if it has a negative element in the objective function row), then we know that increasing this variable's value will increase the objective function. And because this variable can be increased indefinitely, so can the objective function value. Thus, the Simplex algorithm terminates and we can recognize that the problem has an unbounded solution.

The following problem illustrates an unbounded feasible region and unbounded solutions:

$$\text{maximize} \quad z = 5x_1 + 6x_2$$

$$\text{subject to} \quad -x_1 + x_2 \leq 2$$

$$x_2 \leq 10$$

$$x_1, x_2 \geq 0$$

Figure 2.8 shows the feasible region. The initial tableau is given by:

	z	x_1	x_2	s_1	s_2	Solution
z	1	−5	−6	0	0	0
s_1	0	−1	1	1	0	2
s_2	0	0	1	0	1	10

The unboundedness of the feasible region is indicated by the absence of positive elements in the column corresponding to the non-basic variable x_1. The negative coefficient in the top row of this column indicates that x_1 is eligible to increase (from zero) and that, therefore, z can increase indefinitely.

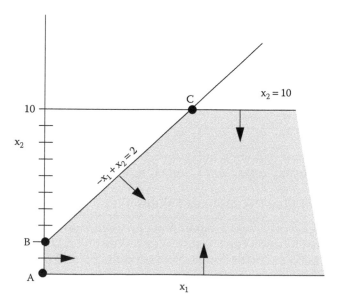

FIGURE 2.8
Unbounded solution.

Our Simplex algorithm, as it is stated, would, in fact, choose x_2 (with the *most* negative coefficient) as the entering variable, and we would move from point A to point B in Figure 2.8, and then subsequently to point C. At that point, we would be faced again with the inevitable: x_1 can be feasibly increased arbitrarily, producing an arbitrarily large value of z.

As noted earlier, a linear programming formulation with an unbounded objective function value undoubtedly represents an invalid model of a real system, since we have no real expectation of achieving unlimited productivity or profitability. Recognizing such a situation, we must reformulate the problem with more careful attention to realistic constraints on the decision variables.

2.7.3 Degenerate Solutions

A solution to a linear programming problem is said to be **degenerate** if one or more of the basic variables has a value of zero. Evidence of the existence of a degenerate solution is found during Step 4 of the Simplex algorithm when there is a tie for the minimum ratio θ, that is, a tie for the leaving variable. In this case, the tie may be broken arbitrarily and one variable is chosen to leave the basis. However, *both* variables participating in the tie will, in fact, become zero, although one of them remains basic.

The presence of a degenerate solution indicates that the linear programming formulation contains at least one redundant constraint. This situation arises in the following problem whose graphical solution is shown in Figure 2.9.

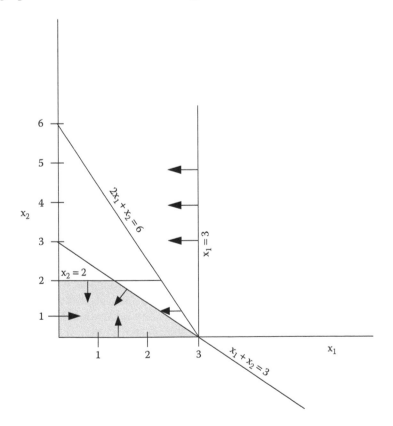

FIGURE 2.9
Degenerate solution.

maximize $\qquad z = 3x_1 + 2x_2$

subject to $\qquad x_1 \leq 3$

$\qquad 2x_1 + x_2 \leq 6$

$\qquad x_2 \leq 2$

$\qquad x_1 + x_2 \leq 3$

$\qquad x_1, x_2 \geq 0$

Note that $x_1 \leq 3$ is redundant, since the constraint $x_1 + x_2 \leq 3$ ensures that $x_1 \leq 3$. Similarly, the constraint $2x_1 + x_2 \leq 6$ is redundant as shown in Figure 2.9. In the initial tableau, x_1 is chosen as the entering variable, and we discover a tie between s_1 and s_2 to leave the basis since $\theta_1 = \theta_2 = 3$.

	z	x_1	x_2	s_1	s_2	s_3	s_4	Solution
	1	-3	-2	0	0	0	0	0
s_1	0	1	0	1	0	0	0	3
s_2	0	2	1	0	1	0	0	6
s_3	0	0	1	0	0	1	0	2
s_4	0	1	1	0	0	0	1	3

Let us arbitrarily select s_1 to leave the basis, and create the next tableau.

	x_1	x_2	s_1	s_2	s_3	s_4	Solution
	0	-2	3	0	0	0	9
x_1	1	0	1	0	0	0	3
s_2	0	1	-2	1	0	0	0
s_3	0	1	0	0	1	0	2
s_4	0	1	-1	0	0	1	0

Notice that the basic variables s_2 and s_4 now have a value of zero. The present solution corresponds to a point where three *redundant* constraints are binding; that is, the slack variables in the first, second, and fourth constraints are zero at this point.

When we now select x_2 to enter the basis, we have a choice between s_2 and s_4 to leave. If we pick s_2, we will discover that the new tableau has a negative cost for s_2, and basic variables x_2 and s_4 are both zero. Since we can now choose x_2 to leave, we could get right back to the tableau where we started. This cycling can continue indefinitely.

Note that, for a two variable problem, degeneracy can occur only when there are redundant constraints. However, in three-variable problems, we could construct four or five constraints such that they all intersect at a common point, and none of them are redundant. (For example, imagine a roof with many sides that all meet at a common peak.) If the problem contains extreme points of this form, and if the Simplex algorithm happens to land on that corner (both rather unlikely in practice), then the algorithm could cycle indefinitely.

Problem degeneracy exposes the only theoretical weakness of the Simplex method: it is possible that the algorithm will cycle indefinitely and fail to converge to an optimal solution. Once a degenerate solution to a problem arises, it is possible that successive iterations of the Simplex method will yield no improvement in the objective function. This phenomenon may be a temporary one, occurring for only one or a few iterations, or it may continue indefinitely, generating the same sequence of non-improving solutions. If it is temporary, then we have merely lost valuable computation time, but we will eventually obtain the desired optimal solution. The more serious possibility, infinite cycling and, therefore, failure of the algorithm, is fortunately not a serious practical problem. Although problems have been constructed that demonstrate this hazard, such cycling in actual problems is so rare that computational modifications to defend against Simplex cycling are not considered to be worthwhile. Therefore, although many practical problems have degenerate solutions, the Simplex algorithm typically cycles only temporarily and reaches the optimal solution without significant degradation in computational efficiency.

2.7.4 Analyzing the Optimal Tableau: Shadow Prices

Once the Simplex method has terminated successfully, we find that the optimal tableau contains not only the solutions for the decision variables, but also auxiliary information that can be of considerable use to the analyst. For example, in the top row of the final tableau, the coefficient of the i-th slack variable is the amount by which the final objective function could be increased for each additional available unit of the resource associated with that slack variable. These values are called **shadow prices**, and represent the **marginal worth** (or incremental value) of making additional units of the various resources available.

By examining the optimal tableau at the end of Section 2.5, we find a coefficient of 4 for slack variable s_3. This means that the final value of z^* could be increased by 4 for each additional unit of the resource associated with the third constraint. Likewise, the coefficient of 1 for slack variable s_2 indicates that z^* could be increased at a rate of 1 for each added unit of the resource associated with the second constraint.

We are not too surprised to find, in this tableau, a zero marginal worth for the first resource (denoted by a zero coefficient for s_1 in the top row). Since $s_1 = 25$ in the final solution, the first inequality constraint is satisfied with a slack of 25; that is, this resource is not being completely consumed in this solution. Therefore, we would not expect any increase in the objective function to result from adding any more units of a resource that is presently already under-utilized.

Decision makers and analysts are usually in a position to know whether the resource limitations (that appear on the right hand sides of the linear system of constraints) are truly fixed or whether resource allocations could be modified by acquiring additional resources. Management can determine the economic advisability of increasing the allotment of the i-th resource by examining the shadow price: the shadow price is the maximum per unit price that should be paid to increase the allotment of that resource by one unit, in order to achieve a net gain in the objective.

Having made the earlier observations about the unit worth of resources, it is important to point out that the increases in resource allocations must be relatively small increases. The economic measure of the value of increasing the availability of any given resource is valid *only* as long as such an increase does not change the optimal *basic* solution. When the right-hand sides of constraints are changed, we do in fact have a different linear programming problem. Analyzing the extent to which resource capacities (or availabilities) can be changed without altering the optimal set of basic variables is one of the topics covered in the following section of this chapter.

2.8 Duality and Sensitivity Analysis

When making an economic interpretation of the objective function of a linear programming problem, an alternative and useful point of view is obtained by computing the collective contributions of all the resources. If we multiply the original availability of each resource (shown in the original tableau) by its marginal worth (taken from the final tableau), and form the sum, we obtain precisely the optimal objective function value. In our example at the beginning of Section 2.5, we have marginal worth values of 0, 1, and 4, and resource availabilities of 150, 250, and 500; therefore, the optimal objective function value can be expressed as

$$z^* = 2250 = 0(150) + 1(250) + 4(500)$$

This apparently equivalent way of viewing the original (or **primal**) linear programming problem is a manifestation of what is called the **dual problem**. The study of **duality** provides the theoretical foundation for practical analysis of optimal solutions obtained with the Simplex method. This topic is especially important because the full and effective use of many linear programming software implementations requires a familiarity with the concepts of duality.

Sensitivity analysis is the study of how a solution to a problem changes when there are slight changes in the problem parameters, without solving the whole problem again from scratch. It is, therefore, an analysis of how sensitive a solution is to small perturbations in the problem data. Objective function coefficients, constraint coefficients, and resource capacities are problem data that may be difficult or costly to obtain. These values may be introduced into the linear programming model as rough estimates or imperfect observations, and they might be values that change over time, as costs fluctuate or resources availabilities vary.

If all problem data were certain and constant over time, there would be no need for sensitivity analysis. Each new problem would be based on exact data, and the solution would be a perfect one. In practice, such is rarely the case. Thus, the problem formulation that is solved initially may not be exactly the *right* problem, that is, the one that is valid at the time resources are actually procured, costs are incurred, or profits are made.

If it could be determined, through the process of sensitivity analysis, which of the problem parameters are the most critical to the optimality of the original problem solution, then analysts could take greatest care in supplying and refining specifically those parameters to which the solution is most sensitive. Sensitivity analysis tools are of great value to management because they can help to provide a thorough understanding of a problem solution, the range of problem parameters over which a solution is valid, and how the solution can be changed by making changes in costs, profits, or resource availability. Duality theory provides the foundation underlying these tools.

2.8.1 The Dual Problem

A linear programming problem and its dual are related in the sense that both problems are based on the same problem data, and an optimal solution to either one of the problems prescribes the optimal solution to the other. These *companion* problems might even be thought of as two different views of the same problem, but with different economic or engineering interpretations, and possibly with different computational implications.

Consider any linear programming formulation that is in the form of a maximization problem with constraints of the less than or equal type or equality constraints. (A constraint in which the inequality is a $\geq$ type can be multiplied by -1 to reverse the direction of the inequality sign, resulting possibly in a negative right-hand-side value.) We will call this the **primal problem**. If all constraints are inequalities and the decision variables are non-negative, the primal problem can be written as:

$$\text{maximize} \quad c_1x_1 + c_2x_2 + \ldots + c_nx_n$$

$$\text{subject to} \quad a_{11}x_1 + a_{12}x_2 + \ldots + a_{1n}x_n \leq b_1$$

$$a_{21}x_1 + a_{22}x_2 + \ldots + a_{2n}x_n \leq b_2$$

$$\vdots \qquad \qquad \vdots$$

$$a_{m1}x_1 + a_{m2}x_2 + \ldots + a_{mn}x_n \leq b_m$$

where the variables $x_1, \ldots, x_n$ are non-negative.

In general, the corresponding **dual problem** is constructed as follows:

- The dual problem is a minimization problem.
- For every variable x_i in the primal problem, there is a constraint in the dual problem.
  If $x_i \geq 0$ in the primal, the constraint is a $\geq$ inequality in the dual.
  If x_i is unrestricted in sign, the i-th constraint is an equality in the dual.
- For every constraint in the primal problem, there is a variable y_i in the dual.
  If the constraint is $\leq$, then $y_i \geq 0$ in the dual problem.
  If the constraint is an equality, then y_i is unrestricted in sign in the dual.
- The right hand sides in the primal are the objective function coefficients in the dual.
- The objective function coefficients in the primal are the right hand sides in the dual.
- The coefficient matrix in the primal is *transposed* to form the coefficient matrix for the dual.

The dual problem corresponding to the earlier primal problem is a problem with m variables and n constraints and can be written as:

$$\text{minimize} \quad b_1y_1 + b_2y_2 + \ldots + b_my_m$$

$$\text{subject to} \quad a_{11}y_1 + a_{21}y_2 + \ldots + a_{m1}y_m \geq c_1$$

$$a_{12}y_1 + a_{22}y_2 + \ldots + a_{m2}y_m \geq c_2$$

$$\vdots \qquad \qquad \vdots$$

$$a_{1n}y_1 + a_{2n}y_2 + \ldots + a_{mn}y_m \geq c_n$$

and the variables $y_1, \ldots, y_m$ are non-negative.

Clearly, the dual of the dual problem is the original primal problem, and in many contexts, it is not necessary to stipulate which one of the companion problems is *the primal one* and which is *the dual one*; each is the dual of the other.

Example 2.8.1

Consider the primal problem:

$$\text{maximize} \qquad 3x_1 + 2x_2 - 6x_3$$

$$\text{Subject to} \qquad 4x_1 + 8x_2 - x_3 \leq 5$$

$$7x_1 - 2x_2 + 2x_3 \geq 4$$

$$\text{and } x_1, x_2, x_3 \geq 0$$

The second constraint can be rewritten as $-7x_1 + 2x_2 - 2x_3 \leq -4$. The dual problem is then

$$\text{minimize} \qquad 5y_1 - 4y_2$$

$$\text{subject to} \qquad 4y_1 - 7y_2 \geq 3$$

$$8y_1 + 2y_2 \geq 2$$

$$-y_1 - 2y_2 \geq -6$$

$$\text{and } y_1, y_2 \geq 0$$

Example 2.8.2

The following primal problem has constraints that include both types of inequalities and an equality constraint:

$$\text{maximize} \qquad 4x_1 - 3x_2$$

$$\text{subject to} \qquad 2x_1 - 4x_2 \leq 5$$

$$5x_1 - 6x_2 \geq 9$$

$$3x_1 + 8x_2 = 2$$

$$x_1 + 2x_2 \leq 1$$

$$\text{and } x_1 \geq 0$$

$$\text{and } x_2 \text{ unrestricted in sign}$$

The dual of this problem is formed by rewriting the second constraint as $-5x_1 + 6x_2 \leq -9$, and then following the guidelines presented earlier to obtain:

$$\text{minimize} \qquad 5y_1 - 9y_2 + 2y_3 + y_4$$

$$\text{subject to} \qquad 2y_1 - 5y_2 + 3y_3 + y_4 \geq 4$$

$$-4y_1 + 6y_2 + 8y_3 + 2y_4 = -3$$

$$\text{and } y_1, y_2, y_4 \geq 0$$

$$\text{and } y_3 \text{ unrestricted in sign}$$

(Recall that the Simplex method requires that all variables be non-negative. When an unrestricted variable arises in a formulation, that variable can be replaced by the difference of two new non-negative variables, as suggested and illustrated in Section 2.4.1.)

There is a very apparent structural similarity between a primal and dual pair of problems, but how are their solutions related? In the course of solving a (primal) maximization problem, the Simplex method generates a series of feasible solutions with successively *larger* objective function values (cx). Solving the corresponding (dual) minimization problem may be thought of as a process of generating a series of feasible solutions with successively *smaller* objective function values (yb). Assuming that an optimal solution does exist, the primal problem will converge to its maximum objective function value from below, and the dual problem will converge to its minimum objective function value from above. The primal objective function evaluated at x never exceeds the dual objective function evaluated at y; and at optimality, the two problems actually have the same objective function value. This can be summarized in the following **duality property**:

Duality property: If x and y are feasible solutions to the primal and dual problems, respectively, then $cx \leq yb$ throughout the optimization process; and finally, at optimality, $cx^* = y^*b$.

It follows from this property that, if feasible objective function values are found for a primal and dual pair of problems, and if these values are equal to each other, then both of the solutions are optimal solutions.

The phenomenon of primal and dual problems sharing the same objective function values is not mere coincidence. In fact, the **shadow prices**, which appear in the top row of the optimal tableau of the primal problem, are precisely the **optimal values of the dual variables**. Similarly, if the dual problem were solved using the Simplex method, the shadow prices in that optimal tableau would be the optimal values of the primal variables.

In the illustrative problem from Section 2.5, the dual objective of minimizing $150y_1 + 250y_2 + 500y_3$ is met when the dual variables (shadow prices) have the values $y_1 = 0$, $y_2 = 1$, $y_3 = 4$. Thus, from the dual point of view,

$$z^* = 150(0) + 250(1) + 500(4) = 2250$$

which is equal to the primal objective value

$$z^* = 8x_1 + 5x_2 = 8\,(125) + 5(250) = 2250$$

for optimal x values of $x_1 = 125$ and $x_2 = 250$.

One further characterization relating primal and dual linear programming problems is known as **complementary slackness**. Because each decision variable in a primal problem is associated with a constraint in the dual problem, each such variable is also associated with a slack or surplus variable in the dual. In any solution, if the primal variable is basic (with value ≥ 0, hence having slack), then the associated dual variable is non-basic (with value = 0, hence having no slack). And if the primal variable is non-basic (with value = 0, hence no slack), then the associated dual variable is basic (with value = 0, hence having slack).

This can be observed even in a problem as simple as the one illustrating the Simplex method in Section 2.5. In the final tableau, the primal basic variables x_1, s_1, and x_2 have positive values, while in the top row we see zero values for their three associated dual variables. The non-basic primal variables s_2 and s_3 have zero values, while their associated dual variables are basic and have non-zero values.

This property is described as follows.

> **Complementary Slackness Property**: If in an optimal solution to a linear programming problem, an inequality constraint is not binding, then the dual variable corresponding to that constraint has a value of zero in any optimal solution to the dual problem.

This is merely a formalization of the intuitive notion that the shadow price of a resource associated with a non-binding constraint is zero. That is, there is a zero marginal worth for a resource that is not being fully utilized.

The properties described earlier were based on an assumption that optimal solutions to both primal and dual problems exist, but, of course, not all linear programming problems have optimal feasible solutions; infeasible problems and problems with unbounded solutions were discussed earlier in this chapter. For corresponding primal and dual problems, exactly one of the following mutually exclusive cases always occurs:

1. Both primal and dual problems are feasible, and both have optimal (and equal) solutions.
2. Both primal and dual problems are infeasible (have no feasible solution).
3. The primal problem is feasible but unbounded, and the dual problem is infeasible.
4. The dual problem is feasible but unbounded, and the primal problem is infeasible.

Because the pertinent parameters and goals of any linear programming problem can be expressed in either a primal or dual form, and because solving either the primal or dual problem yields enough information to easily construct a solution to the other, we might reasonably wonder which problem, primal or dual, should we solve when using the Simplex method.

From the standpoint of computational efficiency, we might wish to choose to solve the problem with the *fewer number of constraints*. As is discussed further in Section 2.10.3, the computation time required for the Simplex method is strongly dependent on the number of constraints, and almost independent of the number of variables. Therefore, in the absence of other identifiable structural characteristics of a problem that might make it amenable to the use of specialized solution methods, we could expect to be able to solve most quickly the problem having the smaller number of constraints. This choice becomes more compelling when the linear programming problem has thousands of constraints, and is of much less importance for more moderate-sized problems of a few hundred or less constraints.

An understanding of duality properties and the relation between primal and dual problems gives an analyst some flexibility in formulating, solving, and interpreting a solution to a linear programming problem. Moreover, duality provides the mathematical basis for analyzing an optimal solution's sensitivity to small changes in problem data. We now turn our attention to the types of analysis that can be made once an optimal solution to a linear programming problem has been obtained.

2.8.2 Postoptimality and Sensitivity Analysis

After an optimal solution to a linear programming problem has been found, the analyst's next step is to review the problem parameters and the solution, in preparation for putting the solution into practice. This process of **postoptimality analysis** includes confirming or updating problem parameters (costs and availability of activities and resources), and if there are any changes to the original problem parameters, assessing the effect of these changes on the optimality of the solution. If the changes are small, it may not be necessary to re-optimize the new problem; instead, some small calculation may suffice to identify simple consequences in the previous optimal scenario. **Sensitivity analysis** is the study of the types, ranges, and magnitude of changes in problem parameters whose effects can be determined relatively easily, without the need for solving a new linear programming problem.

In a linear programming model that is relatively insensitive to changes in problem parameters, the original optimal solution may not change even when several parameters vary widely. Other models may be highly sensitive, and the optimality of the original solution may be seriously undermined by the smallest change in even one parameter. When working with less sensitive models, the expense and effort of acquiring extremely accurate data (through extensive sampling, costly tracking, careful observations, etc.) may not be justified. On the other hand, a successful analyst knows the necessity of making a special effort to obtain the most accurate possible problem data when working with very sensitive models.

Sensitivity analysis addresses several different kinds of changes to a linear programming formulation, including:

- Changes in objective function coefficients
- Increases or decreases in the right hand side of a constraint
- Adding a new variable
- Adding a constraint
- Changes in constraint coefficients

Objective function coefficient range analysis identifies the maximum allowable increase and decrease that can occur for each coefficient without changing the current solution. Under the assumption that all other parameters remain unchanged, a change within the allowable range ensures that the current solution will remain optimal and that the values of the decision variables remain unchanged. The objective function value would, of course, change as the coefficient varies over its range.

Example 2.8.3

Consider a simple two variable example:

$$\text{maximize} \quad z = 4x_1 + 3x_2$$

$$\text{subject to} \quad x_1 + x_2 \leq 4$$

$$2x_1 + x_2 \leq 6$$

$$x_1 + 3x_2 \leq 9$$

$$x_1, \; x_2 \geq 0$$

Using the illustration in Figure 2.10, we can observe that the optimal solution occurs at the point (2, 2) with a function value of $z = 14$. If we change the cost coefficients slightly, the optimal solution will stay at the current point. However, if we add more than 1 to the coefficient of x_2, then the current solution will no longer be optimal. Similarly, if we subtract more than 1 from c_2, the solution will change. (See Exercise 2.45.)

Right-hand-side ranging is performed to determine how much the right-hand side of a constraint can vary (increase or decrease) without causing the original optimal

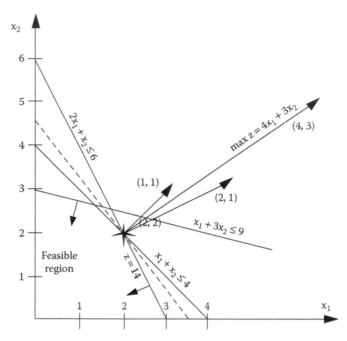

FIGURE 2.10
Illustration of sensitivity analysis.

solution to become infeasible. Changing a constraint alters the feasible region and may affect the shape of the feasible region in the vicinity of the optimal point. (If the original optimal point is no longer a feasible extreme point, a different optimal solution would have to be found.) If a resource is not being completely used (i.e., there is positive slack) in the optimal solution, then clearly the right hand side of the constraint corresponding to that resource can be increased indefinitely. In general, however, possible increases and decreases in right hand sides are measured by analyzing the optimal solution to determine how much slack can be created in the constraint without changing the optimal solution.

In the problem depicted in Figure 2.10, consider what happens when we add 1 to the right hand side of the second constraint, so that the constraint becomes $2x_1 + x_2 \leq 7$. Now, the active constraints at the optimal solution have changed, but the same set of constraints will be active. (The same variables are basic.) As discussed earlier, the objective function will increase by precisely the value of the dual variable corresponding to that constraint. In this example, the objective function will increase by 1.

It is easy to see in the illustration that the right hand side can be increased by 2 without changing the variables in the basis. Beyond that point, the constraint becomes inactive (outside the feasible region). Similarly, the right hand side of constraint 2 can be decreased by 0.5 without changing the basis. At that point, the optimal solution would occur at the intersection of the other two constraints, at (1.5, 2.5), and decreasing beyond that would change the basic variables.

Adding a new variable to a model would require introducing the resource requirements of that new activity or product into a current optimal solution. By analyzing information already in the optimal tableau, it can be determined whether the new variable would be a basic variable in the optimal solution and what would be the value of its coefficient in the objective function. The shadow prices in the optimal solution provide information about the marginal worth of resources, and knowing the resource needs corresponding to the new variable, the value of bringing in the new variable can be computed.

Adding a constraint or **changing constraint coefficients** amounts to rather complicated changes to the original problem. These types of changes to the linear programming model fall logically into the postoptimality analysis framework, but technically these are not changes that can be analyzed or effected by merely using information in the optimal tableau. Such changes are generally best dealt with by solving the modified problem anew.

Almost all commercial software for linear programming, such as the products mentioned in Section 2.10.3, include postoptimality analysis as part of the standard output. Most packages present right-hand-side ranging and objective coefficient ranging information; some also include adding a new variable; rarely are constraint changes included as part of ordinary postoptimality analysis.

The information and insights obtained through sensitivity analysis are especially valuable to management because they provide an indication of the degree of flexibility that is inherent in an operating environment. Such knowledge is helpful in planning, making decisions, and formulating policies for handling fluctuations and imprecision in prices, activities, and resource availabilities used in linear programming models.

2.9 Revised Simplex and Computational Efficiency

The amount of computation required to solve linear programming problems with the Simplex method is indeed arduous; in fact, all but the most trivial problems must be solved with the aid of a computer. Several decades of experience with computer implementations of the Simplex method have led researchers and practitioners to develop various improvements and enhancements to the original Simplex method. The result is a refined version of the standard Simplex, called the **Revised Simplex method**. This method makes much more efficient use of a computer's most valuable resources: CPU computation time and memory space.

Recall that the standard Simplex method performs calculations, at each iteration, to update the entire tableau. Actually, the only data needed at each iteration are the objective function row (to determine the entering variable), the pivot column corresponding to the non-basic entering variable, and the right-hand-side values of the current basic variables (to determine the variable to leave the current basis). Thus, the standard Simplex computes and stores many values that are not needed during the present iteration and that *may* never be needed. The Revised Simplex method performs the same iterations as the standard Simplex, but the details of its computations have specific advantages for computer implementations.

The standard Simplex method generates each new tableau iteratively, based on the previous tableau. However, the Revised Simplex method takes advantage of the fact that all of the information in any tableau can in fact be obtained directly from the original problem equations, if the *inverse* of the matrix of basic columns for that tableau is known. And that inverse can be obtained directly from the original equations if the current basic variables for that tableau are known. Note that the Revised Simplex performs the usual selection of an entering and leaving variable at each iteration, but it carries out only those computations necessary to register that selection and to record the current solution configuration.

Readers acquainted with numerical computation will be aware that matrix inversion is itself a nontrivial task, in terms of both computation time and numerical accuracy.

Therefore, instead of recomputing a basis inverse at each iteration, a **product form** of inverse can be used that allows a new inverse to be computed simply from the previous one. This procedure calls for premultiplying the previous inverse by a matrix that is an identity matrix except in one column. (Only that one column and an indicator of its position in the matrix need be stored explicitly.) Some of the more advanced references listed at the end of this chapter provide a more complete description of product form inverse computation, and of how **re-inversion** can help to maintain accuracy and save storage space.

Although the Revised Simplex method requires some additional bookkeeping that would not be needed if the full tableau were maintained, the method typically requires less computation, uses less storage space, and obtains greater numerical accuracy than the standard Simplex method.

Because only the essential data are computed, Revised Simplex has an advantage, with respect to computation time, over the standard Simplex. This advantage is particularly pronounced when the number of constraints is much less than the number of variables because the size of all the essential data (basic columns and right-hand-side constants) is determined by the number of constraints. (Refer to [Simmons 1972] for a detailed operation-count for the Revised and standard Simplex methods.)

Revised Simplex storage requirements are minimal because it is necessary to store only the basic variables, the basis inverse or its product form, and the constants. The original constraint matrix and objective coefficients can be stored efficiently by the computer's memory manager on conveniently placed storage devices, along with the premultipliers for the product form inverse, if desired.

Perhaps the most attractive advantage offered by the Revised Simplex method is increased numerical accuracy. As discussed in Chapter 1, an algorithm is called numerically unstable if small errors (through round-off in intermediate computations, for example) can lead to very large errors in the final solution. Both the standard and Revised Simplex methods are numerically unstable, but Revised Simplex avoids some of the potential for instability. There is less accumulated round-off error because calculations are performed on a column *only* when it is to enter the basis, not at every iteration. Furthermore, computations are applied to original problem data, not to data that have already undergone (possibly unnecessary) computation.

Typical large linear programming problems have constraint matrices that are very **sparse**, with a large proportion (often in the range of 95%) of zero values. Revised Simplex performs fewer multiplications involving non-zero elements, since Revised Simplex operates on original (sparse) data, whereas standard Simplex operates repeatedly on the entire tableau and quickly creates a dense matrix out of a sparse one. Thus, by taking advantage of sparsity, the Revised Simplex can reduce the amount of computation and therefore maintain numerical accuracy.

The advantages described earlier have been observed so consistently that almost all commercial software for linear programming is based on the Revised Simplex method (with product form inverse) for both phases of the two phase method.

2.10 Software for Linear Programming

Now that we are familiar with linear programming models and a fundamental method for solving these problems, we will turn our attention to some practical considerations necessary for solving large linear programming problems on a computer. Because there is quite a selection of commercially available software for linear programming, anyone in a position to choose a software system for personal use (and certainly anyone contemplating developing their own software) should be aware of the various features to be mentioned in this section. In particular, we will briefly describe some important extensions often found appended to the usual Simplex techniques, and some actual commercial systems that are available. We also include a discussion of **interior methods** that now play an increasingly important role in the practical solution of linear programming problems.

2.10.1 Extensions to General Simplex Methods

The majority of commercial software for linear programming is based on the Revised Simplex method, and most implementations employ the product form inverse. For efficiency and accuracy on a computer, a variety of additional features may also be incorporated. We merely mention a few of them here, and the interested reader can obtain a more thorough understanding using the references cited at the end of the chapter.

The method used for computing and maintaining tableau information has a strong bearing on the size of problem that can be successfully attempted. More complicated implementations require greater skill and effort but operate with greater speed so that larger problems can be solved.

The explicit inverse method is straightforward and can be efficient and useful for problems involving a few hundred rows. The product form inverse allows for problems in the range of 1000 or so rows. For problems with tens of thousands of rows, LU decomposition techniques have been developed, for use both in the iteration phases and during **re-inversion** of the basis. In simple terms, any basis matrix B can be rewritten as the product of two triangular matrices, L and U where L is lower triangular (with zeros above the main diagonal) and U is upper triangular (with zeros below the diagonal). This format enables very efficient inverse computation and solution of the system.

In a linear program with many variables, it is very time consuming to examine every non-basic variable at each iteration to determine the one to enter the basis. Many linear programming implementations do not go to the effort to select the non-basic variable corresponding to the *most negative* top row coefficient, but rather one corresponding to *any* negative coefficient (i.e., any variable that will improve the objective function). Although this strategy may increase the total number of iterations, it is actually a time-saving and very rational approach because the negative top row coefficients only specify a per-unit improvement in z, and not an absolute overall improvement. Thus any *good* entering variable can be quickly selected for the next basis.

In many linear programming models, there are **upper bound constraints** ($x_j \leq u_j$) for some or all of the variables. Constraints such as these, as well as **generalized upper bounds** ($\Sigma x_j \leq u_j$), can be dealt with using a method, introduced by Dantzig and Van Slyke (1967), that handles these constraints implicitly without enlarging the basis. (Recall that for each explicit constraint, there must be a basic variable; therefore, any additional constraints generally contribute to the amount of work and storage required by the Revised Simplex method.) Handling upper bound constraints implicitly does take time, but practice has shown that this is an advantageous trade-off that serves to keep the problem size from increasing.

Very large linear programming models often result in a constraint matrix A in which the non-zero elements appear in patterns or blocks. When a problem exhibits such a high degree of structure, it may be possible to apply a **decomposition** technique (Dantzig and Wolfe 1960). The original model is partitioned, and the subproblems are then solved individually.

Not only do non-zero elements of A often appear in patterns, but more generally, we find the matrix A to be very sparse. A **sparse** matrix is one with a very large proportion of zero elements. A rule of thumb is that large linear programming models typically have only about 10% non-zero elements; some practitioners claim that 1%–5% is a more realistic range. This sparsity is not a surprising phenomenon when we consider that in any large organization, certain sets of products, people, or processes tend to operate in groups, and are therefore subject to *local* constraints. When such a problem is formulated, a sparse matrix results because each variable is involved in a relatively small number of the constraints.

In order to make better use of available memory, sparse matrices should be stored in some type of a compressed format, using methods such as those described by (Murtagh 1981). For example, each non-zero element could be stored along with an encoded form of its row and column indices. The term *super sparse* has been used to describe matrices that are not only sparse but in which many of the non-zero elements are the same. (e.g., in many applications, the vast majority of non-zero coefficients have a value of one.) In that case, each distinct value need be stored only once, and elements are found via a table of addresses into a table of distinct element values. Sparse matrix handling techniques have been shown to be worthwhile even if the coefficient matrix A is stored on a peripheral memory device. Because transfer time is slow relative to computation time, it is prudent to maintain such large data structures in as compact a form as possible.

Round-off error is a natural consequence of using finite precision computing devices. As was pointed out in Chapter 1, this inability to store computed results exactly is particularly pronounced when we perform arithmetic operations on numeric values of very different magnitudes, where we are often unable to record that portion of a result contributed by the smaller value. In an attempt to remove the source of some of these numerical inaccuracies, most commercial linear programming systems apply some kind of **scaling** before beginning the Simplex method. Rows and columns of the matrix A may be multiplied by constants in order to make the largest element of each row and column the same (Murtagh 1981). To improve the **condition** of a matrix (and, therefore, obtain greater accuracy of its inverse), all the elements of A should be kept within a reasonable range, say within a factor of 10^6 or 10^8 of each other (Orchard-Hays 1968). More elaborate and specific mechanisms for scaling have been devised. In general, a healthy awareness of the limitations of computer arithmetic and numerical computation is essential in understanding and interpreting computed results.

In a problem of any practical size, the elimination of artificial variables from an initial solution can take a considerable amount of computation time. The term **crashing** refers generally to any kind of technique that gives the Simplex method a head start and eliminates some of the early iterations. Crashing sometimes consists of choosing a set of (non-artificial) non-basic variables to enter the basis and replace the artificial variables, even at the expense of temporarily degrading the objective function or making the solution infeasible (Cooper and Steinberg 1974). An even better way to give a boost to the Simplex method is to obtain, from the user or analyst, problem specific information about *which* variables are likely to be basic variables in a final solution. Many commercial systems (particularly those for larger powerful computers) provide a means for introducing such information along with other problem data. It may also be possible to restart Simplex iterations using solutions from previous (incomplete) attempts at optimization.

Many commercial systems contain algorithms for **sensitivity analysis** (also called **ranging procedures** or **postoptimality analysis**). These techniques are applied after the

Simplex method has already produced an optimal solution. Sensitivity analysis allows the user to determine the effect that changes in various problem parameters would have on the optimal solution. Changes in the objective (cost/profit) coefficients and in the resource levels (right hand sides of constraints) are commonly dealt with; some systems consider the addition of decision variables to the original model, but most systems do not handle changes in the constraint coefficients or the addition of new constraints.

The relationship between sensitivity analysis and the dual to a linear programming model was described in Section 2.8. It is not uncommon for commercial software to include subroutines embodying a method known as the **dual Simplex** method. During sensitivity analysis, if problem parameters are changed, the current (optimal) solution may become infeasible. However, the problem is then **dual feasible**, and can be reoptimized using the dual Simplex algorithm.

2.10.2 Interior Methods

The complexity of linear programming problems was for many years one of the most important open questions in theoretical computer science. Efforts were made to prove that Dantzig's Simplex method would always stop sooner than $\binom{n}{m}$ iterations, but instead, problems were devised which drive the Simplex method through the combinatorial explosion of basic solutions. On the other hand, the linear programming problem did not seem to be NP-hard either.

The question was first answered in 1979 when the Russian mathematician Leonid B. Khachiyan published an algorithm for solving linear programming problems in polynomial time. Initial confusion over the importance of Khachiyan's discovery arose for two reasons. First, his results appeared in a very short article in a Russian journal and went unnoticed for months because of its obscurity as well as the fact that the report was written in the Russian language. After some time, Eugene Lawler at the University of California at Berkeley brought the article to the attention of the computer science community. The explanation that Khachiyan himself presented was so abbreviated that mathematicians had little inkling of its content. Finally, through Lawler's efforts, Khachiyan's work was expanded upon (and the details of the proof reconstructed) by Gacs and Lovasz (1981), who not only filled in the gaps in the proof but improved on the efficiency of the algorithm. Only then was the new idea available to the general mathematics community for consideration and discussion. Almost nothing was known about Khachiyan himself, and it was generally assumed, even by Gacs and Lovasz, that he had never published any previous works. However, as it turns out, (Aspvall and Stone 1980) cite four publications by Khachiyan prior to his famous one in 1979.

The second misunderstanding arose because Khachiyan's algorithm was designed for linear programming problems in which c, A, and b are *integers*. Careless reporters publicized incorrectly that Khachiyan had developed a polynomial-time algorithm for *integer programming problems* (such as the traveling salesman problem). Because this part of the story was untrue, there was skepticism concerning just what Khachiyan really had done. Major newspapers around the world contributed to the notoriety (but sadly not to the clarification) of this remarkable discovery.

Because linear programming problems had been suspected of having *borderline* complexity—neither being NP-hard nor having a polynomial algorithm—Khachiyan's demonstration of a polynomial-time algorithm was somewhat surprising and of immense importance. Even George Dantzig, who developed the (worst-case exponential-time) Simplex algorithm, graciously offered the comment that, "A lot of people, including myself, spent a lot of time looking for a polynomial-time algorithm for linear programming. I feel stupid that I didn't see it" (Kolata 1979).

Khachiyan's method operates by defining a sequence of ellipsoids (ellipses in a multi-dimensional space), each smaller than the previous ellipsoid, and each containing the feasible region. The method generates a sequence of points $x_0, x_1, x_2, \ldots$, which form the centers of the ellipsoids. At each iteration, if the center x_k of the ellipsoid is infeasible, a hyperplane parallel to a violated constraint and passing through x_k is used to cut the ellipsoid in half. One half is completely infeasible, but the other half contains the feasible region (if it exists), so a smaller ellipsoid is constructed that surrounds this half. Eventually, some x_k will lie in the feasible region.

From a practical standpoint, Khachiyan's **ellipsoid method** lacked the many years of fine-tuning that had been directed toward improving the efficiency of the Simplex method. Therefore, although it was a polynomial-time algorithm, in practice the Simplex method was the preferred method because typically it performed quite well, and software implementations were readily available. It should be noted, however, that whereas the computation time for the Simplex method is most strongly dependent on the number of constraints m, Khachiyan's method is relatively insensitive to m and more strongly dependent on the number of decision variables n. Thus, it was supposed at the time that Khachiyan's ellipsoid method might eventually be superior, in practice, to the Simplex method for problems with numerous constraints. In any case, just five years later in 1984, yet another new method appeared.

Narendra Karmarkar, a young mathematician at AT&T Bell Laboratories, announced an algorithm for solving linear programming problems that was even more efficient than Khachiyan's method. Karmarkar's method is called an **interior point method** since it operates from *within* the polyhedron of feasible points of the linear programming problem. The algorithm uses a series of *projective transformations* in which the polyhedron is first made smoother (*normalized*), then an arbitrary point is selected which is re-mapped to the center, and a sphere is inscribed in the polyhedron. Then a new point is selected, near the edge of the sphere and in the direction of the optimal solution. The space is then transformed or *warped* again so that this new point is in the center. The process is repeated until the selected point is the optimal solution to the linear programming problem. Karmarkar's method of projective transformations demonstrates a polynomial-time complexity bound for linear programming that was better than any previously known bound.

Karmarkar's original announcement claimed that his method was many times faster than the Simplex method. But since AT&T Bell Laboratories' proprietary interests precluded disclosure of the details of its implementation, it was not at first possible to test Karmarkar's claims. In fact, for several years, the scientific community remained somewhat annoyed because no one outside Bell Laboratories was in a position to duplicate Karmarkar's computational experiments—and hence the traditional scientific peer review process could not take place.

Whereas Karmarkar had claimed computation times 50 times faster than Simplex based codes, outside researchers were implementing Karmarkar's method and observing computation times 50 times *worse*. Eventually, however, over the next 10 years, it became evident that by using special data structures, efficient methods for handling sparse matrices, and clever Cholesky factorization techniques, the performance of Karmarkar's method could become quite competitive with Simplex implementations.

An important side effect of the controversy over the validity of Karmarkar's claims is that it sparked a great deal of interest in examining and refining Simplex implementations. Consequently, there are now many very efficient implementations of both approaches.

An overview (Lustig et al. 1994) indicated that small problems, in which the sum of the number of decision variables plus the number of constraints is less than 2000, can generally be solved faster with the Simplex method. For medium sized problems, in which that sum is less than 10,000, Simplex and interior methods compete evenly. And there are several extremely large linear programming problems that have now been solved by **interior point methods** which have never been solved by any Simplex code. An increasing number of commercial software products contain both interior point methods and Simplex methods that can be used together or separately in solving large or difficult problems. Each of these approaches has its advantages, and hybrid software that combines these complementary methods constitutes a powerful computational tool for solving linear programming problems.

As the methods suggested originally by Karmarkar became more widely understood, numerous researchers made their own various contributions to the practical implementation of interior point algorithms. A very thorough summary of theoretical and implementational developments, as well as computational experimentation, may be found in a feature article by (Lustig et al. 1994). Bixby (1994) presents an enlightening description of commercial interior point methods, options, and performance on benchmark problem instances. Saigal (1995) is a comprehensive reference that includes a large section on interior point methods. Mitra et al. (1988) report experimental studies with hybrid interior/ Simplex methods. Thus, the theoretical merits of Karmarkar's new approach, which had never been doubted, have finally been balanced by considerable practical computational experience. As an illustration of this, recall that interior point methods must remain in the interior of the feasible region. Yet computational experience shows that choosing a step length that gets very close to (and nearly outside of) the boundary of the region is actually most efficient. So-called **barrier** parameters are used to control the interior search in the feasible region.

The interior and barrier methods were inspired by (and incorporate) many of the more general methods of nonlinear programming. It should be noted that interior point methods did not originate with Karmarkar; in fact, the approach had been used since the 1950s for *nonlinear* programming problems. However, Karmarkar can be credited with demonstrating that interior point methods could also be practical for solving *linear* programming problems. Therefore, a student who wishes to fully understand these methods might well begin by reading the introductory notions presented in Chapter 5 on Nonlinear Optimization, and then be prepared to embark on a serious study of the mathematics and numerical analysis underlying general optimization procedures.

2.10.3 Software for Solving Linear Programming

The Simplex method is theoretically not an efficient algorithm because its worst case performance is exponential in the size of the problem being solved. However, empirical evidence, observed over many years and many practical problem instances, shows the Simplex method to be consistently very efficient in practice.

The computational effort required for solving a linear program with the Simplex method is strongly dependent on the number of constraints m, and almost independent of the number of variables n. In typical problems, we find that the number of constraints is much less than the number of variables, and in just such cases, the Revised Simplex has great computational advantage over the standard Simplex. In practical experience, the number of Simplex iterations required, on average, to solve a problem with m constraints, is 2m. A practical, although not strict, upper bound on the number of iterations is 2(m + n) (Ravindran et al. 1987). Total computation time has been observed to increase roughly in the order of m^3.

Thus, a 1000-constraint problem may require a million times as much computation time as a 10-constraint problem. In practice, we can generally expect to obtain solutions to linear programming problems very efficiently, despite the lack of any attractive performance guarantees.

To give some perspective to the notion of **problem size** (and to dispel any misperceptions that may have been created by the very small illustrative examples used earlier in this chapter), we should indicate just what is considered a *large* linear programming problem. Problem size is usually expressed in terms of the number of constraints, the number of decision variables (which may or may not include the slack and surplus variables), and perhaps the number of non-zero coefficients in the matrix. In the early 1950s, when the first linear programming software was being developed, an inversion of a matrix of order 100 was considered state of the art in numerical computation. Nowadays, a linear programming problem with thousands of constraints is routine, and problems with tens to hundreds of thousands of constraints are computationally manageable. Advances in hardware technology have delivered dramatically increased processing speeds, and corresponding hardware and software developments in storage capacities and memory management techniques have facilitated computations on the data representing very large problems.

Software for linear programming has been under development for many decades, first using Simplex and related techniques and now including interior point implementations, decomposition, and barrier methods, among other advances, all having evolved together into standard forms. One might think that there is little room, or need, for any significant changes in LP solver technology. But with steady advances in processor speed and storage capabilities, computational mathematics, algorithm engineering, potentials for parallel and distributed computing, and powerful and convenient modeling systems that encourage analysts to attack ever larger and more challenging problems, we are seeing even more remarkable developments in software.

Software vendors typically offer a variety of versions of their packages. The options may be based on the choice of modeling language and the input/output interfaces, the hardware platform and the underlying operating system. Some of these options and characteristics are presented clearly and succinctly in a very useful series of survey articles by (Sharda 1995, 1992) and (Fourer 2015, 2017) that describe many of the most popular software products now available. We mention a few of them here to provide a glimpse of what is currently in use by practitioners who need to solve linear programming problems.

Many advanced modeling languages and systems, such as those mentioned in Chapter 1, provide interfaces with linear programming solvers. For example, AMPL, GAMS, and MPL facilitate linear optimization with advanced features for large-scale problems and parallel simplex methods by offering access to CPLEX, MINOS, and OSL.

IBM ILOG CPLEX Optimizer (commonly referred to as CPLEX) is designed to solve large, difficult linear programming (and other) problems which some other LP solvers cannot solve efficiently. It has been developed to be fast, robust, and reliable, even for poorly scaled or numerically difficult problems. This software uses a modified primal and dual Simplex algorithm, along with interior point methods. CPLEX is currently used to solve some of the largest problems in the world, some with millions of variables, constraints, and non-zeros. Options include a preprocessor for problem reduction, as well as parallel implementations that have demonstrated record-breaking performance. CPLEX is portable across Windows PCs, Unix/Linux, and Mac OS platforms.

MINOS offers numerically stable implementations of primal Simplex, using sparse LU factorization techniques. This system originated with (Murtagh and Saunders 1987) with versions for PCs, Windows, Unix, and mainframe systems.

LINDO (Linear **IN**teractive and **D**iscrete **O**ptimizer), originally developed by Linus Schrage (1991), is one of the oldest and now among the most popular commercial systems for solving linear programming problems. **LINDO API** and the **LINGO** modeling system offer powerful solvers for linear programs, based on methods including primal and dual simplex for speed and robust computations.

SAS provides an integrated package, with capabilities for solving a wide variety of Operations Research problems. SAS/OR subroutines for solving linear programming problems use two phase Revised Simplex, primal and dual simplex, and interior point methods, and employ decomposition algorithms and efficient sparse-matrix techniques.

Gurobi Optimization solves linear programming problems through the use of advanced algorithms taking advantage of various modern powerful hardware architectures.

IMSL has an established reputation in the field of numerical problem-solving software, known for accuracy and dependability. IMSL contains literally thousands of mathematical and statistical library routines including linear programming routines based on the Revised Simplex method. Routines are implemented on a wide variety of platforms.

This selection of commercial software products is by no means exhaustive; we have merely mentioned several representative packages that are in popular use. With new product enhancements constantly under development, our readers should have no trouble finding many additional sources of software for solving linear programming problems.

2.11 Illustrative Applications

2.11.1 Forest Pest Control Program (Rumpf et al. 1985)

The Maine Forest Service operates a program of aerial pesticide spraying to mitigate the destruction of spruce-fir forests by the spruce budworm. Yearly spraying of the 5 million acre infestation takes place in early summer during a specific stage of insect development, and must be done in dry weather under moderate wind conditions. Spraying is done by aircraft teams consisting of a spray aircraft, a guide plane with a pilot and navigator, and a monitor plane with a pilot and observer. The entire program includes analysis of insect damage and danger assessment of treatment requirements, and cost of chemicals, but one third of the total cost of the program is for aircraft and crews. The Forest Service has therefore wisely investigated the use of quantitative methods to maximize the efficiency of aircraft assignments and to reduce aircraft needs.

The aircraft operate out of eight airfields, and preliminary models were developed to partition the infested area into over 300 regions (spray blocks) about each airfield, and to then assign spray blocks to airfields and aircraft to airfields.

This initially seemed like a natural problem to be formulated as a network problem or integer programming model (see Chapters 3 and 4); but some of the realistic elements of this problem could not be incorporated into the network models, and the integer programming formulation turned out to be prohibitively large. Finally, a linear programming formulation was developed that models the problem realistically and that can be solved quite efficiently.

The decision variables are the times allocated to each aircraft team flying out of each airfield to spray each block. The objective function includes only those variables associated with allowable combinations of blocks, aircraft, and airfields; that is, blocks within operating range of the airfield, aircraft capable of spraying the type of pesticide prescribed for a certain block, and the specified type of aircraft team (planes and crew) stationed at the given airfield. The aim is to minimize total spraying cost.

Constraints are imposed to guarantee sufficient time to completely spray each block (and this depends on the geometrical shape of the block, the speed of the aircraft, the pesticide capacity of the plane, and the availability of chemicals at the airfield). A second category of constraints accounts for the time windows during which weather conditions and budworm development are appropriate for effective aerial spray.

The use of this model has saved time and reduced the cost of the aerial spraying program. It has also provided a framework from which to analyze major modifications to the program, such as loss of an airfield or the availability of a new long-range aircraft, and, in response to environmental concerns, to re-evaluate the actual need for spraying certain areas.

2.11.2 Aircraft and Munitions Procurement (Might 1987)

The US Air Force uses a linear programming model to decide annually how much of its procurement budget should be spent on various different aircraft (such as the F-16, A-10, F-111, and F-15E) and on various conventional munitions. It has been argued that quantitative methods are inapplicable for strategic decisions that are highly unstructured. However, senior level decision makers are rotated frequently and often lack long experience and judgment on which to base procurement decisions. For this reason, quantitative analytical decision support has proved to be of great benefit.

The decision involves analyzing the cost-effectiveness of each aircraft carrying each of several possible munitions. The difficulty arises because the attrition of the aircraft is dependent on the munitions being delivered, and an aircraft may be vulnerable to different types of attack, depending on the weapon it is carrying. Likewise, an aircraft must fly at different altitudes with different munitions and thus anti-aircraft weapons vary in effectiveness. And when the loss rate varies only a few percent, there is considerable variation in the number of attacks an aircraft can make during a conflict; thus, the cost-effectiveness of an aircraft-munitions combination is difficult to measure subjectively.

The data used by the linear program include:

- The effectiveness of each aircraft munitions combination against each target type in each of six different weather conditions
- The attrition (probability of loss) of each aircraft for each aforementioned condition
- The number of munitions delivered on each sortie for each condition
- The number of sorties per day for each aircraft munitions combination
- Current inventory of aircraft and munitions
- Number and value of each type of target
- Cost of each new aircraft and munitions type

Thus, the decision variables are the total number of sorties flown by each aircraft munitions combination against each target type in each of six types of weather. The objective is the sum of these variables, each multiplied by the probability of a successful sortie times the value of the target type.

Five categories of constraints are defined for aircraft, munitions, targets, weather, and budget. The current implementation has pre- and post-processors for data formatting, and can be run with different databases. Output includes listings, tables, and graphical displays indicating, for example, trade-offs of funds expended on aircraft versus munitions, target value destroyed versus expenditure on individual munitions or a mixture of munitions. This linear programming approach to procurement has received enthusiastic acceptance within the military procurement community.

2.11.3 Grape Processing: Materials Planning and Production (Schuster and Allen 1998)

Welch's grape processing company has successfully employed linear programming models for optimizing its management of raw materials in its production and distribution of grape juice products. Welch's, Inc. is owned by a cooperative, the National Grape Cooperative Association, involving 1400 growers of Concord and Niagara grapes in the northern United States. Membership in the cooperative is attractive to grape growers because Welch's offers a reliable and consistent market for grapes, despite fluctuations in agricultural productivity.

Welch's plants comprise a vertically integrated industry, handling the acquisition and pressing of raw grapes, the storage of pasteurized grape juice and concentrates, production of jams, jellies, and juice products, and the warehousing and distribution of finished products. The company wishes to maintain consistent taste in its products, although weather and geography account for great variations in grape characteristics (sweetness, color, etc.) from year to year.

Welch's had a comprehensive *materials requirement planning* system to estimate all the resources needs, from juicing raw grapes to the completion of manufactured products. This, along with a minicomputer based cost accounting system have proved useful, but do not provide optimal cost solutions for the very important juice blending operation; and each run of the system takes so much computational time that interactive real-time use of the system is impractical. Furthermore, whereas most industries try to schedule capacities first and then project their materials requirements, the existing system at Welch's did not incorporate any consideration of capacities such as juice concentrations or transportation between plants. Without use of operational constraints such as these, it was not possible to choose effectively from among a large set of feasible product recipes and to efficiently schedule inter-plant transfers. Optimal movement of raw materials among plants and optimal blending of raw materials into products was not supported by any formal system, and was dealt with by trial-and-error and with input from the simple cost-accounting system.

An initial attempt at modeling this problem resulted in a linear programming formulation with 8000 decision variables. Preliminary testing of this *juice logistics* model indicated the workability of the formulation. But management, lacking understanding of the model and fearing major software management problems, did not fully support the use of the model.

In response to this, analysts dealt with the software maintenance difficulty by choosing economical spreadsheet software (What's Best!), which provided convenient interfaces for the model, the analysts, and management. Unfortunately, the 8000 variables overwhelmed this software package. Analysts revised the model by forming aggregate product groups rather than dealing with individual products (e.g., all purple-juice products could be treated as a single aggregate, from a materials standpoint). In this way, the model was streamlined into one having only 324 decision variables. This aggregate view invoked suspicion of yielding misleading and overly simplified inventory projections. Although such concern is probably justified in production planning and disaggregation of end products, it turned out that for purposes of materials planning, this is a perfectly acceptable simplification.

Once this very tractable model was put into regular use, it was realized that the model not only offered a much better structured approach to planning and resulted in significant cost improvements, but it also functioned effectively as a communication tool. Rather than being treated as a piece of special offline data, the optimal solution produced by this linear programming model became a central point of discussion in management meetings and an essential operational tool for the planning committee. The complete acceptance of the model as a legitimate component in decision-making placed Welch's in a position to make key decisions quickly. A profitable decision was made, for example, on whether to purchase raw grapes on the open market (outside the cooperative) during lean crop years; and the system permits rapid decisions on carrying over inventories of grape juice during record-breaking production years (such as happened in 1991 through 1995), and successfully meeting demand after the harsh winter of 1996 by adjusting product recipes.

The analysts at Welch's attribute the acceptance and successful use of the linear programming model to their having reduced the original model to a size compatible with spreadsheet optimization. This alleviated difficulties with software support. Furthermore, the resulting smaller model was more understandable to people having various levels of mathematical interest, ability, and appreciation. Thus, the simpler model proved to be the most workable one in actual practice. Future plans call for development of a more comprehensive model, capable of incorporating changes in material characteristics over time.

2.12 Summary

Linear programming is a special type of mathematical programming, in which the objective function and the constraints can be expressed as linear functions of the decision variables. Once a problem is formulated as a linear program, it is possible to analyze the model and investigate the nature of the solutions to the problem. Graphical solutions for small problems can be illustrative of some of the characteristics of the solutions. In general, linear programming problems may have a unique optimal solution, multiple optimal solutions, or no optimal feasible solution.

For linear programming problems of practical size, the most widely used technique for obtaining solutions is the Simplex method. Applicable to essentially all linear programming models, the Simplex method provides an efficient and effective means of either solving the problem, or discovering that there is no solution.

Every linear programming problem has a dual problem, which often provides a useful alternative interpretation of the solution to the original problem. The theory of duality also suggests ways in which analysts can determine how sensitive a solution is to minor changes in problem parameters.

Relatively recent research has led to the development of new computational approaches, known as barrier methods, or interior point methods. These techniques can in some cases be used effectively to solve the isolated few problems that had never been successfully dealt with using the Simplex method alone. But more importantly, these newer ideas have been integrated skillfully together with older Simplex algorithms to produce new hybrid software that performs better than any one method used independently.

Key Terms

adjacent extreme points
artificial variables
basic solution
basic variables
Big-M method
binding constraints
complementary slackness
constraints
crashing
degeneracy
degenerate solution
degrees of freedom
dual feasible
dual problem
dual Simplex
duality property
extreme point
feasible solution
feasible space
formulation
ellipsoid method
entering variable
graphical solution
infeasible solution
interior point methods
leaving variable
linear programming
marginal worth
multiple optimal solutions
non-basic variable
non-binding constraints
objective function
optimal feasible solution
optimal solution
pivot column
pivot element

pivot operations
pivot row
postoptimality analysis
primal problem
product form
range analysis
re-inversion
Revised Simplex method
right-hand-side ranging
scaling
shadow prices
Simplex method
Simplex tableau
sensitivity analysis
slack variable
solution
standard form
surplus variable
two phase method
unbounded solution
upper bound constraints

Exercises

2.1 An academic computing center receives a large number of jobs from students and faculty to be executed on the computing facilities. Each student job requires six units of space on disk, and three units of time on a printer. Each faculty job requires eight units of space on disk, and two units of time on a printer. A mixture of jobs is to be selected and run as a batch, and the total disk space and printer time available for a batch are 48 units and 60 units, respectively. The computer center is paid three times as much for running a student job as for running a faculty job. Formulate a linear programming problem to determine the mixture of jobs to be run as a batch that will maximize computer center income.

2.2 A tree farm cultivates Virginia pine trees for sale as Christmas trees. Pine trees, being what they are, require extensive pruning during the growing season to shape the trees appropriately for the Christmas tree market. For this purpose, the farm manager can purchase pruning hooks for $16.60 each. He also has a ready supply of spears (at $3 each) that can be bent into pruning hooks. This conversion process requires one hour of labor, whereas final assembly of a purchased pruning hook takes only 15 minutes of labor. Only 10 hours of labor are available to the manager. With labor rates at $8.40 per hour, the farm manager intends to spend no more than $280 on buying or making pruning hooks this year. In all, how many pruning hooks can he acquire (from outright purchase and through conversion), given these limitations? Formulate this as a linear programming problem.

2.3 A plant has five machines, each of which can manufacture the same two models of a certain product. The maximum number of hours available on the five machines during the next production period are, respectively, 60, 85, 65, 90, and 70. The demand for products created during this next production period is expected to be 850 units of model 1 and 960 units of model 2. The profits (in dollars per hour) and production rates (per hour) are given in tabular form:

| | Profit | | | Production Rate | |
| | Model | | | Model | |
Machine	1	2	Machine	1	2
1	2	5	1	7	9
2	8	3	2	5	4
3	3	6	3	6	3
4	5	3	4	4	8
5	4	7	5	5	6

Let x_{ij} be the number of hours machine i is scheduled to manufacture model j, for $i = 1, \ldots, 5$ and $j = 1, 2$. Formulate a linear programming model to maximize profits.

2.4 Metallic alloys A, B, and C are to be made to customer specifications from four different metals (W, X, Y, and Z) that are extracted from two different ores. The cost, maximum available quantity, and constituent parts of these ores are:

| | | Maximum Tons | Percentage of Constituents | | | |
Ore	Cost ($/ton)	Available	W	X	Y	Z
I	150	2800	40	10	15	25
II	95	3100	30	20	10	20

Customer specifications and selling price for the three alloys are:

Alloy	Specifications	Selling Price ($/ton)
A	At least 30% of X At least 50% of W At most 10% of Y	600
B	Between 30% and 40% of Z At least 40% of X At most 70% of W	500
C	At least 40% of Y At most 60% of W	450

Formulate a linear programming model that meets the specified constraints and maximizes the profits from the sale of the alloys. (*Hint*: Let x_{ijk} be the amount of the i-th metal extracted from the j-th ore and used in the k-th alloy.)

2.5 Show graphically the feasible region corresponding to the following set of constraints:

$-2x_1 + x_2 \geq -4$

$x_1 + x_2 \leq 8$

$-x_1 + x_2 \leq 6$

$x_1, x_2 \geq 0$

Give the coordinates of each of the extreme points of the feasible region.

2.6 What is the feasible region corresponding to the following set of constraints?

$x_1 + 3x_2 \leq 24$

$x_1 \leq 6$

$-x_1 + 2x_2 \leq 10$

$x_1, x_2 \geq 0$

Evaluate the objective function $z = 2x_1 + 5x_2$ at each of the extreme points of this feasible region.

2.7 Solve the following linear programming problem graphically.

maximize $z = x_1 - x_2$

subject to $x_1 + x_2 \geq 1$

$3x_2 \leq 9$

$2x_1 + x_2 \leq 4$

$x_1 \leq \dfrac{3}{2}$

$x_1, x_2 \geq 0$

Give the optimal value of z and the optimal solution (x_1, x_2).

2.8 Solve the following linear programming problem graphically:

maximize $z = -2x_1 + x_2$

subject to $x_1 - x_2 \leq 5$

$x_1 \leq 7$

$x_2 \leq 6$

$x_1 - x_2 \geq -4$

$x_1, x_2 \geq 0$

Outline the feasible region, and give the optimal values of z, x_1, and x_2.

2.9 Examine the following formulation, and comment on the nature of its solution:

maximize $z = 3x_1 - 2x_2$

subject to $x_1 \leq 2$

$x_2 \leq 3$

$3x_1 - 2x_2 \geq 8$

$x_1, x_2 \geq 0$

2.10 Examine the next formulation, and comment on the nature of its solution:

maximize $z = 3x_1 + 4x_2$

subject to $6x_1 + 8x_2 \leq 10$

$x_1 + x_2 \geq 1$

$x_1, x_2 \geq 0$

2.11 Examine the following formulation, and comment on the nature of its solution:

maximize $z = 5x_1 + 4x_2$

subject to $x_2 \leq 10$

$x_1 - 2x_2 \geq 3$

$x_1, x_2 \geq 0$

2.12 Place the following linear programming model in standard form:

maximize $z = 16x_1 + 2x_2 - 3x_3$

subject to $(1)\ x_1 - 6x_2 \geq 4$

$(2)\ 3x_2 + 7x_3 \leq -5$

$(3)\ x_1 + x_2 + x_3 = 10$

$(4)\ x_1, x_2, x_3 \geq 0$

2.13 Place the following linear programming model in standard form:

maximize $z = 5x_1 + 6x_2 + 3x_3$

subject to   (1) $|x_1 - x_3| \leq 10$

                  (2) $10x_1 + 7x_2 + 4x_3 \leq 50$

                  (3) $2x_1 - 11x_3 \geq 15$

                  $x_1, x_3 \geq 0$

                  x_2 unrestricted in sign

2.14 Give all of the basic solutions and basic feasible solutions of the problem in Exercise 2.9.

2.15 Give the coordinates of all of the basic solutions and basic feasible solutions of the problem in Exercise 2.10.

2.16 Use the Simplex algorithm to solve the linear programming formulation from Exercise 2.1. What is the percentage utilization of the disk and printer resources at optimality? Comment on how the university community is likely to react to the optimal solution to this problem.

2.17 Solve the following problem using the Simplex method:

maximize $z = x_1 + 2x_2$

subject to   (1) $x_1 + x_2 \geq 6$

                  (2) $x_2 \leq 6$

                  (3) $x_1 \leq 8$

                  $x_1, x_2 \geq 0$

2.18 Solve the following problem using the Simplex method:

maximize $z = 4x_1 + x_2$

subject to   (1) $3x_1 + x_2 = 3$

                  (2) $4x_1 + 3x_2 \geq 6$

                  (3) $x_1 + 2x_2 \leq 3$

                  $x_1, x_2 \geq 0$

2.19 Apply the Simplex algorithm to each of the following problems. Observe the behavior of the Simplex method and indicate which problems display degeneracy, multiple optima, infeasibility, or an unbounded solution.

a. maximize $3x_1 + x_2$

    subject to   (1) $x_1 \leq 8$

                      (2) $2x_1 - 3x_2 \leq 5$

                      (3) $x_1, x_2 \geq 0$

b. maximize $3x_1 + 4x_2$

    subject to   (1) $x_1 + x_2 \geq 5$

                      (2) $2x_1 + x_2 \leq 4$

                      (3) $x_1, x_2 \geq 0$

c. maximize $x_1 + 2x_2$

    subject to   (1) $x_1 + 2x_2 \leq 10$

                      (2) $x_1, x_2 \geq 0$

d. maximize $3x_1 + 9x_2$

    subject to   (1) $x_1 + 4x_2 \leq 8$

                      (2) $x_1 + 2x_2 \leq 4$

                      (3) $x_1, x_2 \geq 0$

2.20 Create a linear programming problem formulation that has unbounded solutions but in which no evidence of unboundedness appears in the initial Simplex tableau.

2.21 Perform as many Simplex iterations as possible on the example problem in Section 2.7.2. Observe that the algorithm terminates when there are no ratios θ_i from which to choose a variable to leave the basis.

2.22 Solve the following linear programming problem using the Two Phase Simplex method.

maximize $z = 4x_1 + x_2$

subject to $3x_1 + x_2 = 3$

$4x_1 + 3x_2 \geq 6$

$x_1 + 2x_2 \leq 3$

$x_1, x_2 \geq 0$

2.23 Examine this linear programming formulation:

maximize $x_1 + 2x_2$

subject to $x_1 + 2x_2 \leq 10$

$x_1, x_2 \geq 0$

Comment on the nature of its solution(s). How does this change if the first constraint is removed from the problem?

2.24 Solve the following linear programming problem graphically.

maximize $x_1 - x_2$

subject to $x_1 + x_2 \geq 1$

$3x_2 \geq 9$

$2x_1 + x_2 \leq 4$

$x_1 \leq 1.5$

$x_1, x_2 \geq 0$

2.25 What determines the number of basic variables in a linear programming problem solution?

2.26 What is the value of a non-basic variable in a feasible solution of a linear programming problem?

2.27 In an optimal Simplex tableau, what is the economic interpretation of the objective function row entry corresponding to the i-th slack variable?

2.28 In a Simplex tableau, what is the interpretation of the entries in the right-hand-side column?

2.29 What is the consequence of a tie for the entering basic variable?

2.30 What if there is a tie for the leaving basic variable?

2.31 What if, in the objective function row of a final tableau, there is a zero in a column corresponding to a non-basic variable?

2.32 What happens in the Simplex algorithm if you choose, as the entering variable, a variable with a negative objective row coefficient but not the *most* negative coefficient?

2.33 Solve the following problem using the Simplex method:

maximize $\quad z = x_1 + 9x_2 + x_3$

subject to $\quad x_1 + 2x_2 + 3x_3 \le 9$

$\qquad\qquad 3x_1 + 2x_2 + 2x_3 \le 15$

$\qquad\qquad x_1, x_2, x_3 \ge 0$

2.34 Use the Two Phase Simplex method to solve the following problem:

minimize $\quad z = 16x_1 + 2x_2 - 3x_3$

subject to $\quad x_1 - 6x_2 \ge 4$

$\qquad\qquad 3x_2 + 7x_3 \le -5$

$\qquad\qquad x_1 + x_2 + x_3 = 10$

$\qquad\qquad x_1, x_2, x_3, \ge 0$

2.35 A business executive has the option of investing money in two plans. Plan A guarantees that each dollar invested will earn 70 cents a year hence, and plan B guarantees that each dollar invested will earn $2 two years hence. Plan A allows yearly investments, while in plan B, only investments for periods that are multiples of two years are allowed. How should the executive invest $100,000 to maximize the earnings at the end of three years? Formulate this problem as a linear programming problem.

2.36 An investment portfolio management firm wishes to develop a mathematical model to help decide how to invest $1 million for one year. Municipal bonds are to be bought in combinations that balance risk and profit. Three types of bonds are being considered:

- AAA rated bonds yielding 6% annually and which must be purchased in units of $5000
- A rated bonds yielding 8% annually and which must be purchased in units of $1000, and
- J rated (junk) bonds yielding 10% annually and which must be purchased in units of $10,000.

The Board of Directors has specified that no more than 25% of the portfolio should be invested in (risky) junk bonds, and at least 40% should be invested in AAA rated bonds. Bonds are to be purchased with the objective of maximizing earnings at the end of the year. It may be assumed that the stated yield dividend is paid at the end of the year, and that no other distributions are made during the year. Formulate this problem as a linear programming problem.

2.37 A philanthropist wishes to develop a mathematical model to help him decide how to donate his spare cash to several worthy causes. He has $10 million to distribute among the recipients, and he would like to donate in units of thousands of dollars.

Three organizations would like to receive funds: Our Great State University, the Friends of the Grand Opera, and the Save the Humuhumunukunukuapua'a Society. The philanthropist wants to give at most 50% of his cash to any one organization. The desirability of the philanthropist's giving to any particular recipient is to be measured in terms of the number of tax credits he will receive. The value of giving to an educational institution is rated at 10 credits for every $1000 donation, while the value of $1000 donation to the music lovers is rated at 8 credits, and each $1000 donation to the wildlife conservation is rated at 6 credits. Write a linear programming model to help this philanthropist maximize the number of tax credits that can be achieved by contributing among these three groups.

2.38 Solve the following problem graphically:

maximize $\quad$ $z = -2x_1 + x_2$

subject to $\quad$ $x_1 - x_2 \leq 5$

$\qquad\qquad$ $x_1 \leq 7$

$\qquad\qquad$ $x_2 \leq 6$

$\qquad\qquad$ $x_1 - x_2 \geq -4$

$\qquad\qquad$ $x_1, x_2 \geq 0$

2.39 Write the dual of the primal linear programming problem in Exercise 2.7.

2.40 Write the dual of the primal problem in Exercise 2.8. Solve the dual problem, and identify the shadow prices.

2.41 Solve the dual problem corresponding to the primal problem in Exercise 2.12. Determine whether optimal solutions exist. If so, describe the relation between the primal shadow prices and dual variables at optimality.

2.42 Describe the nature of the solutions of the primal problem in Exercise 2.10 and its dual problem.

2.43 Each of the following statements refers to the Simplex algorithm. Fill in the blanks with an appropriate letter from the following choices:

1. If all slack and surplus variables are zero in an optimal solution, then _____.
2. If a basic variable has the value zero in an optimal solution, then _____.
3. If an artificial variable is non-zero in an optimal solution, then _____.
4. If a non-basic variable has zero coefficient in the top row of an optimal tableau, then _____.

Completion alternatives:

A. There are multiple optimal solutions.
B. The current solution is degenerate.
C. All constraints are equalities at optimality.
D. The shadow prices are inverses of the dual variables.
E. No feasible solution exists.
F. The solution is unbounded.

2.44 The following statements are intended to describe the relationship between primal and dual linear programming problems. For each statement, fill in the blank to indicate the most appropriate choice from the alternatives shown in the following list.

   1. The optimal objective function value in the primal problem corresponds to _____.

   2. The shadow prices in the optimal primal tableau correspond to _____.

   3. Basic variables in the optimal primal tableau correspond to _____.

   4. The variables in the primal problem correspond to _____.

   5. Shadow prices in the optimal dual tableau correspond to _____.

      Completion alternatives:

      A. The primal non-basic variables

      B. The dual non-basic variables

      C. The primal constraints

      D. Optimal basic variables in the dual problem

      E. The optimal objective function value in the dual

      F. The shadow prices in the dual

      G. Basic variables in the optimal primal problem

      H. The constraints in the dual problem

2.45 Recall Example 2.8.3 and verify the range within which changes in objective function coefficient c_2 can vary without affecting the optimal solution.

2.46 What was the theoretical significance of the algorithm developed by Khachiyan for solving linear programming problems?

2.47 What is the practical significance of the *interior point* methods, as originated by Karmarkar, for solving linear programming problems? How do these methods compare in practice with the traditional Simplex-based methods?

References and Suggested Readings

Albers, D. J., and C. Reid. 1986. An interview with George B. Dantzig: The father of linear programming. *The College Mathematics Journal* 17 (4): 293–314.

Arbel, A. 1993. *Exploring Interior-Point Linear Programming Algorithms and Software*. Cambridge, MA: MIT Press.

Aronofsky, J. S., and A. C. Williams. 1962. The use of linear programming and mathematical models in underground oil production. *Management Science* 8: 394–402.

Aspvall, B., and R. E. Stone. 1980. Khachiyan's linear programming algorithm. *Journal of Algorithms* 1: 1–13.

Bartels, R. H. 1971. A stabilization of the simplex method. *Numerische Mathematik* 16: 414–434.

Benichou, M., J. M. Gauthier, G. Hentges, and G. Ribiere. 1977. The efficient solution of large-scale linear programming problems—Some algorithms techniques and computational results. *Mathematical Programming* 13: 280–322.

Bixby, R. E. 1994. Progress in linear programming. *ORSA Journal of Computing* 6 (1): 15–22.

Bland, R. G., D. Goldfarb, and M. J. Todd. 1981. The ellipsoid method: A survey. *Operations Research* 26 (9): 1039–1091.

Borgwardt, K. H. 1980. *The Simplex Method: A Probabilistic Analysis*. New York: Springer-Verlag.

Calvert, J. E., and W. L. Voxman. 1989. *Linear Programming*. Orlando, FL: Harcourt Brace Jovanovich.

Charnes, A., and W. W. Cooper. 1961. *Management Models and Industrial Applications of Linear Programming*, Vol. I and II. New York: John Wiley & Sons.

Charnes, A., W. W. Cooper, and A. Henderson. 1953. *An Introduction to Linear Programming*. New York: John Wiley & Sons.

Cooper, L., and D. Steinberg. 1974. *Methods and Applications of Linear Programming*. Philadelphia, PA: W.B. Saunders.

Dantzig, G. B. 1963. *Linear Programming and Extensions*. Princeton, NJ: Princeton University Press.

Dantzig, G. B., and M. N. Thapa. 1996. *Linear Programming*. New York: Springer-Verlag.

Dantzig, G. B., and R. M. Van Slyke. 1967. Generalized upper bounding techniques for linear programming. *Journal of Computer and System Sciences* 1 (3): 213–226.

Dantzig, G. B., and P. Wolfe. 1960. Decomposition principle for linear programs. *Operations Research* 8: 101–111.

Dantzig, G., A. Orden, and P. Wolfe. 1955. The generalized Simplex Method for minimizing a linear form under linear inequality restraints. *Pacific Journal of Mathematics* 5 (2): 183–195.

Emmett, A. 1985. Karmarkar's algorithm: A threat to simplex? *IEEE Spectrum* 22: 54–55.

Fieldhouse, M. 1986. *Commercial Linear Programming Codes on Microcomputers*. In J. D. Coelho, and L.V. Tavares (Eds.), *OR Models on Microcomputers*. New York: North-Holland, Elsevier.

Fourer, R. 2015. Software survey: Linear programming. *OR/MS Today* 42 (3): 52–63.

Fourer, R. 2017. Software survey: Linear programming. *OR/MS Today* 44 (3): 48–59.

Gacs, P., and L. Lovasz. 1981. Khachiyan's algorithm for linear programming. *Mathematical Programming Study* 14: 61–68.

Gal, T. 1992. Putting the LP survey into perspective. *OR/MS Today* 19 (6): 93.

Gass, S., H. Greenberg, K. Hoffman, and R. W. Langley (Eds.). 1986. *Impacts of Microcomputers on Operations Research*. New York: North-Holland.

Gass, S. I. 1985. *Linear Programming*, 5th ed. New York: McGraw-Hill.

Gill, P. E., and W. Murray. 1973. A numerically stable form of the simplex method. *Linear Algebra and its Applications* 7: 99–138.

Gill, P. E., W. Murray, M. A. Saunders, and M. H. Wright. 1981. *A Numerical Investigation of Ellipsoid Algorithms for Large-scale Linear Programming, in Large-Scale Linear Programming*, Vol. 1. Laxenburg, Austria: IIASA.

Gill, P. E., W. Murray, and M. H. Wright. 1981. *Practical Optimization*. New York: Academic Press.

Hadley, G. 1962. *Linear Programming*. Reading, MA: Addison-Wesley.

Harvey, C. M. 1979. *Operations Research: An Introduction to Linear Optimization*. New York: North-Holland.

Higle, J. L., and S. W. Wallace. 2003. Sensitivity analysis and uncertainty in linear programming. *Interfaces* 33 (4): 53–60.

Hillier, F. S., and G. J. Lieberman. 2010. *Introduction to Operations Research*, 9th ed. Boston, MA: McGraw-Hill.

Hooker, J. N. 1986. Karmarkar's linear programming algorithm. *Interfaces* 16 (4): 75–90.

IBM Mathematical Programming System Extended/370 (MPSX/370) Logic Manual #LY19-1024-0 and Primer #GH19-1091-1.

IBM Mathematical Programming System/360 Version 2. Linear and Separable Programming, User's Manual #H20-0476-2.

Karmarkar, N. 1984. A new polynomial-time algorithm for linear programming. *Combinatorica* 4 (4): 373–395.

Katta, G. M. 1976. *Linear and Combinatorial Programming*. New York: John Wiley & Sons.

Khachiyan, L. G. 1979. A polynomial algorithm in linear programming. *Soviet Mathematics Doklady* 20 (1): 191–194.

Kolata, G. B. 1979. Mathematicians amazed by Russian's discovery. *Science* 206: 545–546.

Loomba, N. P. 1976. *Linear Programming: A Managerial Perspective*. New York: Macmillan.

Luenberger, D., and Y. Ye. 2015. *Linear and Nonlinear Programming*, 4th ed. New York: Springer.

Lustig, I. J., R. E. Marsten, and D. F. Shanno. 1994. Interior point methods for linear programming: Computational state of the art. *ORSA Journal on Computing* 6 (1): 1–14.

McCall, E. H. 1982. Performance results of the simplex algorithm for a set of real-world linear programming models. *Communications of the ACM* 25 (3): 20–212.

Might, R. J. 1987. Decision support for aircraft munitions procurement. *Interfaces* 17 (5): 55–63.

Miller, C. E. 1963. *The Simplex Method for Local Separable Programming, in Recent Advances in Mathematical Programming*. New York: McGraw-Hill.

Mitra, G., M. Tamiz, and J. Yadegar. 1988. Investigation of an interior search method within a simplex framework. *Communications of the ACM* 31 (12): 1474–1482.

Müller-Merbach, H. 1970. *On Round-Off Errors in Linear Programming*. New York: Springer-Verlag.

Murtagh, B. A. 1981. *Advanced Linear Programming: Computation and Practice*. New York: McGraw-Hill.

Murtagh, B. A., and M. A. Saunders. 1987. MINOS 5.1 User's Guide: Technical Report SOL. In *Systems Optimization Laboratory*. Stanford, CA: Stanford University.

Murty, K. G. 1989. *Linear Complementarity, Linear and Nonlinear Programming*. Berlin, Germany: Heldermann Verlag.

Nash, S. G., and A. Sofer. 1996. *Linear and Nonlinear Programming*. New York: McGraw-Hill.

Nering, E., and A. Tucker. 1992. *Linear Programming and Related Problems*. Boston, MA: Academic Press.

Orchard-Hays, W. 1968. *Advanced Linear Programming Computing Techniques*. New York: McGraw-Hill.

Ravindran, A., D. T. Phillips, and J. J. Solberg. 1987. *Operations Research: Principles and Practice*. New York: John Wiley & Sons.

Rumpf, D. L., E. Melachrinoudis, and T. Rumpf. 1985. Improving efficiency in a forest pest control spray program. *Interfaces* 15 (5): 1–11.

Saigal, R. 1995. *Linear Programming: A Modern Integrated Analysis*. Boston, MA: Kluwer Academic.

Saunders, M. A. 1976. *A Fast, Stable Implementation of the Simplex Method using Bartels-Golub Updating, in Sparse Matrix Computations*. New York: Academic Press.

Schrage, L. 1991. *LINDO: An Optimization Modeling System*, 4th ed., Text and Software. Danvers, MA: Boyd and Fraser.

Schrage, L. 1986. *Integer and Quadratic Programming with LINDO*. Palo Alto, CA: The Scientific Press.

Schrijver, A. 1986. *Theory of Linear and Integer Programming*. New York: John Wiley & Sons.

Schuster, E. W., and S. J. Allen. 1998. Raw material management at Welch's, Inc. *Interfaces* 28 (5): 13–24.

Shanno, D. F. 1985. Computing Karmarkar Projections Quickly. University of California, Davis Graduate School of Administration.

Sharda, R. 1992. Linear programming software for personal computers: 1992 Survey. *OR/MS Today* 19 (3): 44–60.

Sharda, R. 1995. Linear programming solver software for personal computers: 1995 report. *OR/MS Today* 22 (5): 49–57.

Simmons, D. M. 1972. *Linear Programming for Operations Research*. San Francisco, CA: Holden-Day.

Taha, H. A. 2011. *Operations Research: An Introduction*, 9th ed. Upper Saddle River, NJ: Pearson.

Vanderbei, R. J. 2013. *Linear Programming: Foundations and Extensions*, 4th ed. New York: Springer.

Wilson, D. G. 1992. A brief introduction to the IBM optimization subroutine library. *IBM Systems Journal* 31 (1): 9–10.

Wolfe, P. 1980. A bibliography for the ellipsoid algorithm. Yorktown Heights, NY: IBM Research Center.

XMP Software. 1991. *User's Manual for the OB1 Linear Programming System*. Incline Village, NV.

Ye, Y. 1997. *Interior-Point Algorithms: Theory and Analysis*. New York: John Wiley & Sons.

3

Network Analysis

Network analysis provides a framework for the study of a special class of linear programming problems that can be modeled as network programs. Because such a vast array of problems can be viewed as networks, this is one of the most significant classes of applications in the field of Operations Research. Some of these problems correspond to a physical or geographical network of elements within a system, while others correspond more abstractly to a graphical approach to planning or grouping or arranging the elements of a system.

The diversity of problems that fall quite naturally into the network model is striking. Networks can be used to represent systems of highways, railroads, shipping lanes, or aviation patterns, where some supply of a commodity is transported or distributed to satisfy a demand. Pipeline systems or utility grids can be viewed as fluid flow or power flow networks, while computer communication networks represent the flow of information, and an economic system may represent the flow of wealth. In some cases, the problem may call for routing a vehicle or a commodity between certain specified points in the network; other applications may require that some entity be circulated throughout the network.

By using the network model more abstractly, we can solve problems that require assigning jobs to machines, or matching workers with jobs for maximum efficiency. Network methods can also be applied to project planning and project management, where various activities must be scheduled in order to minimize the duration of a project or to meet specified completion dates, subject to the availability of resources.

All of these apparently different problems have underlying similarities: all consist of a set of centers, junctions, or nodes that are interconnected (logically or physically) by links, channels, or conveyances. Because of this, a study of general network models and techniques will provide us with tools that can be applied to a variety of applications. As we study these models, we will see that it is the mathematical structure or form of the problem that is important and not necessarily the application. Furthermore, the successful use of network models is largely dependent on a skillful analyst's ability to perceive the structure of a problem and to assess whether the network framework is an appropriate approach to a solution. We will see examples in which there is more than one way to represent the problem as a network model, and one formulation may be superior to others.

This chapter begins with some basic definitions and properties of graphs and networks. Algorithms are then presented for finding the maximum flow in a network, optimally transporting a commodity from supply points to demand points, matching or pairing appropriate elements in a system, and efficiently designing a network such that every pair of points has some connecting path. Methods are described for finding the shortest route between points in a network, and then these methods are applied to multistage decision-making processes and project-planning problems.

3.1 Graphs and Networks: Preliminary Definitions

A **graph** is a structure consisting of a set of **nodes** (vertices, points, or junctions) and a set of connections called **arcs** (edges, links, or branches). Each connection is associated with a pair of nodes and is usually drawn as a line joining two points. If there is an orientation or direction on the arcs, then the graph is said to be **directed**, otherwise it is **undirected**. The **degree** of a node is the number of arcs attached to it. An **isolated** node in a graph is one that has no arc attaching it to any other node, and therefore such a node is of degree zero.

In a directed graph, if there is an arc *from* node A *to* node B, then node A is said to be a **predecessor** of node B, and node B is a **successor** of node A. The arc is often designated by the ordered pair (A, B).

For certain applications, it is useful to refer to a **path** from some given node to another. Let $x_1, x_2, \ldots, x_n$ be a sequence of distinct nodes, such that nodes adjacent to each other in the sequence are connected to each other by an arc in the graph. That is, if the sequence contains x_i, x_{i+1}, then either the arc (x_i, x_{i+1}) or the arc (x_{i+1}, x_i) exists in the graph. Then we say there is a path from x_1 to x_n, that consists of the nodes and their connecting links. In Figure 3.1, there is a path from node A to node G that can be described by A, (A, B), B, (B, C), C, (E, C), E, (E, G), G. When the arc connecting nodes x_i and x_{i+1} in a path is (x_i, x_{i+1}), it is called a **forward arc**; if the connecting arc is (x_{i+1}, x_i), it is a **backward arc**.

In the illustration, the path contains the three forward arcs (A, B), (B, C), and (E, G) and one backward arc (E, C). If all the arcs in a path are forward arcs, then the path is called a **directed chain** or simply a **chain**. If the graph is undirected, then the terms **path** and **chain** are synonymous. If $x_1 = x_n$ in the path, then the path is called a **cycle** or a **cyclic path**. In the illustration, we see the cyclic path

$$B, (B, C), C, (E, C), E, (E, G), G, (G, B), B$$

although this is *not* a cyclic chain because it includes the backward arc (E, C). A **connected graph** is a graph that has at least one path connecting every pair of nodes.

A graph is a **bipartite** graph if the nodes can be partitioned into two subsets S and T, such that each node is in exactly one of the subsets, and every arc in the graph connects a node in set S with a node in set T. Such a graph is a **complete bipartite** graph if each node in S is connected to every node in T. The graph in Figure 3.2 is a complete bipartite graph in which nodes A and B are in one subset, and nodes C, D, and E are in the other.

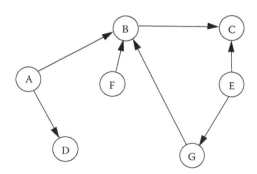

FIGURE 3.1
Paths in a graph.

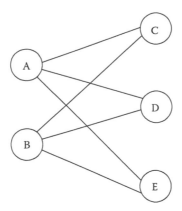

FIGURE 3.2
A complete bipartite graph.

A **tree** is a directed connected graph in which each node has at most one predecessor, and one node (the root node) has no predecessor. In an undirected graph, we have a tree if the graph is connected and contains no cycles. (If there are n nodes, there will be n − 1 arcs in the tree.) Figure 3.3 contains illustrations.

A **network** is a directed connected graph that is used to represent or model a system or a process. The arcs in a network are typically assigned weights that may represent a cost or value or capacity corresponding to each link in the network.

A node in a network may be designated as a **source** (or origin), and some other node may be designated as a **sink** (or destination). A network may have multiple sources and sinks. A **cut set** (or simply a **cut**) is any set of arcs which, if removed from the network, would disconnect the source(s) from the sink(s). Because networks are commonly used to represent the transmission of some entity from a source node to a sink node, we introduce the concept of **flow** through a network. Flow can be thought of as the total amount of an entity that originates at the source, makes its way along the various arcs and passes through intermediate nodes, and finally arrives at (or is consumed by) the destination (sink) node. The study of network flow is the subject of the next section.

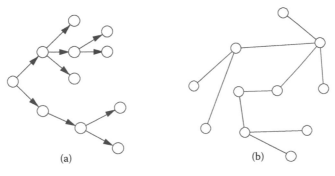

FIGURE 3.3
Trees (n = 11): (a) directed tree and (b) undirected tree.

3.2 Maximum Flow in Networks

Maximum flow problems arise in networks where there is a source and a sink connected by a system of directional links, each having a given capacity. The problem is to determine the greatest possible flow that can be routed through the various network links, from source to sink, without violating the capacity constraints. The commodity flowing in the network is generated only at the source and is consumed only at the sink. The source node has only arcs directed out of it, and the sink node has only arcs directed into it. Intermediate nodes neither contribute to nor diminish the flow passing through them.

As an example, consider a data communication network in which processing nodes are connected by data links. In Figure 3.4, data being collected or generated at site A must be transmitted through the network as quickly as possible to a destination processor at site G where the data can be archived or processed. Each data link has a capacity (probably some function of baud rate and availability or band width) that effectively limits the flow of data through that link. Alternatively, one can envision a power generation and distribution system as a network flow model in which power is generated at the source and conducted through transform stations to end users. Capacities are shown as labels on the arcs.

The **maximum flow problem** can be stated precisely as a linear programming formulation. Let n be the number of nodes, and let nodes 1 and n be designated as source and sink, respectively. The decision variables x_{ij} denote the amount of flow along the arc from node i to node j (i, j = 1, ..., n). The **capacity** of the arc from node i to node j is the upper limit on the flow through this arc, and is denoted u_{ij}. If we let f denote the total flow through the network, then to maximize the total flow, we would want to

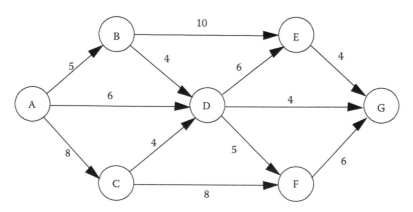

FIGURE 3.4
Data communications network.

$$\text{maximize} \qquad z = f$$

$$\text{subject to} \qquad \sum_{i=2}^{n} x_{1i} = f \qquad\qquad (1)$$

$$\sum_{i=1}^{n-1} x_{in} = f \qquad\qquad (2)$$

$$\sum_{i=1}^{n} x_{ij} = \sum_{k=1}^{n} x_{jk} \text{ for } j = 2, 3, \ldots, n-1 \qquad (3)$$

$$x_{ij} \le u_{ij} \text{ for all } i, j = 1, \ldots n \qquad\qquad (4)$$

Constraints (1) and (2) state that all the flow is generated at the source and consumed at the sink. Constraint (1) ensures that a flow of f leaves the source, and because of conservation of flow, that flow stops only at the sink. Constraints (3) are the *flow conservation equations* for all the intermediate nodes; nothing is generated or consumed at these nodes. Constraints (4) enforce arc capacity restrictions. All flow amounts x_{ij} must be non-negative. Actually, constraint (2) is redundant.

As with all of the network models in this chapter, this problem could be solved using the Simplex method. However, we can take advantage of the special network structure to solve this problem much more efficiently. One of the most commonly used methods is an iterative-improvement method known as the **Ford-Fulkerson labeling algorithm**. An initial feasible flow can always be found by letting the flow through the network be zero (all $x_{ij} = 0$). The algorithm then operates through a sequence of iterations, each iteration consisting of two phases: (1) first we look for a way to increase the current flow, by finding a path of arcs from source to sink whose current flow is less than capacity (this is called a *flow augmenting path*); and then (2) we increase the current flow, as much as possible, along that path. If in phase (1) it is not possible to find a flow augmenting path, then the current flow is optimal. We will first outline the basic algorithm, and then fill in the details.

3.2.1 Maximum Flow Algorithm

Initialization: Establish an initial feasible flow.

Phase 1: Use a labeling procedure to look for a flow augmenting path. If none can be found, stop; the current flow is optimal.

Phase 2: Increase the current flow as much as possible in the flow augmenting path (until some arc reaches its capacity). Go to Phase 1.

The search for a flow augmenting path in Phase 1 is facilitated by a labeling procedure that begins by labeling the source node. We will use a check mark (✓) on our figures to indicate that a node has been labeled. From any labeled node i, we must examine outgoing arcs (i, j) and incoming arcs (j, i), for unlabeled nodes j. We label (✓) node j if the current flow in outgoing arc (i, j) is less than its capacity u_{ij}, or if the current flow in incoming arc (j, i) is greater than zero. Labeling a node i means that we could increase the total flow in the network from the source as far as node i. If the sink node eventually can be labeled, then a flow augmenting path has been found. If more than one flow augmenting path exists, choose any one arbitrarily.

In Phase 2, the arcs in the flow augmenting path are first identified. Then by examining the differences in current flow and capacity flow on all forward arcs in the path, and the current flow in all backward arcs, we determine the greatest feasible amount by which the total flow through this path can be increased. Increase the flow in all forward arcs by this amount, and decrease the flow in all backward arcs by this amount.

We will now illustrate the maximum flow algorithm by applying it to the network pictured in Figure 3.4. Let us assume initially that the flow in all arcs is zero, $x_{ij} = 0$ and f = 0. In the first iteration, we label nodes A, B, C, D, and G, and discover the flow augmenting path (A, D) and (D, G), across which we can increase the flow by 4. So now, $x_{AD} = 4$, $x_{DG} = 4$, and f = 4.

In the second iteration, we label nodes A, B, C, then nodes E, D, and F, and finally node G. A flow augmenting path consists of links (A, B), (B, D), (D, E), and (E, G) and flow on this path can be increased by 4. Now $x_{AB} = 4$, $x_{BD} = 4$, $x_{DE} = 4$, $x_{EG} = 4$, and f = 8.

In the third iteration, we see that there remains some unused capacity on link (A, B), so we can label nodes A, B, and E, but not G. It appears we cannot use the full capacity of link (A, B). However, we can also label nodes C, D, F, and G, and augment the flow along the links (A, D), (D, F), and (F, G) by 2, the amount of remaining capacity in (A, D). Now $x_{AD} = 6$, $x_{DF} = 2$, $x_{FG} = 2$, and f = 10.

In the fourth iteration, we can label nodes A, B, C, D, F, and G. Along the path from A, C, D, F, to G, we can add a flow of 4, the remaining capacity in (F, G). So $x_{AC} = 4$, $x_{CF} = 4$, $x_{FG} = 6$, and f = 14.

In the fifth iteration, we can label all nodes except G. Therefore, there is no flow augmenting path, and the current flow of 14 is optimal.

Notice that in any network, there is always a bottleneck that in some sense impedes the flow through the network. The total capacity of the bottleneck is an upper bound on the total flow in the network. Cut sets are, by definition, essential in order for there to be a flow from source to sink, since removal of the cut set links would render the sink unreachable from the source. The capacities on the links in *any* cut set *potentially* limit the total flow. One of the fundamental theorems of Ford and Fulkerson states that the minimum cut (i.e., the cut set with minimum total capacity) is in fact the bottleneck that precisely determines the maximum possible flow in the network. This *Max-Flow Min-Cut* Theorem provides the foundation for the maximum flow labeling algorithm presented earlier. During Phase 1 of the algorithm, if a flow augmenting path cannot be found, then we can be assured that the capacity of some cut is being fully used by the current flow. This minimum cut is the set of links that separate the nodes that are labeled (✓) from those that are not labeled. Observe that, by definition of the labeling algorithm, every **forward** arc in the cut set (from a labeled to an unlabeled node) must be at capacity. Similarly, every **reverse** arc in the cut set (from an unlabeled to a labeled node) must have zero flow. Therefore, the capacity of the cut is precisely equal to the current flow and this flow is optimal. In other words, a saturated cut defines the maximum flow.

In the final iteration of the example earlier, the cut set that separates the labeled nodes from the unlabeled nodes is the set of links (E, G), (D, G), and (F, G). The capacity of this cut set is 4 + 4 + 6 = 14, which is just exactly the value of the optimal flow through this network.

If all of the arcs in a network are forward arcs, it is easy to identify a flow augmenting path. Each edge in the path is below capacity and we can increase the flow until some edge reaches capacity. To appreciate the idea of reverse arc labeling, consider the situation shown in Figure 3.5a. In the diagram, each arc from node i to node j is labeled with (x_{ij}, u_{ij}).

Suppose our initial path is from node 1 to 2 to 4 to 6, with a flow of 4. At this point, shown in Figure 3.5b, there is no direct path from the source node 1 to the sink node 6 that allows an increase in flow. However, the algorithm will find the path

$$
\begin{array}{ccccc}
(0,4) & (0,4) & (4,4) & (0,4) & (0,4) \\
1 \longrightarrow 3 \longrightarrow 4 \longleftarrow 2 \longrightarrow 5 \longrightarrow 6
\end{array}
$$

Increase the flow on each forward arc by 4, and decrease the flow on the reverse arc. The resulting flow is shown in Figure 3.5c with a total flow of 8. Notice that the net effect,

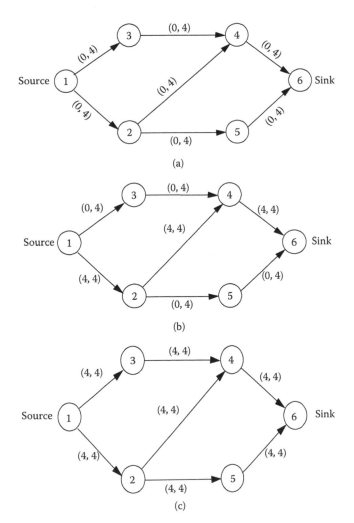

FIGURE 3.5
Maximum flow example: (a) original network, (b) path augmenting, and (c) optimal maximum flow.

with respect to the reverse arc, is that we decided to take the flow out of node 2 and send it somewhere else (namely to nodes 5 and 6). Similarly, we decided to use the *new* flow at node 4, coming from node 3, instead of the flow from node 2. Therefore, if we can label node 4, we can effectively divert the flow at node 2 to *create* additional flow through the entire network.

3.2.2 Extensions to the Maximum Flow Problem

There are several interesting extensions to the maximum flow problem. The existence of **multiple sources** and **multiple sinks** requires only a minor change in our original network model. Suppose, for example, nodes 1A, 1B, and 1C are sources, and nodes nA, nB, nC, and nD are sinks, as shown in Figure 3.6a. This network can be modified to include a *super-source* node (which we will call 1S) and a *super-sink* node (nS). The super source is connected to the multiple sources via links unrestricted in capacity, as in Figure 3.6b; and likewise, the multiple sinks are connected to the super sink by uncapacitated links, as in Figure 3.6c.

Because none of the new uncapacitated links could possibly contribute to any minimum cut, the maximum flow from the super-source node 1S to the super-sink node nS will also be the maximum flow in the multiple-source multiple-sink problem.

We can use this same construction to handle the situation in which some or all of the sources have a limited capacity by simply placing a capacity on the arc from the super-source to the capacitated source node. Capacities on the sinks can be handled in the same way.

The basic maximum flow algorithm is normally used to solve a part of a more complex problem. For example, in the next section, we will encounter almost the same problem, but where there is a per-unit cost associated with each arc in the network, and we want to minimize total cost. There are, however, some direct applications of the maximum flow algorithm. One of these occurs in network capacity planning. For example, an electric utility company may use network flow to determine the capacity of its present system. By identifying the cut sets, it can easily determine where additional lines must be installed in order to increase the capacity of the existing grid.

The complexity of maximum flow algorithms is dependent on the method used for selecting the flow augmenting paths. Because network flow algorithms are used so often in practical applications, efforts have been made to develop faster versions. A shortest path augmentation method developed by (Edmonds and Karp 1972) is used in an algorithm having complexity $O(ne^2)$, where n is the number of nodes and e is the number of edges.

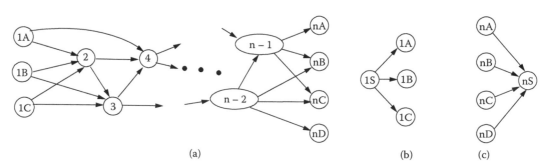

(a)   (b)   (c)

FIGURE 3.6
Multiple sources and sinks: (a) original network, (b) super source, and (c) super sink.

Dinic's method (Dinic 1970) of using so-called *blocking flows* requires $O(n^2e)$ computation time, while Karzanov's method (Karzanov 1974) based on the idea of *preflows* is dependent solely on the number of nodes, and requires $O(n^3)$ time.

Extensions to the maximal flow problem include multi-commodity problems, maximal dynamic flow problems, and cost effective increases in network capacity. These topics are discussed fully in the references by Battersby (1970), Hu (1970), and Price (1971).

3.3 Minimum Cost Network Flow Problems

When there are costs associated with shipping or transporting a flow through a network, the goal might be to establish a *minimum cost* flow in the network, subject to capacity constraints on the links. The minimum cost flow problem is interesting not only because the general model is so comprehensive in its applicability, but also because special cases of the model can be interpreted and applied to quite a variety of resource distribution and allocation problems.

3.3.1 Transportation Problem

One of the simplest minimum cost network flow problems is one in which every node is either a source (supply) or a sink (demand). For example, we could imagine a distributor with several warehouses and a group of customers. There is a cost associated with serving each customer from any given warehouse.

In this model, we have m supply nodes, each with an available supply s_i, and n demand nodes, each with a demand of d_j. And we assume that the total supply in the network is equal to the total demand:

$$\sum_{i=1}^{m} s_i = \sum_{j=1}^{n} d_j$$

The objective is to satisfy all the demands, using the available supply, and to accomplish this distribution using minimum cost routes. The formulation of the problem is as follows:

$$\text{minimize} \quad z = \sum_{i=1}^{m} \sum_{j=1}^{n} c_{ij} x_{ij}$$

$$\text{subject to} \quad \sum_{j=1}^{n} x_{ij} = s_i \qquad \text{for } i = 1, \ldots, m \qquad (1)$$

$$\sum_{i=1}^{m} x_{ij} = d_j \qquad \text{for } j = 1, \ldots, n \qquad (2)$$

$$x_{ij} \geq 0 \qquad \text{for all i and j} \qquad (3)$$

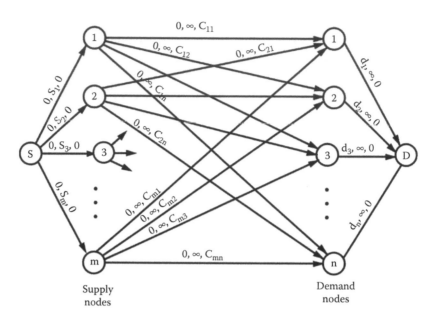

FIGURE 3.7
Transportation problem as minimum cost network flow problem.

Because the set of supply nodes is distinct from the set of demand nodes, and all nodes in the network belong to one of these sets, this transportation model can be pictured as a bipartite graph, with the addition of a super-source node S and a super-sink node D. In Figure 3.7, arcs connecting supply nodes to demand nodes represent the actual distribution routes. Each arc in the drawing is labeled with a triple, indicating a lower bound on the flow, an upper bound on the flow, and a per unit cost for the flow along the arc. Arcs from the super source S impose the (upper bound) supply limits, and, of course, carry no cost. Similarly, arcs to the super sink D enforce the (lower bound) demand requirements. It should be clear that finding a minimum cost flow from node S to node D in this network precisely solves the **transportation problem** that we have formulated, and the resulting minimum cost is the cost of the optimal distribution of the commodity through the transportation network.

To illustrate the solution approach, we will use a simple example of a distributor with three warehouses and five customers. Because of the simple structure of the transportation problem, it is probably easier to visualize the problem in matrix form, as shown in Table 3.1.

In the table, c_{ij} in row i and column j of the matrix represents the cost of sending one unit of product from source i to sink j. Similarly, x_{ij} represents the number of units sent from source i to sink j, the current *flow* solution.

Consider the example problem in Table 3.2. Observe that the total demand of 65 units is equal to the total supply. Because most of the x_{ij} values will be zero, we will write them in only when they are positive.

We will describe how to solve this problem using the Simplex method. After all, this is a linear programming problem. However, the special structure of the transportation problem will allow us to take a number of shortcuts. The Simplex method says that we should first find any basic feasible solution, and then look for a simple *pivot* to improve the solution. If no such improvement can be found, the current solution must be optimal.

TABLE 3.1

Transportation Problem

Sources	Sinks (Customers)					
(Warehouses)	1	2	3	4	5	Supply
1	c_{11} x_{11}	c_{12} x_{12}	c_{13} x_{13}	c_{14} x_{14}	c_{15} x_{15}	s_1
2	c_{21} x_{21}	c_{22} x_{22}	c_{23} x_{23}	c_{24} x_{24}	c_{25} x_{25}	s_2
3	c_{31} x_{31}	c_{32} x_{32}	c_{33} x_{33}	c_{34} x_{34}	c_{35} x_{35}	s_3
Demand	d_1	d_2	d_3	d_4	d_5	

TABLE 3.2

Transportation Problem Example

Sources	Sinks (Customers)					
(Warehouses)	1	2	3	4	5	Supply
1	28	7	16	2	30	20
2	18	8	14	4	20	20
3	10	12	13	5	28	25
Demand	12	14	12	18	9	65

The first simplification to the basic Simplex method is that we do not need a complex two phase method to find a basic feasible solution. Instead, we present three fast and commonly used techniques for obtaining an initial solution.

3.3.1.1 Northwest Corner Rule

If we ignore the total cost, it is trivial to find an initial feasible solution. We simply assign the first group of customers to the first warehouse until the capacity is exhausted, and then start assigning customers to the second warehouse until it too is at its capacity, and so on.

We begin at the upper left corner of the tableau, the *northwest* corner. Increase the flow in this cell as much as possible until the flow is equal to the supply in this row or the demand in this column. Reduce the demand and the supply in this row and column by the amount of the flow, since the requirement has now been satisfied. Draw a line through the row or column that has zero remaining required. (If both are zero, select either one arbitrarily.) Repeat the northwest corner rule on the reduced matrix.

Consider the example in Table 3.2. Begin with row 1 and column 1. Since the demand is 12 and the supply is 20, the flow can be at most 12. Reduce the limit on row 1 and column 1 by 12, and draw a line through column 1. The reduced problem is shown in Table 3.3. The reduced problem (without column 1) has x_{12} (row 1, column 2) in the northwest corner. We let $x_{12} = 8$ because the remaining supply in row 1 is 8. This time, we delete row 1, and subtract 8 from supply s_1 and demand d_2.

TABLE 3.3

Northwest Corner Rule

Sources (Warehouses)	Sinks (Customers) 1		2		3		4		5		Supply
1	12	28		7		16		2		30	~~20~~ 8
2		18		8		14		4		20	20
3		10		12		13		5		28	25
Demand	~~12~~ 0		14		12		18		9		

TABLE 3.4

Initial Northwest Corner Solution

Sources (Warehouses)	Sinks (Customers) 1		2		3		4		5		Supply
1	12	28	8	7		16		2		30	20
2		18	6	8	12	14	2	4		20	20
3		10		12		13	16	5	9	28	25
Demand	12		14		12		18		9		65

The final solution is presented in Table 3.4. The reader should verify this result. The total cost of this solution is given by $(12 \cdot 28) + (8 \cdot 7) + (6 \cdot 8) + (12 \cdot 14) + (2 \cdot 4) + (16 \cdot 5) + (9 \cdot 28) = 948$.

There are several features of this solution that we should notice. First, it should be clear that the procedure always produces a *feasible* solution. For a solution to be feasible, every customer must be receiving all of the necessary demand from some warehouses, and no warehouse may exceed its supply. In fact, all of the rows and columns will be satisfied at equality. Because this method never transports more than the remaining supply or demand, we have only to verify that no customer gets less than what it asked for.

Suppose the last customer did not get all its required demand; then that row will not be deleted. Moreover, there must be some excess supply at one of the warehouses, so *that column* has not been deleted. Therefore, there is still one cell left for the northwest corner rule to work in. (The technique stops only when every cell in the matrix has been deleted.)

The second thing to notice is that we must always *start* at x_{11} and we must *finish* at x_{mn} (for m warehouses and n customers). Moreover, at each step, the algorithm will delete one row or one column. In the last cell, the remaining demand in column n and the supply in row m must be identical. Because there are m rows and n columns, the solution will use exactly $(m + n - 1)$ cells and therefore $(m + n - 1)$ of the x_{ij} will have a positive value. In our example, we have $3 + 5 - 1 = 7$ cells that are selected for a positive flow.

In the Chapter 2 presentation of the Simplex method, it was stated that the number of basic variables is precisely equal to the number of constraints. In the linear programming formulation of the transportation problem, there are m equality constraints for the supply at the m warehouses, and n constraints for the demands of the n customers. Therefore, one would expect (m + n) non-zero (basic) variables. All other (non-basic) variables are zero. The apparent discrepancy can be explained by observing that the linear programming constraints are not independent. If the last constraint were deleted, and we solved that problem, we would find that the solution will have all warehouse supply satisfied at equality, and the first (n − 1) customers will have their demand satisfied at equality. All remaining demand must be assigned to customer n. Because total supply equals total demand, the demand for customer n will automatically be satisfied exactly. In other words, when the corresponding linear programming problem is solved with (m + n − 1) constraints, there will be exactly (m + n − 1) basic variables, and introducing the additional constraint will not change this.

3.3.1.2 Minimum Cost Method

The northwest corner rule is a quick way to find a feasible solution. However, the method ignores any cost information; hence, it is unlikely that the initial solution will be a very good one.

The same approach can be extended in an obvious way to search for a basic feasible solution while attempting to minimize the total cost.

Step 1: Select the cell in the matrix that has the smallest cost, breaking ties arbitrarily.

Step 2: Increase the flow in this cell as much as possible until the flow is equal to the supply in the row or the demand in the column.

Step 3: Reduce the supply and the demand in this row and column because the requirement has now been satisfied.

Step 4: Draw a line through the row or column that has zero remaining supply or demand. If both are zero, select either one arbitrarily. Repeat the procedure from Step 1 on the reduced matrix.

This method is very similar to the northwest corner rule in that it selects one cell, saturates it, and deletes a row or column. It is also guaranteed to find a basic feasible solution with precisely (m + n − 1) selected flow cells. However, unlike the northwest corner rule, this method tries to match customers and warehouses, with some consideration of costs.

The method is illustrated in Table 3.5, where we first find the minimum cost cell to be $c_{14} = 2$. Therefore, we satisfy as much of the customer 4 demand as possible from warehouse 1. In this case, all 18 units can be supplied. We reduce the remaining supply at warehouse 1 to 2 units, and delete customer 4.

In the next iteration, the minimum (undeleted) entry is $c_{12} = 7$, and we will satisfy as much of the demand of customer 2 as possible from warehouse 1. In this case, warehouse 1 has only 2 units left, so the flow x_{12} is set at 2, row 1 is deleted, and the demand of customer 2 is reduced to 12.

The final solution is presented in Table 3.6. The total cost of this solution is given by

$$(2 \cdot 7) + (18 \cdot 2) + (12 \cdot 8) + (8 \cdot 20) + (12 \cdot 10) + (12 \cdot 13) + (1 \cdot 28) = 610$$

TABLE 3.5

Iteration 1: Minimum Cost

Sources (Warehouses)	Sinks (Customers)					Supply
	1	2	3	4	5	
1	28	7	16	2 18	30	~~20~~ 2
2	18	8	14	4	20	20
3	10	12	13	5	28	25
Demand	12	14	12	0	9	65

TABLE 3.6

Minimum Cost Final Solution

Sources (Warehouses)	Sinks (Customers)					Supply
	1	2	3	4	5	
1	28	7 2	16	2 18	30	20
2	18 12	8	14	4 8	20	20
3	10 12	12	13 12	5 1	28	25
Demand	12	14	12	18	9	65

As before, this solution is a basic feasible solution with precisely seven basic variables. However, the total cost is considerably lower than the cost of the solution obtained with the northwest corner rule. It is important to realize that obtaining the improved initial feasible solution did require more computation time. At the first step of the northwest corner rule, the single cell in the top left corner is selected. In the corresponding first step of the minimum cost algorithm, it is necessary to search all of the $m \cdot n$ cells in the matrix to find the one having the least cost. (When m is 100 and n is 10,000, this additional work takes a considerable amount of time.)

There are a wide variety of other algorithms available for finding an initial feasible solution. Typically, they all exhibit the property that better initial solutions require more computation time. The value of spending a lot of effort searching for better initial solutions is somewhat questionable; the Simplex method will enable us to derive the optimal solution from *any* initial solution. The only advantage of using good initial solutions is that it should reduce the number of pivot operations required later.

3.3.1.3 Minimum "Row" Cost Method

The computational requirements of the minimum cost method can be reduced significantly without completely sacrificing the spirit. In Step 1, instead of looking for the minimum cost element in the whole matrix, we simply look for the minimum cost element in

TABLE 3.7

Minimum Row Cost Final Solution

Sources	Sinks (Customers)					
(Warehouses)	1	2	3	4	5	Supply
1	28	7	16	2	30	20
	2		18			
2	18	8	14	4	20	20
	8	12				
3	10	12	13	5	28	25
	4		12		9	
Demand	12	14	12	18	9	65

the first row. We continue to do this until warehouse 1 is saturated. Step 1 will now require scanning n elements instead of m · n elements. However, by assigning the best possible customer to warehouse 1, the method still tends to find low cost solutions.

Table 3.7 illustrates the final solution using the minimum row cost method. It has a total cost of

$$(2 \cdot 7) + (18 \cdot 2) + (8 \cdot 18) + (12 \cdot 8) + (4 \cdot 10) + (12 \cdot 13) + (9 \cdot 28) = 638$$

This solution has only a slightly higher cost than the cost of 610 that was obtained with the minimum cost method, and it required less work. In general, this is representative of the performance one would expect of the two methods, although, of course, it would be possible to construct simple examples in which the minimum row cost method produced better solutions.

3.3.1.4 Transportation Simplex Method

Before we explain the procedure for finding the optimal solution, it will be useful to describe a simple modification that transforms the original problem into an equivalent new problem. Consider our example from Table 3.7, which shows the initial basic feasible solution obtained using the minimum row cost method. Observe what happens if we subtract $1 from every cost element in the first row. Because warehouse 1 has a supply of 20 units, *every* feasible solution will have a total of 20 units in row 1. Reducing the cost of each element by $1 will reduce the cost of every feasible solution by exactly $20. In particular, the cost of the optimal solution will decrease by $20.

The optimal solution to the new *reduced* problem (in terms of the flow variables x_{ij}) is exactly the same as the optimal solution of the original problem. We simply solve the new problem and then add $20 to the optimal objective function value. Furthermore, if we reduce all of the costs in the first row by 2 or 3 or 4, we will not change the problem; we will simply reduce the total cost of every solution by $40 or $60 or $80, respectively.

Similarly, consider the first column of the matrix corresponding to customer 1. Clearly, every feasible solution will have a total of 12 units distributed somewhere in column 1. If every cost element in column 1 were reduced by 1 or 2 or 3, then the total cost of every feasible solution would decrease by $12 or $24 or $36, respectively. The new reduced problem is identical to the original one with respect to the optimal flow values x_{ij}.

TABLE 3.8

Reduced Cost Solution

Sources (Warehouses)	Sinks (Customers) 1	2	3	4	5	Supply	u_i
1	11 2	0	−4 18	0	−5	20	0
2	0 8	0 12	−7	1	−16	20	1
3	0 4	12 12	0	0 9	0	25	−7
Demand	12	14	12	18	9	65	
v_j	17	7	20	2	35		

Now consider our example problem. We will construct an equivalent problem by subtracting constants from the costs in the rows and columns. The reduced problem will have the property that the reduced cost corresponding to every basic variable cell will be precisely zero. This is illustrated, for our example, in Table 3.8. We let u_i denote the amount subtracted from every cost element in row i and v_j represent the amount subtracted from every element in column j.

The reader should verify that all the *reduced costs* c'_{ij} in this table obey the relationship:

$$c'_{ij} = c_{ij} - u_i - v_j$$

where c_{ij} is the original cost. As discussed earlier, finding the optimal solution to this problem is exactly the same as solving the original problem. Note that $u_3 = -7$. This indicates that we *added* 7 to row 3 instead of *subtracting* 7. Clearly, we can add a constant to a row or column as well as subtract a constant without changing the problem.

The reduced problem has several interesting features. In particular, the total cost of the current solution, in terms of c'_{ij}, is precisely zero. The reader should verify that we have reduced the total cost of the solution by 638. In addition, some of the reduced costs corresponding to the non-basic cells are negative. Consider the cell x_{23}, with $c'_{23} = -7$. If we could increase the number of units of flow, from warehouse 2 to customer 3, by one unit, we could reduce the total cost by 7. That is, the total cost now is zero, and it would become −7. If we increase the flow of units from warehouse 2 to customer 3, it will also be necessary to decrease the flow to some other customer from warehouse 2, and from some other warehouse to customer 3. At all times, the total supply and demand constraints must be maintained. In the example, if we increase x_{23} by 1, decrease x_{21} by 1, increase x_{31} by 1, and decrease x_{33} by 1, we will maintain all supply and demand equalities, and the total cost will be reduced by 7. Moreover, if we restrict ourselves to using only basic variable cells, this solution is unique.

If we continue to increase the flow on x_{23}, we will further decrease the cost of the solution by 7 per unit. However, we cannot continue to do this indefinitely. Specifically, for every unit that we increase x_{23}, it is necessary to decrease x_{21} and x_{33} by 1. Because x_{ij} must be non-negative, we can decrease x_{21} by 8 and x_{33} by 12. Therefore, the maximum increase for x_{23} is 8, giving a decrease of \$56 in the cost. When $x_{23} = 8$, x_{21} becomes zero and we remove x_{21} from the basis to let x_{23} enter. The new solution is illustrated in Table 3.9.

TABLE 3.9

Transportation Simplex

Sources (Warehouses)	Sinks (Customers) 1	2	3	4	5	Supply	u_i
1	11	0 — 2	-4	0 — 18	-5	20	0
2	0 — 0	0 — 12	-7 — 8	1	-16	20	1
3	0 — 12	12	0 — 4	0	0 — 9	25	-7
Demand	12	14	12	18	9	65	
v_j	17	7	20	2	35		

The cost of this solution is -56. If we put these same flows in the original table, we would discover that the total cost is $582, precisely 56 less than the cost of our initial solution.

In Table 3.9, the reduced cost is no longer zero for all basic variable cells. The new cell x_{23} has $c'_{23} = -7$. In order to make this zero again, we can either add 7 to row 2 or add 7 to column 3. (It does not matter which we select.) Suppose we add 7 to column 3 (decrease v_3 by 7). Then, we will also be forced to subtract 7 from row 3 (in order to keep c'_{33} at zero) and then add 7 to columns 1 and 5 (in order to keep c'_{31} and c'_{35} at zero). The new reduced cost solution is shown in Table 3.10.

Once again, this new problem is identical to the original. The current basic feasible solution has a value of zero, and there is an opportunity to further reduce the cost if we can increase the flow from warehouse 2 to customer 5.

Before doing this, let's make one observation: it will be useful for us to depict the problem in a slightly different way. In Figure 3.8, the problem has been drawn as a network with only the basic flow edges shown. Observe that the basic edges form a tree. In other words, if we ignore the directions of the edges, there are no circuits.

The network also has the property that there are exactly $(m + n - 1)$ edges. If we had a basic solution that had less than this number of edges, then we would arbitrarily add extra basic cells *with a zero flow* to keep the total at $(m + n - 1)$. Because the additional flow is zero, the extra basic variables do not affect the total cost.

TABLE 3.10

New Reduction Cost Solution

Sources (Warehouses)	Sinks (Customers) 1	2	3	4	5	Supply	u_i
1	18	0 — 2	3	0 — 18	2	20	0
2	7	0 — 12	0 — 8	1	-9	20	1
3	0 — 12	5	0 — 4	3	0 — 9	25	0
Demand	12	14	12	18	9	65	
v_j	10	7	13	2	28		

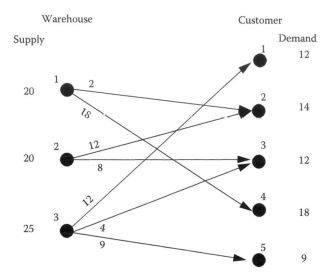

FIGURE 3.8
Basic flow tree.

However, because of these properties, introducing a new edge into the basis *will always create a single circuit*. For example, when we try to introduce the variable x_{25} into the basis, we get the network shown in Figure 3.9. This produces a unique circuit on the variables x_{25}, x_{35}, x_{33}, and x_{23}.

If we want to increase the flow on x_{25} by an amount Δ, and still maintain equality at the supply and demand nodes, we must decrease x_{35}, increase x_{33}, and decrease x_{23}, all by the amount Δ. In order to maintain feasibility, x_{35} and x_{23} must remain non-negative, and hence the maximum value of Δ is 8. We set $x_{25} = 8$, $x_{35} = 1$, $x_{33} = 12$, and $x_{23} = 0$, thus adding x_{25} to the basis and removing x_{23}. (If two variables become zero simultaneously, we can

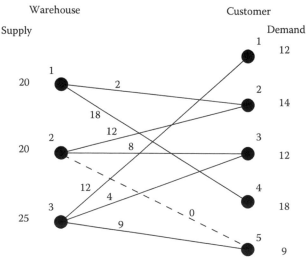

FIGURE 3.9
Transportation network with circuit.

TABLE 3.11

Transportation Simplex Continued

Sources	Sinks (Customers)						
(Warehouses)	1	2	3	4	5	Supply	u_i
1	27	0 / 2	12	0 / 18	11	20	−9
2	16 / 0	0 / 12	9 / 0	1	0 / 8	20	−8
3	0 / 12	−4	0 / 12	−6	0 / 1	25	0
Demand	12	14	12	18	9	65	
v_j	10	16	13	11	28		

arbitrarily select one to leave the basis.) The new solution, with the new reduced costs computed, is shown in Table 3.11. In order to remove $c'_{25} = -9$, we increased row 2 by 9.

The resulting cost should have decreased by 8 (the new flow in x_{25}) times −9 (the reduced cost) = −\$72. When we substitute the new flow into the original problem, we discover that the new total cost is \$510, a reduction of \$72 from the previous basic feasible solution cost of \$582. The following steps summarize the Transportation Simplex method.

3.3.1.5 Transportation Simplex

I. Compute the reduced costs c'_{ij} such that every basic cell has a zero reduced cost. (Initially, assume $c'_{ij} = c_{ij}$, and the u_i and v_j are all zero.)

  a. Construct the basic variable network (tree) as in Figure 3.9. Select any u_i and assign to it any arbitrary fixed value.

  b. For each unfixed v_j that is adjacent to a fixed u_i, adjust v_j such that c'_{ij} is zero, and then call v_j fixed.

  c. For each unfixed u_i that is adjacent to a fixed v_j, adjust u_i such that c'_{ij} is zero, and call u_i fixed.

  d. Repeat steps 2 and 3 until all u_i and v_j are fixed.

  e. Compute all non-basic costs as $c'_{ij} = c_{ij} - u_i - v_j$.

II. If any non-basic c'_{ij} is negative, let x_{ij} enter the basis. (As in the ordinary Simplex method, we can choose any negative c'_{ij}.)

  a. Identify the unique even cycle defined by the edge x_{ij} and other basic variable edges.

  b. Alternately increase and decrease the flow in the edges in this circuit until at least one basic variable has a zero flow. Remove that variable from the basis.

  c. Repeat the algorithm completely from the beginning (Part I) by recomputing the reduced costs.

Continuing with our example, in Table 3.11, for Part II of the algorithm, we find $c'_{34} = -6$. Therefore, x_{34} can enter the basis. The unique basic cycle is (x_{34}, x_{35}, x_{25}, x_{22}, x_{12}, x_{24}). The increase of the flow in this alternating circuit is limited by a decrease of 1 in the flow on x_{35}.

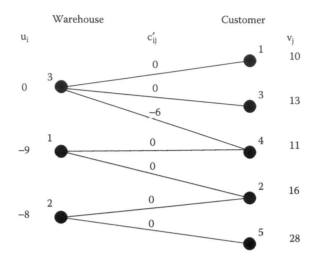

FIGURE 3.10
Basic network tree.

Therefore, x_{34}, x_{25}, and x_{12} increase by 1, and x_{35}, x_{22}, and x_{24} decrease by 1. The variable x_{34} enters the basis and x_{35} leaves the basis.

When we now return to Part I of the algorithm, we can select any basic cell. There are some small computational savings to be obtained if we choose the basic variable that just entered. Consider the new basic network tree in Figure 3.10. (In the figure, we have reordered the warehouse and customer numbers to eliminate crossing lines.)

Notice the edge corresponding to $c'_{34} = -6$. In order to get $c'_{34} = 0$, we must decrease either u_3 or v_4. Suppose we decrease v_4 by 6. Then, in order to keep all other $c'_{ij} = 0$ for basic edges, we must increase u_1 and u_2 by 6, and decrease v_2 and v_5 by 6. The new reduced cost matrix is shown in Table 3.12.

The total cost is $6 lower for a total of $604 in the original problem. Moreover, all of the reduced costs are now non-negative. Just as in the Simplex method, when the reduced costs are all non-negative, the current solution must be optimal.

TABLE 3.12

Optimal Solution

Sources	Sinks (Customers)						
(Warehouses)	1	2	3	4	5	Supply	u_i
1	21	0	6	0	11		−3
		3		17		20	
2	10	0	3	1	0		−2
	11				9	20	
3	0	2	0	0	6		0
	12		12	1	0	25	
Demand	12	14	12	18	9	65	
v_j	10	10	13	5	22		

3.3.2 Assignment Problem and Stable Matching

Our discussion of transportation models has dealt with the flow of some entity or material between nodes of a network. By imposing a few simple assumptions on the transportation model, we find that we have an apparently new kind of optimization problem.

Suppose, for example, that we wish to assign n people to n jobs; that is, we wish to associate each person with exactly one job, and vice versa. Cost parameters c_{ij} denote the cost of assigning person i to job j. Decision variables now have a completely new meaning, representing an association or bond between two entities rather than the flow of a commodity between two nodes. Specifically, each variable x_{ij} is to have a value of either zero or one:

$$x_{ij} = \begin{cases} 1, & \text{if person i is assigned to job j} \\ 0, & \text{otherwise} \end{cases}$$

If in the transportation model we require m = n, and assign all the supply and demand parameters a value of 1, then we have the following formulation for the **assignment problem**.

$$\text{minimize} \quad z = \sum_{i=1}^{n} \sum_{j=1}^{n} c_{ij} x_{ij}$$

$$\text{subject to} \quad \sum_{j=1}^{n} x_{ij} = 1 \qquad \text{for } i = 1, \ldots, n \qquad (1)$$

$$\sum_{i=1}^{n} x_{ij} = 1 \qquad \text{for } j = 1, \ldots, n \qquad (2)$$

$$\text{and } x_{ij} = 0 \text{ or } 1 \qquad \text{for all i and j}$$

Re-examining Figure 3.7 under the current assumptions, we see that we are establishing a *flow* of 1 unit *out of* each *person* node and a *flow* of 1 unit *into* each *job* node. The constraints corresponding to supply and demand constraints in the transportation model enforce the one to one association between persons and jobs. The aforementioned constraints (1) specify that each person be assigned to exactly one job, while constraints (2) specify that each job have exactly one person assigned to it.

Because network problems with integer parameters can be solved using the Simplex method to obtain integer solutions, we might simply replace the 0–1 constraint by the constraints $x_{ij} \geq 0$ and $x_{ij} \leq 1$, and treat this problem as an ordinary linear programming problem. The difficulty here lies in the inefficiency that may result from problem degeneracy. (Notice that we have 2n constraints, and only n of the decision variables are allowed to have a value greater than zero. Therefore, in any feasible solution, n − 1 basic variables are zero; that is, *any* feasible solution to the assignment problem is degenerate.) Fortunately, the highly specialized structure of the assignment model can be exploited in an efficient algorithm designed specifically for this problem. The algorithm is known as the **Hungarian Method**, named in honor of the Hungarian mathematicians König and Egervary who established the fundamentals upon which the algorithm is based.

The simple structure of the assignment model leads to a solution that is intuitively easy to follow. The key to this method lies in the fact that a constant may be added to or subtracted from any row or column in the cost matrix without affecting the optimal solution. Suppose we add a constant k to row p of the cost matrix. Then the new objective function

$$z' = \sum_{j=1}^{n} (c_{pj} + k) x_{pj} + \sum_{\substack{i=1 \\ i \neq p}}^{n} \sum_{j=1}^{n} c_{ij} x_{ij}$$

$$= \sum_{i=1}^{n} \sum_{j=1}^{n} c_{ij} x_{ij} + k \sum_{j=1}^{n} x_{pj}$$

= original objective function plus a constant

Similarly, if we add a constant k to column q, then

$$z' = \sum_{i=1}^{n} (c_{iq} + k) x_{iq} + \sum_{i=1}^{n} \sum_{\substack{j=1 \\ j \neq q}}^{n} c_{ij} x_{ij}$$

$$= \sum_{i=1}^{n} \sum_{j=1}^{n} c_{ij} x_{ij} + k \sum_{i=1}^{n} x_{iq}$$

= original objective function plus a constant

We will use this property of the assignment model to modify (repeatedly, if necessary) the cost matrix, and thereby create a new matrix in which the location of zero elements indicates an optimal feasible solution.

In order to do this, we wish to create a cost matrix with a zero in every row and every column. If we can do this, then our modified objective function value is zero; and since the cost cannot be negative, we know a zero value is optimal.

As an example, consider the cost matrix

$$\begin{bmatrix} 4 & 9 & 8 \\ 6 & 7 & 5 \\ 4 & 6 & 9 \end{bmatrix}$$

To obtain zero elements, we subtract the smallest element from each row. Subtracting 4, 5, and 4 from rows 1, 2, and 3, respectively, we obtain the modified cost matrix

$$\begin{bmatrix} 0 & 5 & 4 \\ 1 & 2 & 0 \\ 0 & 2 & 5 \end{bmatrix}$$

This does not yet identify for us a feasible solution, but if we subtract 2 (the smallest element) from the second column, we obtain

$$\begin{bmatrix} 0 & 3 & 4 \\ 1 & 0 & 0 \\ 0 & 0 & 5 \end{bmatrix}$$

From this we can make an optimal feasible assignment using the zero elements marked with squares.

$$\begin{bmatrix} \boxed{0} & 3 & 4 \\ 1 & 0 & \boxed{0} \\ 0 & \boxed{0} & 5 \end{bmatrix}$$

Assignment variables $x_{11} = x_{23} = x_{32} = 1$, and all the others are zero. The actual objective function cost, based on the original cost matrix, is $4 + 6 + 5 = 15$.

Now look at a problem in which the solution is not revealed quite so readily. The cost matrix

$$\begin{bmatrix} 2 & 11 & 2 & 6 \\ 3 & 10 & 9 & 4 \\ 8 & 6 & 6 & 6 \\ 10 & 13 & 15 & 13 \end{bmatrix}$$

can be immediately reduced to the matrix

$$\begin{bmatrix} 0 & 9 & 0 & 4 \\ 0 & 7 & 6 & 1 \\ 2 & 0 & 0 & 0 \\ 0 & 3 & 5 & 3 \end{bmatrix}$$

Now, every row and column contains a zero element, so we cannot subtract any more constants in the obvious way. However, we can make only three feasible assignments. At this point, the Hungarian method prescribes that we draw the minimum possible number of horizontal and vertical lines so that all zero elements are covered by a line. (The number of such lines that will be necessary is just exactly the number of feasible job assignments that can be made using the current cost matrix.)

A simple procedure for obtaining the minimum number of lines can be summarized as follows. Suppose you have made as many assignments as possible (to zero entries in the matrix), but there are less than n assignments:

1. *Mark* every row that has no assignment.
2. *Mark* every column that has a zero in a marked row.

3. *Mark* every row that has an assigned zero in a marked column.

4. Repeat from Step (2) until no new columns can be marked.

In Step (2), if we ever *mark* a column that has *not* been assigned yet, we can construct a new solution with one additional assignment. Column j was marked because row i was marked. Shift the assignment in row i to column j. This frees up another marked column. Assign this new marked column in a similar way until, eventually, we can assign a marked row that previously had no assignment.

Otherwise, draw a line through every *unmarked* row and every *marked* column. It is easy to verify that these lines cover every zero and that the number of lines equals the number of current assignments. For example, in the modified cost matrix

$$\begin{bmatrix} 0 & 9 & \boxed{0} & 4 \\ \boxed{0} & 7 & 6 & 1 \\ 2 & \boxed{0} & 0 & 0 \\ 0 & 3 & 5 & 3 \end{bmatrix}$$

mark row 4, mark column 1, and mark row 2. After drawing the three lines, select the minimum uncovered element, subtract this value from all the uncovered elements, and add it to all elements at the intersection of two lines. In this case, we select the value 1, subtract it from uncovered elements on rows 2 and 4, and add it to the intersection elements in the first column. (Although the Hungarian method is popularly described in terms of drawing lines and manipulating covered and uncovered elements, observe that these operations are just equivalent to subtracting and adding a constant to entire rows and columns. In our example, we are subtracting the constant value 1 from rows 2 and 4 and adding 1 to column 1.) The result is the further modified cost matrix

$$\begin{bmatrix} 1 & 9 & \boxed{0} & 4 \\ 0 & 6 & 5 & \boxed{0} \\ 3 & \boxed{0} & 0 & 0 \\ \boxed{0} & 2 & 4 & 2 \end{bmatrix}$$

from which we can make four feasible assignments: $x_{13} = x_{24} = x_{32} = x_{41} = 1$. The cost of this assignment is obtained from the original cost matrix as $c_{13} + c_{24} + c_{32} + c_{41} = 2 + 4 + 6 + 20 = 22$.

This process ensures that at least *one new zero* entry will be generated at each iteration, but the *number of assignments* does not necessarily increase. However, the Hungarian method *is* guaranteed to solve the problem; this iterative procedure will be repeated as many times as necessary so that a complete feasible assignment is finally obtained.

The Hungarian method is relatively efficient for solving large problems. However, there are more efficient commercial codes available that can dramatically reduce computation time. This can be very important when an application requires, for example, that several thousand assignment problems be solved as subroutines in a larger problem.

In case there is a mismatch between the number of people and the number of jobs, the problem is brought into balance by adding either *dummy* people or *dummy* jobs, as needed. For example, if there are m people and n jobs, and m > n, then there are not enough jobs so we add m − n dummy jobs, and a set of zero-valued cost coefficients for each. Once the balanced problem is solved, any person assigned to a dummy job actually has no job. Similarly, if m < n, the problem is balanced with dummy people; and in the final solution, n − m jobs actually have no one assigned to them.

3.3.2.1 Stable Matching

While the classical assignment problem seeks to find an association of objects that is optimal from a collective, or global, point of view, it does not necessarily consider *individual* preferences or affinities. Suppose the entries in the cost matrix actually represent rankings, so that finding a minimum cost assignment actually associates objects according to their preferences. Now if the objects being associated with each other are people being assigned to machines, the people probably have preferences, while the machines do not. But if we have employees (people) being assigned to employers (also people), then most likely there are preferences on both sides. Similar situations arise, for example, when medical residents are being assigned to hospitals, or when graduate students become associated with certain doctoral programs, because in all these cases there are mutual preferences involved, which certainly might be different on the side of the *employer* than on the side of the *employee*. The workers could probably rank their preferences for employers *and* the employers could probably rank their preferences among the pool of potential employees. In this case, there are two cost matrices, reflecting the preferences of both groups.

If we wish to treat this as an ordinary assignment model, a single cost matrix can be constructed by simply adding corresponding elements of the rank matrices (Exercise 3.9 at the end of this chapter). Remember, however, that the (i, j)-th element of the employee rankings does not get added to the (i, j)-th element of the employer rankings, but rather to the (j, i)-th element. Information about employee i and employer j is in the (i, j)-th position in the first matrix but in the (j, i)-th position in the second matrix.

The Hungarian method, when applied to this problem, yields a solution that is in some sense for the collective good of both employees and employers. But what about the individuals or employers who do not get their first or even second choices? The behavioral reaction of these people is dealt with by using a model that is known as the *stable marriage problem* (so-called because this model hypothetically could be used to represent the preferences of groups of people who are to be matched for marriage) (Knuth 1976).

For this example, we will use a group of four men and a group of four women. Consider the following preference matrices, and the corresponding cost matrix composed in the way we described earlier.

		Woman							Man			
		W	X	Y	Z			A	B	C	D
Man	A	2	1	3	4	Woman	W	1	3	4	2
	B	1	2	3	4		X	3	2	4	1
	C	4	1	2	3		Y	1	3	4	2
	D	1	3	2	4		Z	4	2	1	3

		W	X	Y	Z
Cost =	A	3	4	4	8
	B	4	4	6	6
	C	8	5	6	4
	D	3	4	4	7

The matching A–Y, B–X, C–Z, D–W has a cost of $4 + 4 + 4 + 3 = 15$, and is optimal when viewed as an ordinary assignment problem; but from an individual perspective, that matching leaves something to be desired. Notice that man A and woman W both prefer each other over the one they are matched to. A matching is called **unstable** if two people who are not married prefer each other to their spouses. In our example, A and W acting according to their preferences would leave Y and D, respectively, for each other. Then there would be little choice for Y and D but to get together with each other—a disappointment for each, since now each is paired with a second-ranked choice, whereas previously both had been matched with their first-ranked choices. (Observe that this rearrangement A–W, B–X, C–Z, D–Y has the same cost, $z^* = 15$, as the previous matching when viewed as a simple assignment problem.)

Finding stable matchings is a difficult problem, both from a sociological and a computational standpoint. Even the problem of determining *whether* a matching is stable is difficult; and the process of removing instabilities one at a time is not only slow but can lead to circularities that prevent the algorithm from terminating.

A better approach seems to be to construct stable matchings from the outset. In fact, algorithms exist to construct a stable matching efficiently. However, the overall quality (cost) of the assignment may be quite poor (everyone may be unhappy but stable), and all known algorithms for this tend to be biased in favor of one group or the other (men over women, employers over employees, etc.). A well-known *propose and reject* algorithm constructs a stable assignment in $O(n^2)$ time, but unfortunately the matching is done from a *man-optimal* point of view, and in fact a consequence of the method is that each woman obtains the worst partner that she can have in any stable matching. The only remedy is to create a stable matching from a *woman-optimal* point of view, with the corresponding consequence to each man. We can clearly see here that there are important economic and sociological effects involving employment stability and worker satisfaction for which we currently have no good solutions (Ahuja et al. 1993).

3.3.3 Capacitated Transshipment Problem

The most general form of the minimum cost network flow problem arises when some commodity is to be distributed from sources to destinations. Each node can create a certain **supply** or absorb some **demand** of the commodity. It is not necessary for each unit of the commodity to be shipped directly from a source to a destination; instead, it may be **transshipped** indirectly through intermediate nodes on its way to its destination. In fact, the total supply could conceivably be routed through any node in transit. Links can have upper and lower bounds on the flow that may be assigned to them. The object then is to meet the demands without exceeding the available supply, and to do so at minimum cost. This model is known as a *minimum cost flow problem* or as a **capacitated transshipment problem**. We let x_{ij} represent the number of units shipped along the arc from node i to node j, and c_{ij} denote the per unit cost of that shipment. Capacities are specified by lower bounds ℓ_{ij} and upper bounds u_{ij} on each arc from node i to node j. Flow balance equations enforce the constraint for a net supply s_i at each node i. The net supply at a node is expressed as total flow out minus total flow in. (If s_i is negative, it will be interpreted as a net demand constraint.) The formulation is as follows:

$$\text{minimize} \quad z = \sum_{i=1}^{n} \sum_{j=1}^{n} c_{ij} x_{ij}$$

$$\text{subject to} \quad \sum_{j=1}^{n} x_{ij} - \sum_{k=1}^{n} x_{ki} = s_i \qquad \text{for } i = 1, \dots, n \qquad (1)$$

$$\ell_{ij} \leq x_{ij} \leq u_{ij} \qquad \qquad \text{for all } i \text{ and } j \qquad (2)$$

Summations are taken over all index values for which the corresponding arcs exist in the network. To keep the notation simple, we assume that $\ell_{ij} = u_{ij} = 0$ for all non-existent arcs.

Most introductory textbooks that describe the transshipment problem, explain how it can be modeled as an expanded transportation problem with dummy demands and supplies for each intermediate node. The two models *are*, in fact, equivalent. And although that approach will work for small problems, it is not recommended for any applications of practical size.

The minimum cost network flow problem could also be solved using the Simplex method presented in Chapter 2. However, the special structure in the formulation makes the problem amenable to more efficient solution techniques. The structure is apparent in the flow balance equations (constraints [1] in our previous formulation). The variables x_{ij} appear with coefficients of only 0, +1, and −1 in each equation. And because each arc flows into exactly one node and out of exactly one node, each variable appears in exactly two of the flow balance equations. This matrix of coefficients is known as a **node-arc incidence matrix** and is fundamental to the methods that have been tailored for use on this problem.

One efficient technique for solving the minimum cost flow problem is a specialization of Dantzig's Simplex algorithm, and has been called the *Simplex on a graph* algorithm (Kennington 1980). One implementation of this method is reported to be over 100 times faster than a general linear programming code applied to the minimum cost flow problem.

Another method, developed by Fulkerson specifically for the minimum cost flow problem, is called the *out-of-kilter* algorithm. Each arc is either *in kilter* or *out of kilter*, indicating whether that arc could be in a minimum cost solution. *Kilter numbers* specify how far an arc is from being *in kilter*. Beginning with any maximum feasible flow, the algorithm repeatedly selects an out-of-kilter arc, and adjusts the flow in the network so as to reduce the kilter number of the chosen arc, while not increasing the kilter number of any other arc, and maintaining feasible flow. When all arcs are in kilter, the current solution is the minimum cost flow. Clear and complete descriptions of this method may be found in several of the references cited at the end of this chapter, including Kennington (1980), Price (1971), Battersby (1967, 1970), Hu (1970), and Tarjan (1983).

The following example, from Glover and Klingman (1992), illustrates the creative use of the transshipment model for production planning and distribution decisions. A major U.S. car manufacturer must determine the number of cars of each of three models M1, M2, and M3 to produce at the Atlanta and Los Angeles plants, and how many of each model to ship from each plant to distribution centers in Pittsburgh and Chicago. Subject to bounds on production capacities, demands, and shipment capacities, the objective is to identify a minimum cost production-distribution plan. A network model for this problem is given in which arcs from plant locations to plant/model nodes are labeled with upper and lower bounds on production levels, and with production costs for each model at each plant. Similarly, arcs from distribution/model nodes to distribution point nodes are labeled to indicate bounds on demands. Links from plant/model nodes to distribution/model nodes are labeled with the appropriate transportation costs, and with capacity restriction limits, if any.

A solution to this problem determines the production and distribution decision for the car manufacturer; but, moreover, it solves a multi-commodity problem with a straightforward transshipment model. By having distinct nodes for each model type, the production and distribution plan for each model is established.

3.4 Network Connectivity

3.4.1 Minimum Spanning Trees

Now consider a network problem in which we wish to select the fewest possible arcs in the network that will keep the graph connected. Recall that a graph is connected if there is at least one path between every pair of nodes. We furthermore want to select just those arcs with the smallest weights or costs. This is called the **minimum spanning tree** problem.

A typical application for a minimum spanning tree may arise in the design of a data communications network that includes processor nodes and numerous (possibly redundant) data links connecting the nodes in various ways. We would like to determine the set of data links, with the lowest total cost, that will maintain connectivity, so that there is some way to route data between any pair of nodes. Similarly, in any type of utility distribution network or transportation network, it may be desirable to identify the minimum set of connections to **span** the nodes.

Such a minimal set of arcs always forms a tree. Clearly, the inclusion of any arc resulting in a cycle would be a redundant arc, and this could not be a *minimum* spanning tree. To see this, suppose that the optimal solution contains a cycle. Select any arc (i, j) in the cycle, and delete it. Notice that any two nodes that were connected using arc (i, j) are *still* connected because nodes i and j are still connected by moving the other way around the cycle. Therefore, the solution could not have been optimal because we easily constructed a better (less costly) one.

We present two algorithms for solving this problem. The choice of which one to use for a particular application depends on the density or sparsity of the network in question. The two algorithms are quite simple, and are sometimes called *greedy algorithms* because at each stage we make the decision that appears locally to be the best; and in so doing, we finally arrive at an overall solution that is optimal. (As has already been suggested, it is a rare and wonderful thing when we are able to solve combinatorial problems using simple greedy algorithms.)

Our first solution to the minimum spanning tree problem is **Prim's algorithm**, which operates by iteratively building a set of connected nodes as follows:

1. Arbitrarily select any node initially. Identify the node that is connected to the first node by the lowest cost arc. These two nodes now comprise the connected set, and the arc connecting them is a part of the minimum spanning tree.

2. Determine an isolated node that is *closest* (connected by the lowest cost arc) to some node in the connected set. (Break ties arbitrarily.) Add this node to the connected set of nodes and include the arc in the spanning tree. Repeat this step until no nodes remain isolated.

Prim's algorithm is illustrated by the example shown in Figure 3.11a, where the sequence of pictures (b) through (e) shows the iterative construction of the minimum spanning tree. Node B is arbitrarily chosen as the initial node. Node C is its *closest* neighbor. Then node E is attached, followed by node D and finally node A. In the figure, nodes are outlined boldly as they become connected.

The arcs in the spanning tree have weights 1, 2, 4, and 5, yielding a cost of 12 for the minimal spanning tree. Note that the choice of initial node B is arbitrary, and any choice for the initial node would have yielded a tree whose cost is 12.

The complexity of Prim's algorithm is $O(n^2)$ for an n-node network. If the network is sparse (with much less than n^2 arcs), the performance of this algorithm on large networks is unnecessarily slow. For such cases, we have an alternative algorithm, known as **Kruskal's algorithm**, whose performance is $O(e \log e)$ where e is the number of arcs. Thus, in a sparse network where e is much less than n^2, Kruskal's algorithm is superior; whereas in dense networks, Prim's algorithm is preferred.

Kruskal's algorithm operates by iteratively building up a set of arcs. We examine all the arcs, in increasing order of arc cost. For each arc, if the arc connects two nodes that are currently not connected (directly or indirectly) to each other, then the arc is included in the spanning tree. Otherwise, inclusion of the arc would cause a cycle and therefore could not be a part of a minimum spanning tree. This algorithm is another example of a greedy method. With Kruskal's algorithm, we ensure a minimum cost tree by examining and choosing the lowest cost spanning arcs first. Figure 3.12 shows the sequence of arcs chosen for a minimum spanning tree for the network in Figure 3.12a.

Tarjan provides a historical perspective on solutions to spanning tree problems, and describes several efficient variations to Prim's and Kruskal's algorithms. In such implementations, the improved complexity hinges on the use of specialized data structures (such as heaps and priority queues). Tarjan also discusses mechanisms for sensitivity analysis (Tarjan 1982): an algorithm is available for testing whether a given spanning tree is minimal, and it is also possible to determine how much the cost on each arc can be changed without affecting the minimality of the current spanning tree.

(a)   (b)   (c)   (d)   (e)

FIGURE 3.11
Prim's algorithm: (a) original network, (b) first iteration, (c) second iteration, (d) third iteration, and (e) last iteration.

(a)	(b)	(c)	(d)	(e)

FIGURE 3.12

Kruskal's algorithm: (a) original network, (b) first iteration, (c) second iteration, (d) third iteration, and (e) last iteration.

It is interesting to note how difficult the minimum spanning tree problem becomes when certain constraints are added. If we place limits on the degree of all the nodes in the spanning tree, then the minimum spanning tree problem becomes NP-hard. Such restrictions might reasonably apply in an actual application, for example, where we could have a limited number of I/O ports on each microprocessor in a multiprocessor network.

3.4.2 Shortest Network Problem: A Variation on Minimum Spanning Trees

In the minimum spanning tree problem, we choose a minimum cost subset of arcs that connect the vertices. But suppose that, instead of choosing a set of arcs from among those *already* in the network, we allow ourselves to introduce new connections in addition to the original arcs. Consider the following common problem. An electrician has decided where to place the outlets in a home, and now wants to connect the outlets back to the circuit box using the minimum amount of wire. Note that *any* circuit is a spanning tree. But, as any electrician will tell you, to minimize the total length of cable, you should in fact introduce *new* nodes (junction boxes) in the network, and *then* find the minimum spanning tree.

Consider the simple network in Figure 3.13 in which the nodes are the vertices of an equilateral triangle and the arcs connect each pair of nodes. The length (or weight) of each of the arcs is four units. A minimum spanning tree has a length of 8, and is obtained by choosing any two of the three arcs as shown in Figure 3.13a–c. But if instead of choosing a subset of the given arcs, we judiciously introduce a *new* node or junction point, we find that we are able to span the three nodes with line segments whose total length is only about 6.928. This is the **shortest network** that spans the three original vertices, and is illustrated in Figure 3.13d.

Clearly, this could represent a substantial saving in the cost of links if we were designing the connections in communication networks, circuit board layouts, or highway or utility distribution networks. This example is an instance of what is called the **Steiner tree problem**: where should we introduce new nodes in the network to minimize the corresponding spanning tree?

The difficulty of the Steiner tree problem lies in selecting the *location* of the extra *junction* points. Geometric intuition probably tells us that the solution in Figure 3.13d is better than

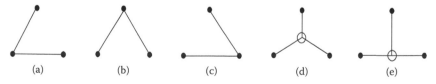

(a)	(b)	(c)	(d)	(e)

FIGURE 3.13

Shortest network problem: (a) a minimum spanning tree, (b) another minimum spanning tree, (c) a third minimum spanning tree, (d) shortest network, and (e) sub-optimal junction point.

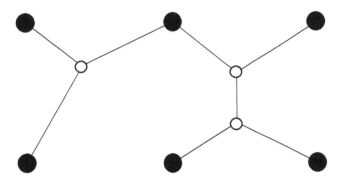

FIGURE 3.14
Steiner tree problem.

the one in Figure 3.13e. However, consider a slightly larger problem, such as the graph with six nodes arranged in a grid in Figure 3.14. Is this an optimal Steiner tree? In fact, there is a slightly better set of junction points and connections than the ones shown in the figure, but how would we know this? And what about solving much larger problems?

The best known algorithms for solving the Steiner tree problem are based on an algorithm of Melzak (1961); and although numerous modifications to that algorithm have improved its efficiency, the algorithms still require exponential computation time.

Although the Steiner tree problem is NP-hard, we still have practical algorithms that yield approximations to the solutions that we want. In fact, we even have the guarantee that a Steiner tree is at most 17.6% shorter than a minimum spanning tree. Thus, we can use an efficient greedy algorithm (such as Prim's or Kruskal's) and obtain a spanning tree whose length is at most only about 21% greater than that of a Steiner tree whose calculation may require exponential effort. Here again, the analyst is faced with the choice of accepting a possibly suboptimal solution that can be obtained easily, versus a provably optimal solution that is obtainable only at enormous computational expense. Of course, the household electrician is probably inserting a few extra junctions at obvious locations and very likely feels that his solution is convenient and satisfactory from a practical standpoint. See Bern and Graham (1989) for an interesting historical perspective on Steiner problems, exact and approximate algorithms.

3.5 Shortest Path Problems

We will now consider a class of network problems in which we try to determine the shortest (or least costly) route between two nodes. The chosen route need not necessarily pass through all other nodes. An obvious application of this type of problem is represented by a vehicle traveling from a departure point to a final destination passing through different points via the shortest route. Similarly, a distributed computer network that must route data along the shortest path between designated pairs of processing nodes. We will also see other, less obvious applications that can be solved with shortest path algorithms (see exercises) or with methods reminiscent of shortest path algorithms (Sections 3.6 and 3.7).

The **shortest path problem** can be viewed as a transshipment problem having a single source and a single destination. The supply at the source and the demand at the destination are each considered to be one unit, and the cost of sending this unit between any two adjacent nodes is indicated by the cost (weight or distance) on the arc connecting the two nodes. By finding a minimum cost transshipment, we are in fact determining the shortest route by which the unit can travel from the source to the destination. Although the shortest path problem could be dealt with by using the more general transshipment model, the structure of the shortest path problem makes it amenable to much more specialized and efficient algorithms.

3.5.1 Shortest Path through an Acyclic Network

There are several well-known algorithms for finding the shortest path between certain pairs of nodes in a network. We will concentrate first on a particularly simple algorithm that is based on the use of recursive computations. This approach to shortest path problems will also provide us with a foundation for the study of dynamic programming and project management in the next two sections of this chapter.

As an illustration, consider the acyclic network in Figure 3.15, where arc labels d_{ij} denote distance from node i to node j. Notice that in an acyclic graph, it is always possible to name the nodes in such a way that an arc is oriented *from* a lower-numbered node *to* a higher numbered node. (A consequence of this property is that such a network can be represented by an adjacency matrix that is upper triangular, requiring only $(n^2 + n)/2$ storage locations in computer memory instead of n^2.) We wish to determine the shortest path from the lowest-numbered node to the highest-numbered node.

The algorithm operates by assigning a label to each node, indicating the shortest distance from that node to the destination. A node is *eligible* for labeling if all its successors have been labeled.

1. Initially, the destination node is given a label of zero, indicating that there is no cost or distance associated with going from that node to itself.
2. Choose any *eligible* node k, and assign it a label p_k as follows:

   $p_k = \min \{d_{kj} + p_j\}$, the minimum taken over all successors j of node k
3. Repeat Step 2 until the source node is labeled. The label on the source *is* the shortest distance from the source to the destination.

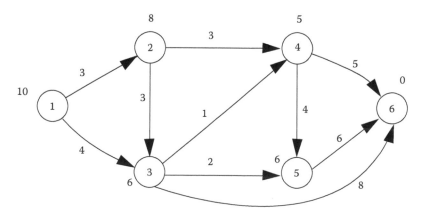

FIGURE 3.15
Acyclic network with node labels.

In the illustration in Figure 3.15, initially $p_6 = 0$. Next, node 5 is eligible and $p_5 = 6 + 0 = 6$. The label for node 4 is computed as $p_4 = \min \{5 + 0, 4 + 6\} = 5$. Node 3 is now eligible, and $p_3 = \min \{1 + 5, 2 + 6, 8 + 0\} = 6$. The label on node 2 is $p_2 = \min \{3 + 5, 3 + 6\} = 8$, and finally $p_1 = \min \{3 + 8, 4 + 6\} = 10$. Thus, the length of the shortest path is 10, and the path itself is obtained by tracing back through the computations to find the path containing the arcs (1,3), (3,4), (4,6).

This backward labeling procedure has an intuitive appeal when the problem is small enough that the labels can be shown in a diagram. For larger problems, we may obtain better insight by examining the recursive structure of the computations. For this, we will again use the illustrative network from Figure 3.15. We wish to determine a label for node 1; but in order to compute p_1, we require the labels for nodes 2 and 3. Obtaining these labels involves the recursive labeling procedure (twice). Each of these recursive computations in turn requires further recursion. The pattern of recursive *calls* to obtain the label on the first node is illustrated as follows, where L(i) denotes p_i:

$$L(1) = 4 + L(3)$$
$$= 4 + \left[1 + L(4)\right]$$
$$= 4 + \left[1 + \left[5 + L(6)\right]\right]$$
$$= 4 + \left[1 + \left[5 + 0\right]\right]$$
$$= 10$$

Observe that the label on each node summarizes information on higher-numbered nodes. In fact, the value of the label on any node *is* actually the length of the shortest path from that node to the destination.

3.5.2 Shortest Paths from Source to All Other Nodes

A more general algorithm that can be applied to any network having all arc labels non-negative is known as **Dijkstra's algorithm**. This algorithm begins with the source node and determines the shortest paths from the source to every other node. During the operation of Dijkstra's algorithm, the nodes are partitioned into two sets: a set, which we shall call S, to contain nodes for which the shortest distance from the source is known, and another set T to contain nodes for which this shortest distance is not yet known. A label p_i is associated with every node i and specifies the length of the shortest path known *so far* from the source (node 1) to node i. Again, we let d_{ij} denote the direct distance from node i to node j.

1. Initially, only the source node is placed in set S, and this node is labeled zero, indicating that there is zero distance from the source to itself.

2. Initialize all other labels as follows:

$$p_i = d_{1i} \qquad \text{for } i \neq \text{source node 1}$$

and $p_i = \infty$ if node i is not connected to the source

3. Choose a node w, not in set S, whose label p_w is minimum over all nodes not in S, add node w to S, and adjust the labels for all nodes v, not in set S, as follows:

$$p_v = \min \{p_v, \; p_w + d_{wv}\}$$

4. Repeat Step 3 until all nodes belong to set S.

In step 3, we assume that p_v is the shortest distance from the source to node v directly through nodes in S. When we add node w to S, we check whether or not the new distance through w is shorter, and update if necessary. We will use the network shown in Figure 3.16 to illustrate Dijkstra's algorithm.

Initially S = {1}, and $p_1 = 0$, $p_2 = 5$, $p_3 = 3$, $p_4 = 8$, $p_5 = \infty$, and $p_6 = \infty$. We then choose the minimum label 3 on node 3, and S = {1, 3}. Labels are now

$$p_2 = \min \{5, 3 + \infty\} = 5$$

$$p_4 = \min \{8, 3 + \infty\} = 8$$

$$p_5 = \min \{\infty, 3 + 4\} = 7$$

$$p_6 = \min \{\infty, 3 + 8\} = 11$$

In the next iteration, we select the label 5 on node 2, so that S = {1, 3, 2} and new labels are

$$p_4 = \min \{8, 5 + 2\} = 7$$

$$p_5 = \min \{7, 5 + \infty\} = 7$$

$$p_6 = \min \{11, 5 + \infty\} = 11$$

From these labels, we break a tie arbitrarily and select the minimum label 7 on node 5. Now S = {1, 3, 2, 5} and

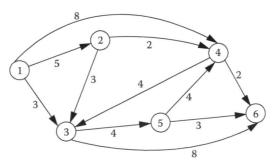

FIGURE 3.16
Shortest path with Dijkstra's algorithm.

$$p_4 = \min \{7, 7+4\} = 7$$

$$p_6 = \min \{11, 7+3\} = 10$$

Now we choose node 4 and $S = \{1, 3, 2, 5, 4\}$, and

$$p_6 = \min \{10, 7+2\} = 9$$

Finally, node 6 is added to set S. The final labels are $p_1 = 0$, $p_2 = 5$, $p_3 = 3$, $p_4 = 7$, $p_5 = 7$, and $p_6 = 9$, and the values of these labels indicate the lengths of the shortest paths from node 1 to each of the other nodes.

On a dense graph of n nodes and e arcs, represented by an adjacency matrix, Dijkstra's algorithm executes in time $O(n^2)$. In a sparse network where e is much less than n^2, it is worthwhile to represent the graph as an adjacency *list*, and to manage the node partitions using a priority queue implemented as a partially ordered tree (Aho and Hopcroft 1974). In that case, the running time is $O(e \log n)$.

The proof of optimality of Dijkstra's algorithm requires that all the arcs have positive labels. But consider a network in which arcs represent stages of a journey. Along certain arcs a cost is incurred (positive cost), while on other arcs it is possible to turn a profit (negative costs). Our objective would be to find a minimum cost path from source to destination and, if possible, a path with negative cost (i.e., a profitable path). An algorithm developed by Bellman (1958) and Ford Jr. (1956) will solve this problem as long as there is no cycle in which the sum of the arc lengths is negative. (Observe that, if there were a cycle with a negative total length, then we could simply travel around the cycle indefinitely reducing our cost with no lower bound.)

Suppose we have a network for which we would like to know the shortest distance between *any* two nodes. This is called the *all-pairs* shortest path problem. For this problem, Dijkstra's algorithm could be applied n times (using a different node each time as the source) to obtain the desired result in time $O(n^3)$. Another algorithm known as Floyd's algorithm provides the solution in a more direct way, also in time $O(n^3)$ but with a much lower constant factor than Dijkstra's algorithm. However, for large sparse graphs, clever use of data structures will allow Dijkstra's algorithm to operate in $O(n e \log n)$ time. Algorithms for the second shortest path through a network, the n-th shortest path, and for *all* possible paths between two specified nodes, are described and illustrated in Price (1971).

3.5.3 Problems Solvable with Shortest Path Methods

We have shown how shortest path methods can be used to determine the shortest (fastest, or least costly) route between two locations in a network. A couple of additional illustrations should indicate the great variety of problems that can be modeled and solved in this way.

A frequently cited example is one in which we wish to determine the most cost-effective schedule for the replacement of equipment over a period of time. Let us suppose circuit boards for A/D conversion in a navigation computer are to be replaced at intervals over a period of 6 months. Ideally, replacement should occur *before* an actual breakdown in order to maintain an operational system. Frequent replacement incurs capital expenses and costs of labor for installation. But infrequent replacement may lead to increased maintenance

TABLE 3.13

Equipment Replacement Costs

		Circuit Board Replaced				
		Feb	Mar	Apr	May	June
Circuit Board Installed	Jan	5.00	6.75	8.25	12.50	16.80
	Feb		5.25	6.25	9.50	11.50
	Mar			5.25	7.25	9.00
	Apr				5.50	8.20
	May					5.80

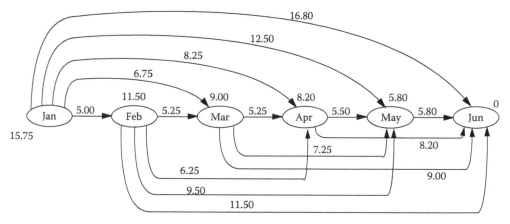

FIGURE 3.17
Equipment replacement schedule.

costs and unacceptably high rates of system downtime. If we collect data on the costs of purchase, installation, and maintenance, cost of expected downtime, and salvage value of replaced boards, we can arrive at a tabularized summary of these expenses, such as shown in Table 3.13.

Any circuit board becomes a candidate for replacement after one month. This problem can be represented as a network (Figure 3.17) with nodes representing the months, and arcs labeled with the costs shown in the table. By finding the shortest path between node Jan and node Jun, we obtain the optimal (least costly) replacement policy. The route Jan → Mar → Jun, with minimal cost 6.75 + 9.00 = 15.75, indicates that circuit boards installed in January should be replaced in March and again in June.

This approach is often used for practical situations. However, observe that if we add a node for July, or August, the optimal solution will change. We can overcome this problem by using a *rolling horizon*. For example, in January, we might use a 24-month formulation to decide when to perform the *first* replacement. That is, we use just the *first* shortest path. When we get to that month chosen for replacement, we formulate a *new* shortest path problem for the *next* 24 months. Many other practical problems have a similar structure.

An apparently unrelated set of problems is often illustrated in the form of riddles or puzzles. The context may involve ferrying missionaries and cannibals, foxes and chickens, monkeys and bananas; or separating a volume of some liquid by using an apparently inappropriate set of containers or measuring devices; or rearranging the elements of a plastic puzzle. In each of these problems, there is some initial configuration, and a sequence of simple one-step moves or operations, concluding eventually in some desired goal configuration. Each of these problems can be solved in the following way. Create a set of nodes in which each node represents a possible configuration of the system. Place a directed arc to indicate where a transition can be made from one configuration node to another through *one* simple move. Assign a cost of 1 to each arc in the network. If there are multiple goal configurations, join those nodes to a common sink node and label these new arcs zero. The shortest path from the initial configuration node to the sink or goal configuration node represents a solution to the problem, and moreover, this path describes the solution obtainable in the smallest number of steps.

3.6 Dynamic Programming

Dynamic Programming is an approach to solving mathematical programming problems by decomposing a problem into simpler, interdependent, subproblems, and then finding solutions to the subproblems in stages, in such a way that eventually an optimal solution to the original problem emerges. Because this approach has been used particularly for applications that require decisions to be made dynamically over time, the descriptive name *dynamic programming* has come into common use. However, this procedure is applicable to any problem that can be dealt with as a staged decision-making process.

In most of the optimization problems that we have seen thus far, all of the decision variables have been dealt with simultaneously. Arbitrarily complex interactions among decision variables are precisely what make general mathematical programming problems difficult. However, many problems have a structure that allows us to break the problem into smaller problems that can be dealt with *somewhat* independently. As long as we are able to preserve the original relationship among the subproblems, we may find that the total computational effort required to solve the problem as a sequence of subproblems is much less than the effort that would be required to attack all components of the problem simultaneously.

Unlike linear programming and other specialized mathematical programming formulations, dynamic programming does not represent any certain class of *problems*, but rather an approach to *solving* optimization problems of various types. Because the procedure must be tailored to the problem, the successful application of dynamic programming principles depends strongly on the intuition and talent of the analyst. Insight and experience are required in order for a problem-solver to perceive just how (or whether) a problem can be decomposed into subproblems, and to state mathematically how each stage is to be solved and how the stages are related to one another. Exposure to a large number of illustrative dynamic programming applications, including discrete and continuous variables, probabilistic systems, and a variety of objective function forms, would be required in order to provide truly useful and comprehensive insights into the craft of dynamic programming. Even then, it must be admitted that many problems simply do not lend themselves efficiently to the dynamic programming framework.

We will examine some examples, and in the process we will also describe some of the unifying themes and notations of the dynamic programming approach. For further exposure to this problem-solving tool, refer to the discussions by Bellman (1957), Nemhauser (1966), Beightler (1976), and White (1969).

3.6.1 Labeling Method for Multi-Stage Decision Making

Our first example of the use of the dynamic programming approach involves a choice of transportation routes. Figure 3.18 shows a system of roads connecting three sources H_i that generate hazardous by-products with two sites D_j designated for the disposal of hazardous waste materials. Three political borders (shown by dashed-lines) must be crossed in transit. Each straight-line section of road requires one day's travel time, so it is a four-day drive from any H_i to any D_j. However, at each border crossing, regulations require container inspection and possible recontainerization, and this can cause delays at each checkpoint. The number of days delay that can be anticipated is shown in the circle drawn at each checkpoint. The problem is to determine the route from generation sites to disposal sites that involves the minimum delays.

The **stages** in this multi-stage decision process correspond to the three borders that must be crossed. In the terminology of dynamic programming, the various checkpoints at each stage are called **states**. Thus, there are four states in the first stage, and three states in each of the second and third stages.

To solve this problem, we take an approach that is similar to the backward labeling method for shortest path through an acyclic graph. Our decisions will be made, beginning with the final stage, Stage 3, and moving backward (to the left) through the earlier stages. At each stage, we phrase our decision in the following way: for each possible *state* in the current stage, if this state is ever reached, what would be the minimum delay from here to the dump sites? If this question can be answered at every stage, then eventually at the first stage, we will have established our minimum delay route, as desired.

The mechanism that we will use is a backward node-labeling scheme. When we arrive at Stage 3, the delay to the dump site is just the delay at the third border crossing. We label each checkpoint node accordingly, as shown in Figure 3.19a.

At stage 2, the delay at the top node is 5 plus either four or three additional days. We choose the minimum 3 and label that node with $5 + 3 = 8$. The other two nodes are labeled in the same way, as shown in Figure 3.19b.

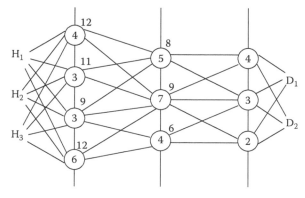

FIGURE 3.18
Hazardous waste disposal routes.

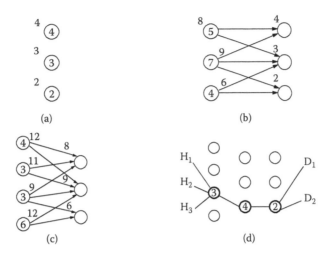

FIGURE 3.19
Minimum delay path: (a) stage 3, (b) stage 2, (c) stage 1, and (d) optimal path.

Backing up to Stage 1, we similarly compute four labels, as shown in Figure 3.19c. Since all four checkpoints at Stage 1 are uniformly accessible from each of the generation sites, we can conclude that the minimum delay path goes through the node labeled 9 at the first border crossing (with a delay of 3). The optimal path is highlighted in Figure 3.19d, where the total delay of 9 is obtained by crossing the second border at the bottom node (where delay is 4), and from there crossing the third border at its bottom node (with a delay of 2).

3.6.2 Tabular Method

Dynamic programming problems can usually be represented more succinctly in tabular form rather than as a graph. Consider the following problem. A Director of Computing Facilities must decide how to allocate five computer systems among three locations: the Library, the University Computer Center, and the Computer Science Lab. The number of users who can be accommodated through various allocations is shown in Table 3.14.

By viewing this problem as a staged decision process, we can determine the optimal allocation that will provide computer access to the greatest number of users. Let Stage 1

TABLE 3.14

Computer Allocation Problem

Number of Computers Allocated	Number of Users Served at Each Location		
	Library	University Computer Center	CS Lab
0	0	0	0
1	3	5	8
2	6	10	12
3	7	11	13
4	15	12	13
5	20	24	18

denote the decision of how many computers to place in the Library, Stage 2 denote the decision for the Computer Center, and Stage 3 for the Computer Science Lab. As before, we will begin with the last stage, and work backward.

At the third stage, we do not know how allocations may be made at earlier stages, but regardless of what earlier allocations may have been decided, we wish to determine the optimal allocation for the remaining available computers. Since this is the last stage, we clearly should allocate all remaining computers (i.e., the ones that were not allocated in Stage 1 and Stage 2) to the Lab, as shown in Table 3.15.

At the second stage, the alternatives are somewhat more interesting. Again, we do not know what allocations may be made at earlier stages (Stage 1); but since this is not the last stage, we must consider the possibility of allocating only a portion of what is available, leaving some computers for allocation in Stage 3. The various possible allocations in Stage 2 are shown in Table 3.16. Each entry represented by a sum includes the number of users that can be served by placing some computers here at this stage, plus the optimal number that could be served by saving the remaining available computers for later stages.

We can conclude the solution to this problem now by solving Stage 1. In this case, we do not have to consider different numbers of available computers: we know that all five are available because there are no preceding stages (during which any could be allocated). We do, however, have the option to allocate any number of them, as shown in Table 3.17.

TABLE 3.15

Allocation to Computer Science Lab

	Computer Science Lab	
Number Available	Number to Allocate	Optimal Number of Users Served
0	0	0
1	1	8
2	2	12
3	3	13
4	4	13
5	5	18

TABLE 3.16

Allocation to University Computer Center

Number Available	Payoff for the Number Allocated to the University Computing Center						Optimal Number of Users Served	By Allocating
	0	1	2	3	4	5		
0	0						0	0
1	0 + 8	5 + 0					8	0
2	0 + 12	5 + 8	10 + 0				13	1
3	0 + 13	5 + 12	10 + 8	11 + 0			18	2
4	0 + 13	5 + 13	10 + 12	11 + 8	12 + 0		22	2
5	0 + 18	5 + 13	10 + 13	11 + 12	12 + 8	24 + 0	24	5

TABLE 3.17

Allocation to the Library

| Number | Payoff for the Number Allocated to the Library | | | | | | Optimal Number of | |
Available	0	1	2	3	4	5	Users Served	By Allocating
5	0 + 24	3 + 22	6 + 18	7 + 13	15 + 8	20 + 0	25	1

The problem is now solved. The optimal number of users, 25, that can be served is obtained by allocating one computer to the Library. That leaves 4 available for Stage 2, and from the table for Stage 2, we know that the optimal decision is to allocate 2 to the Computer Center, leaving 2 for Stage 3, the Computer Science Lab. At Stage 3, we allocate both available computers. Thus, by placing 1, 2, and 2 computers, respectively, in the Library, Computer Center, and Lab, we can serve 3 + 10 + 12 = 25 computer users.

Notice that we could have used a graphical representation of this problem as shown in Figure 3.20, and the backward labeling technique, to find the optimal alloca-tion. However, even in a problem of this size, the number of arcs becomes large and

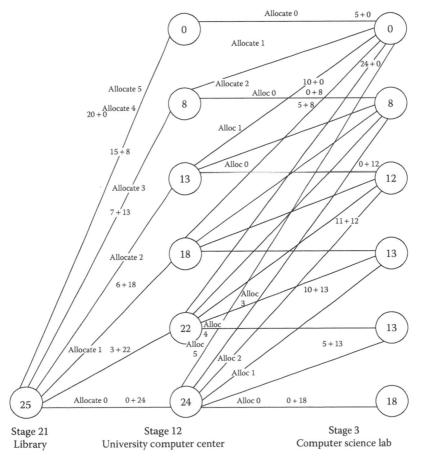

FIGURE 3.20
Graphical representation of computer allocation problem.

awkward to display. We accomplish exactly the same thing conceptually using the more convenient tabular representation.

3.6.3 General Recursive Method

Using dynamic programming, we have now solved two problems—waste-disposal routing and computer allocation—as staged-decision problems. Each point where a decision is made is referred to as a **stage** of the decision process. In some problems, these stages correspond to stages in time; in other cases, they refer to geographical stages; and in others, the stages may reflect a more abstract logical decomposition of the larger problem. The structuring of a complex problem into simpler stages of decision-making is the fundamental characteristic of the dynamic programming approach.

Within each stage, **states** are defined in such a way as to embody all the information needed in order to make the current decision and to fully define the ramifications of any current decision on future decisions. The specification of states is a critical performance factor in any dynamic programming solution. In practical problems, the number of possible states can quickly become unmanageable. Successful applications usually require considerable skill in the definition of the states.

In our illustrative examples, each stage has only one state variable (to specify *which* check-point on a border crossing, or *how many* computers are available for allocation to the current location). Some problems require more than one state variable, and each *state* of the system is represented by each possible combination of state variable values. Clearly, the number of possible states increases exponentially as the number of state variables grows, and the computational effort involved in solving the problem may become prohibitively expensive.

Decision variables in a dynamic programming model define the decisions made at each stage. Each decision yields some payoff (or return) that contributes to the **objective function**. Because of the staged structure of this method of problem-solving, determining the optimal value of a decision variable is a process that cannot be based on the entire problem but rather on only those stages of the problem that have already been dealt with. After identifying a final stage, and associating a payoff with each state in that stage, we then repeatedly move backward to preceding stages using a backward recursive relation, until we have finally arrived at an initial stage, and have thus sequentially arrived at a solution to the entire problem. Decisions at each stage must be made in accordance with the **dynamic programming principle of optimality**, which is stated as follows: regardless of the decisions made to arrive at a particular state, the remaining decisions must constitute an optimal policy for all successive stages, with respect to the current decision.

Suppose that our problem has N stages, and we are currently trying to compute stage n. Let s_n denote the state and d_n denote the decision made when there are n stages remaining in the solution process. Let $f_n(s_n, d_n)$ denote the total payoff or return for the last $N - n$ stages, given current state s_n and current decision d_n. The optimal return for these $N - n$ stages is then written as $f_n^*(s_n) = f_n(s_n, d_n^*)$, meaning that d_n^* is the optimal decision for this state, regardless of how we arrive at this state. Clearly, if we can work backward to an initial stage, then $f_1^*(s_1)$ is the optimal objective function value for an N-stage problem.

The return function for any state is written in terms of the return obtained from succeeding stages:

$$f_n^*(s_n) = \max_{d_n}\{r(s_n, d_n) + f_{n+1}^*(s_{n+1})\}$$

where $r(s_n, d_n)$ is the return resulting from making decision d_n while in state s_n at the current stage, and s_{n+1} is the new state that we will be in at stage $n + 1$ if we are in s_n now, and make decision d_n. Observe that we have previously computed the optimal cost for completing the solution process from *all* possible states s_{n+1}. This recursive relation identifies the optimal policy for each state with $N - n$ stages remaining, based on the optimal policy for each state with $(N - n) - 1$ stages remaining.

In the computer allocation example earlier, the Computer Science Lab location represents Stage 3, the University Computer Center is Stage 2, and the Library is Stage 1. States represent the number of computers available in a stage, and the decision variable specifies how many to allocate in this stage. Therefore, to find the optimal allocation, we must compute

$$f_1^*(\text{Library}) = \max_{d_1}\{r(s_1, d_1) + f_2^*(s_2)\}$$

where $s_2 = s_1 - d_1$. For this we need to have computed

$$f_2^*(s_2) = \max_{d_2}\{r(s_2, d_2) + f_3^*(s_3)\}$$

where $s_3 = s_2 - d_2$. Finally, f_3^* is trivial to compute for all states in Stage 3 because all remaining available computers should be used. The recursive computations for this example are shown for Stage 3 in Table 3.18, for Stage 2 in Table 3.19, and for Stage 1 in Table 3.20.

After the backward recursion is applied, the optimal objective function value is known, but the sequence of decisions leading to that optimum must be retrieved by tracing forward

TABLE 3.18

Stage Three

s_3	$f_3(s_3, d_3)$						d_3^*	$f_3^*(s_3, d_3)$
	$d_3 = 0$	1	2	3	4	5		
0	0						0	0
1		8					1	8
2			12				2	12
3				13			3	13
4					13		4	13
5						18	5	18

TABLE 3.19

Stage Two

s_2	$f_2(s_2, d_2)$						d_2^*	$f_2^*(s_2, d_2)$
	$d_2 = 0$	1	2	3	4	5		
0	0						0	0
1	0 + 8	5 + 0					0	8
2	0 + 12	5 + 8	10 + 0				1	13
3	0 + 13	5 + 12	10 + 8	11 + 0			2	18
4	0 + 13	5 + 13	10 + 12	11 + 8	12 + 0		2	22
5	0 + 18	5 + 13	10 + 13	11 + 12	12 + 8	24 + 0	5	24

TABLE 3.20

Stage One

	$f_1(s_1, d_1)$							
s_1	$d_1 = 0$	1	2	3	4	5	d_1^*	$f_1^*(s_1, d_1)$
5	0 + 24	3 + 22	6 + 18	7 + 13	15 + 8	20 + 0	1	25

to identify, at each stage, the decision that was chosen during the backward recursion. In the example, $s_1 = 5$ and $d_1^* = 1$. Therefore, $s_2 = 4$. When $s_2 = 4$, $d_2^* = 2$, and hence $s_3 = 2$. When $s_3 = 2$, $d_3^* = 2$.

Our discussion of dynamic programming has addressed only the most essential features of the method, and we should now mention some variations to this problem-solving approach. In our two examples, there were a finite number of states at each stage, representing discrete roads to choose or whole items to allocate. Applications involving arbitrary allocations of money or weight, for example, may be modeled with a continuous state-space. In this case, the graphical and tabular methods are useless, but the recursive relations readily apply.

In each of our sample problems, the return at any stage was *added* to cumulative returns from succeeding stages. This was appropriate because the time delays and the number of users served were additive in nature. Different applications may involve costs that are compounded together in arbitrary mathematical ways. For example, in the hazardous waste disposal problem, if the checkpoints introduced probabilities of contamination or spillage, then the probabilities (of *no* contamination) at successive stages should be multiplied, rather than added, to find the *safest* route. In that case,

$$f_n(s_n, d_n) = r(s_n, d_n) \cdot f_{n+1}^*(s_{n+1})$$

where $f_{n+1}^*(s_{n+1})$ is the minimum probability of contamination from stage $n + 1$ in state s_{n+1}, and s_{n+1} is the state that we would be in if we were in state s_n at stage n and made decision d_n.

Our recursive relations have been expressed in the form of *backward* recursion, based on the stages *remaining* in the decision process. For most problems, it would be equally valid to define *forward* recursive relations, based on *completed* decision stages. The final result will be the same. For example, in the computer allocation problem, our state variables could represent *the number of machines left* in backward recursion, or we could define a forward recursive model based on *the number of machines allocated so far*. However, the definition of the state variables is often more intuitively appealing in one direction for a particular application.

3.7 Project Management

The planning and coordination of large complex projects, consisting of many tasks or activities, is often viewed as less of an *optimization* problem and more of a *management* procedure aimed at completing a project under certain resource constraints and with attention to various cost-time trade-offs. However, certain aspects of project management can be dealt with conveniently by using network optimization methods that were discussed earlier in this chapter.

During the 1950s, two methodologies were developed—independently and simultaneously—for project management, and both approaches were based on network models. One method, called the **Critical Path Method (CPM)**, was developed for the management of construction and production activities; while the other, called the **Program Evaluation and Review Technique (PERT)**, was developed for the U.S. Navy in scheduling research and development activities for the Polaris missile program. CPM is based on deterministic specifications of task durations, and is therefore appropriate for production projects in which previous experience with the subtasks allows management to make reliable time estimates. PERT, on the other hand, is based on probabilistic estimates of task durations, and thus is most useful in a research and development environment where task completion times cannot be known in advance. Because both PERT and CPM approach project scheduling using similar network models and methods, the terms PERT and CPM are sometimes used interchangeably or collectively as *PERT-CPM methods*.

Large scale projects generally consist of a set of tasks or activities whose completion times are known or can be estimated (using a range of values, for example), and for which precedence constraints are specified, indicating that certain activities must be completed before others can begin. Simply identifying the distinct activities, and determining their durations and interdependencies, is an important part of the *planning* of any large project. PERT-CPM methods then provide for the construction of a network diagram, from which we can determine the minimum overall project duration and identify those tasks whose timely completion is critical or essential to the minimum project completion time. The purpose of this phase is to construct a *schedule* or time chart with start and finish times for each activity. Information may also be available that will allow us to evaluate the effect of putting extra money, people, or machines into a particular task in order to shorten the project duration. Thus, we can use the network to evaluate cost-time trade-offs. Finally, once the project is underway, the network diagram can be used in monitoring or *controlling* the project, to follow the progress of the various activities, and to make adjustments where appropriate. These three phases—planning, scheduling, and controlling—are essential to the effective management of any large project. In the following sections, we will see how the network methods underlying PERT and CPM help to support these phases of management.

3.7.1 Project Networks and Critical Paths

A project network provides a graphical representation of the precedence relations among all the activities in a project. Each **activity** is represented by an arc in the network. The nodes in the network denote **events** corresponding to points in time when one or more activities are completed. Directions on the arcs indicate the sequence in which events must occur. Additionally, a node is added at the beginning of the network to represent the *start* event for the entire project. Similarly, a final node is introduced to denote the *finish* event for the project.

As an illustration, we will build a project network for a set of six activities with the following precedence constraints:

1. A precedes D
2. A and B precede C
3. C and D precede F
4. E precedes F

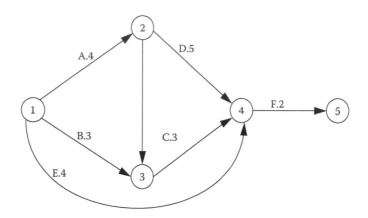

FIGURE 3.21
Project network.

The project network diagram is shown in Figure 3.21. Solid arcs denote activities A through E. Activities C, D, and E must all precede activity F. Therefore, we use event 4 to represent the time at which activities C, D, and E are all finished, and activity F can begin. We cannot combine events 2 and 3. We want event 2 to represent that A has finished and D can begin. Event 3 represents that A and B are finished and C can begin. To do this, we introduce a dummy activity from event 2 to event 3 with zero duration. The sole purpose of this is to ensure that event C does not start until event A has finished.

We let the variable t_i represent the time at which event i occurs, and d_{ij} denote the duration of the activity represented by the arc between nodes i and j. In this example, suppose $d_{12} = 4$, $d_{13} = 3$, $d_{14} = 4$, $d_{23} = 0$, $d_{24} = 5$, $d_{34} = 3$, and $d_{45} = 2$. These individual activity lengths are shown in Figure 3.21 along the appropriate arcs. Since t_1 and t_5 are the *start* and *finish* times, total project length is $(t_5 - t_1)$.

Now that the activities have been identified and described in the diagram, our next objective is to determine a minimum length project schedule; that is, to determine *when* each activity should begin so that precedence constraints are met and so that the entire project completes as quickly as possible. We can write the formulation as a linear programming problem, with constraints to assure that successive events i and j are separated from one another by at least the required duration of the event on the arc (i, j):

$$\text{minimize} \quad z = t_5 \notin - t_1$$

$$\text{subject to} \quad t_2 - t_1 \geq 4$$

$$t_3 - t_1 \geq 3$$

$$t_4 - t_1 \geq 4$$

$$t_3 - t_2 \geq 0$$

$$t_4 - t_2 \geq 5$$

$$t_4 - t_3 \geq 3$$

$$t_5 - t_4 \geq 2$$

$$\text{and } t_i \geq 0 \quad \text{for all } i = 1, 2, \ldots, 5$$

Note that this formulation could be solved with the ordinary Simplex method, but clearly there is a special network structure to the problem.

In order to minimize the project duration, we have to realize that actually we must find the longest sequence of linearly ordered activities; that is, we must find the **longest path** through the network. This insight gives us a slightly different perspective on the problem.

Consider the following linear programming problem. Let $x_{ij} = 1$ if activity (i, j) is in the longest path, and $x_{ij} = 0$ otherwise. This problem can be written as:

$$\text{maximize} \qquad 4x_{12} + 3x_{13} + 4x_{14} + 5x_{24} + 3x_{34} + 2x_{45}$$

subject to

$$
\begin{array}{llll}
-x_{12} - x_{13} - x_{14} & & & = -1 \\
x_{12} & - x_{23} - x_{24} & & = 0 \\
x_{13} & + x_{23} & - x_{34} & = 0 \\
x_{14} + & x_{24} & + x_{34} - x_{45} & = 0 \\
& & x_{45} & = 1
\end{array}
$$

$$\text{all } x_{ij} = 0 \text{ or } 1$$

The objective function adds up the total length of the longest path, while the constraints ensure that the solution represents a path from event 1 to event 5. The first constraint states that only one edge can leave node 1. The last constraint states that only one edge can enter node 5. The other constraints specify that the number of incoming arcs equals the number of outgoing arcs in each of the interior nodes. The only feasible solution to this problem is a path, and the optimal solution is the longest path.

These two problems are in fact equivalent. The second one is called the dual problem of the first. (Recall from the discussion in Section 2.8 that every linear programming problem has a dual problem, and typically the two versions represent a different view or interpretation of the same problem parameters.) Notice that the first problem has one constraint for each activity and one variable for each event, while the second formulation has a constraint for each event and a variable for each activity.

If we inspect the previous dual formulation, we can see that the constraints require that one *unit of flow* is to be routed from node 1 to node 5. We now recognize that this is the specialized form of the transshipment model that we dealt with in Section 3.5 to find the shortest path through a network. In our project management application, however, we *minimize* project duration by *maximizing* the path length. We can therefore treat our project scheduling problem as a *longest path* problem.

By finding the longest path through the project network, we are also finding what is known as the *critical path*. A **critical path** is a path from the start node to the finish node, with the property that any delay in completing activities along this path will cause a delay in overall project completion. The activities along the critical path are called **critical activities**.

To describe the PERT-CPM method for identifying critical activities in a project, we need two definitions. The **earliest time** for a node j, denoted E_j, is the time at which event j will occur if all previous activities are started as early as possible. The *start* node 1 has $E_1 = 0$ since there are no predecessors. Then any other node's earliest time can be determined as long as all its predecessors' earliest times have been calculated. We can make a **forward pass** through the network, calculating E_j for each event j as

$$E_j = \max_i \{E_i + d_{ij}\}$$

where (i, j) are all the arcs entering node j, and d_{ij} is the duration of the activity represented by arc (i, j). Once we have the earliest time for the *finish* event, we know the earliest possible completion time for the entire project.

The **latest time** for a node i, denoted L_i, is the latest time that event i can occur without causing delay in the completion of the project beyond its earliest possible time. Once we have made the forward pass to determine the earliest project completion time, we make a **backward pass** through the network. For a network of n nodes, $L_n = E_n$, then L_i can be determined for any node i as long as all of that node's successors' latest times have been calculated. The general formula is

$$L_i = \min_j \{L_j - d_{ij}\}$$

where (i, j) are all the arcs leaving node i.

The **slack time for an event** is the difference between the latest time and the earliest time for that event. Events having a slack time of zero are called **critical events**. The **slack time of an activity** (i, j) is $L_j - E_i - d_{ij}$. Activities with slack time zero are the **critical activities**, which must be completed without delay if the minimum feasible project duration is to be achieved.

Now re-examine the project network in Figure 3.21 to determine a critical path and construct a time chart. During the forward pass, we obtain the following earliest times:

$$E_1 = 0$$

$$E_2 = \max_1 \{0 + 4\} = 4$$

$$E_3 = \max_{1,2} \{0 + 3, 4 + 0\} = 4$$

$$E_4 = \max_{1,2,3} \{4 + 5, 4 + 3, 0 + 4\} = 9$$

$$E_5 = \max_4 \{9 + 2\} = 11$$

Therefore, the minimum completion time for the project is 11 time units. In a backward pass, we obtain latest times for each event as follows:

$$L_5 = E_5 = 11$$

$$L_4 = \min_5 \{11 - 2\} = 9$$

$$L_3 = \min_4 \{9 - 3\} = 6$$

$$L_2 = \min_{3,4} \{6 - 0, 9 - 5\} = 4$$

$$L_1 = \min_{2,3,4} \{4 - 4, 6 - 3, 9 - 4\} = 0$$

From these results, we can determine the critical path. Since $E_1 = L_1$, $E_2 = L_2$, $E_4 = L_4$, and $E_5 = L_5$, the critical events are at nodes 1, 2, 4, and 5; and therefore the critical activities are

TABLE 3.21

Project Time Chart

Activity	Duration	Earliest Start	Latest Start	Earliest Finish	Latest Finish	Slack Time
A	4	0	0	4	4	0
B	3	0	1	3	4	3
C	3	4	6	7	9	2
D	5	4	4	9	9	0
E	4	0	5	4	9	5
F	2	9	9	11	11	0

activities A, D, and F (the activities along the critical path). We also notice that the slack times for the activities are

$$A: L_2 - E_1 - 4 = 4 - 0 - 4 = 0$$

$$B: L_3 - E_1 - 3 = 6 - 0 - 3 = 3$$

$$C: L_4 - E_3 - 3 = 9 - 4 - 3 = 2$$

$$D: L_4 - E_2 - 5 = 9 - 4 - 5 = 0$$

$$E: L_4 - E_1 - 4 = 9 - 0 - 4 = 5$$

$$F: L_5 - E_4 - 2 = 11 - 9 - 2 = 0$$

and the activities with zero slack time are the critical activities. The noncritical activities B, C, and E could be delayed as much as 3, 2, and 5 time units, respectively, without extending the duration of the project.

All of this information can be summarized in the time chart shown in Table 3.21. This layout provides a clear and convenient tool for management to use in scheduling noncritical activities, considering possible improvements in the project schedule, or in evaluating the effects of delays along the critical path.

3.7.2 Cost versus Time Trade-Offs

The methods presented thus far have dealt solely with scheduling activities in order to achieve a minimum project duration, and no consideration has been given to the cost of the project. In addition to *direct* costs associated with each individual activity, there are typically *indirect* costs that may be viewed as overhead costs and that are proportional to the duration of the entire project. These costs may include such expenses as administrative or supervisory costs, equipment and facilities rental, and interest on capital. A financially realistic project manager may be willing to add resources, involving some direct expense, to certain activities in order to reduce the duration of those activities, and thereby to reduce the project duration and the attendant indirect costs. CPM provides a mechanism for minimizing the *total* (direct plus indirect) costs of a project.

Suppose that for every activity, we know the normal duration and the budgeted cost associated with completing the activity under normal circumstances. Suppose also that, through additional expenditures, the duration of each activity can be reduced. This is known as **crashing**. For each activity then, we know the **crash completion time** and the

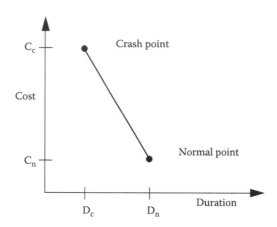

FIGURE 3.22
Cost versus time trade-off.

crash cost. By crashing critical jobs, total project length can be reduced. If the cost of crashing is less than the indirect costs that can be saved, then we not only reduce total costs but we can also enjoy various subjective benefits associated with completing a project ahead of schedule.

Figure 3.22 shows a straight-line relationship that is typically assumed, describing crash costs and durations and normal costs and durations. Each activity has its own cost vs. time trade-off, represented by the *slope* of the straight line, and its own crash point (or **crash limit**) beyond which no amount of added resources can reduce the activity's duration.

We take advantage of cost vs. time trade-offs in the following way. Using normal costs and durations for all activities, we first determine a critical path, as before. Then we consider reducing the duration of critical activities.

If we crash all the critical activities simultaneously, then almost certainly the network's critical path will have changed, and we suddenly find that we are working on the wrong problem. Instead, we should choose *one* of the critical activities to crash; in particular, we should choose the one that will yield the greatest reduction in schedule length per unit of added costs. This choice is easily made by simply selecting the activity having the smallest cost vs. time slope.

Having now chosen *which* critical activity to crash, we must still proceed with caution. As the duration of a critical activity is reduced, the activity may cease to be critical (there is now a new critical path in the network). At this point, it is useless to further reduce this activity, and instead we should be investing in the reduction of some *currently* critical activity. It has been suggested that the least-slope critical activity be crashed by only one time unit, then a possibly new critical path found. This process is repeated until all critical activities are at their crash limits.

Another consideration in deciding how far to crash an activity is the reduction in indirect costs that can be achieved. Since the aim is presumably to minimize the *sum* of activity costs and indirect costs, every crash operation should be undertaken only if it can be justified with respect to total project costs.

As an example, consider again the project network of Figure 3.21. The normal and crash points for each activity are given in Table 3.22, where D_n denotes the normal duration of the activity, C_n denotes the normal cost, D_c denotes the crash limit, and C_c denotes the

TABLE 3.22

Crash Costs

Activity	Normal		Crash		Crashing Cost per Day
	D_n	C_n	D_c	C_c	
A	4	400	2	820	210
B	3	500	3	500	—
C	3	350	2	500	150
D	5	300	3	700	200
E	4	100	3	125	25
F	2	200	1	300	100

crash cost. The cost versus time slopes for each activity are computed as $(C_c - C_n)/(D_n - D_c)$, and are shown in the far right column.

Suppose that indirect costs amount to \$220/day; therefore, the total project cost under a normal schedule is (400 + 500 + 350 + 300 + 100 + 200) plus (\$220/day · 11 days) = 1850 + 2420 = \$4270. If all activities were at their crash point, then the project duration would be 7 days, and the total project cost would be (820 + 500 + 500 + 700 + 125 + 300) + (\$220/day · 7 days) = 2945 + 1540 = \$4485. Clearly in this case, we are paying crash costs for activities that do not contribute to the reduction in project length. So, we would expect the optimal schedule to fall somewhere between these two extremes.

Beginning with the normal schedule, where the critical activities are A, D, and F, we find that we can crash activity F at a cost of only \$100/day; and by crashing activity F to its limit, we can reduce total overhead by \$220, for a net savings of \$120. The total project cost would then be \$4150, and the project duration is 10 days.

The critical path has not changed, so we now consider critical activities A and D. The daily reduction at the least cost is obtained by crashing activity D. Crashing D by one day costs \$200, but saves \$220; therefore, the total cost is now \$4130, and project duration is nine days. Since the critical path still includes activity D, we can crash it by one additional day, to obtain an eight-day project at a total cost of \$4110.

Activity A is now the only critical activity that is not at its crash limit, and we can save \$220 − \$210 = \$10 by crashing A to three days for a total project cost of \$4100. At this point, activities A and B are on parallel critical paths; therefore, any crashing must be applied simultaneously to both projects. In our case, project B cannot be crashed, and therefore the project duration cannot be reduced to less than seven days. (Notice that if project B could have been reduced but if the combined cost of crashing activities A and B exceeded \$220, then crashing them would not have been economical.)

Since critical activities A, B, D, and F are all crashed as far as possible to reduce the project duration, the current schedule is optimal. The durations of activities A, B, C, D, E, and F, respectively, are 3, 3, 3, 3, 4, and 1. The project cost is (610 + 500 + 350 + 700 + 100 + 300) + 7(220) = 2560 + 1540 = \$4100.

3.7.3 Probabilistic Project Scheduling

For certain types of projects, there may be no previous experience from which to determine the duration of the individual activities. PERT provides a means of handling such uncertainties through the use of probabilities for the completion times of the activities.

The project manager is required to provide three time estimates for each activity: an optimistic duration, denoted as a, specifying the minimum reasonable completion time if all goes well; a pessimistic duration, denoted as b, specifying the maximum duration if things go badly; and a most probable duration, denoted as m.

To apply critical path methods to a project layout based on probabilistic completion time estimates, we need to know two statistics for each activity. The **expected time** to complete each activity can be used as the actual time in order to find a critical path (as in the deterministic case), and the **variance** will give an indication of the amount by which the project might deviate from its expected project duration. These statistics are obtained, in PERT, by assuming that activity durations follow a Beta distribution.

Based on this assumption, the expected time μ for an activity is approximated as

$$\mu = \frac{(a+b+4m)}{6}$$

because the midpoint (a + b)/2 is given about half the weight of the mode m. Illustrative distributions are shown in Figure 3.23. In many probability distributions, the tails (a and b in our case) are considered to lie about three standard deviations from the mean μ; therefore, the standard deviation σ = (b − a)/6, and the variance σ² = [(b − a)/6]².

These statistics are now used in the following straightforward way. The activity means μ are used as activity durations, and the critical path method is used to determine the critical activities. The **expected project duration** D is the sum of all the means of the activities on the critical path. Likewise, the variance V of the project duration is the sum of the variances of the activities on the critical path.

Under PERT assumptions, the Central Limit Theorem implies that the project duration (being the sum of independent random variables) is normally distributed with mean D and variance V. Using tables for a normal distribution, we can, for example, determine the probability that the actual project duration will fall in a certain range, or the probability of meeting certain specified deadlines. For a more detailed discussion of probabilistic project scheduling, refer to the textbook by Ravindran et al. (1987).

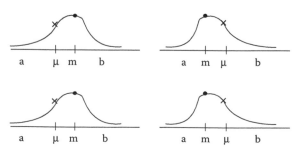

FIGURE 3.23
Expected time for activity.

3.8 Software for Network Analysis

Many network problems can be solved with software developed for ordinary linear programming problems. But more specialized software for network problems has been developed that takes advantage of the distinctive structure of network formulations, and can be used to solve network problems very efficiently. Real network problems may involve hundreds of thousands of nodes and millions of arcs, and fortunately there is software available for solving such large problems on a variety of hardware platforms. Some of the more noteworthy ones are mentioned here.

IBM CPLEX Optimization Studio has a network optimizer available through a callable library for various platforms.

SAS/OR OPTNET is a system for analyzing various characteristics of networks and solving network optimization problems and related models having network-structured data. This software handles general assignment problems, performs critical path analysis, determines minimum cost network flows, finds shortest paths, and solves transportation problems. It performs cycle detection and analyzes connectivity in networks, and does project scheduling and resource-constrained scheduling. OPTNET interfaces with OPTMODEL, described earlier in Chapter 1.

TransCAD is an integrated system of Geographic Information System (GIS) and transportation modeling capabilities, designed to help transportation professionals plan, organize, manage, and analyze transportation data. It offers a complete toolbox of analytical methods for mapping, assignment, site location, minimum cost distribution, transportation, vehicle routing and scheduling, planning, logistics, and marketing. TransCAD supplies state-of-the-art data collection tools for accessing data from the Global Positioning System (GPS).

COIN-OR, the open source OR software website (www.coin-or.org), offers software tools for network optimization. Sifaleras (2015) summarized these tool into Coin Graph Classes (Cgc), the Efficient Modeling and Optimization in Networks (LEMON), and VRPH. Cgc is a collection of network representations and algorithms aiming to facilitate the development and implementation of network algorithms; LEMON is a C++ template library providing efficient implementations of network optimization algorithms and common graph structures; and VRPH constitutes an open source C++ package in a software library containing tools to create metaheuristic algorithms for the Vehicle Routing Problem.

Mascopt (Mascotte Optimization) is an open source project that provides a set of Java-based tools for network optimization problems to help implementing solutions to network problems by providing a data model of the network and the demands, libraries to handle networks and graphs, and ready to use implementation of existing algorithms or linear programs. It also provides some graphical tools to display graph results. For more detail, Lalande et al. (2004).

Google-OR Tools provides open source solvers for network flow problems in its graph libraries.

Finally, More and Wright (1993) published a guide to optimization software that included descriptions of older computer programs including network optimization software such as GENOS, LNOS, LSNNO, NETFLO, and NETSOLVE.

3.9 Illustrative Applications

3.9.1 DNA Sequence Comparison Using a Shortest Path Algorithm (Waterman 1988; Wagner and Fischer 1974)

A problem that arises frequently in the field of cell biology is the comparison of DNA sequences and the analysis of how one sequence is transformed into another. A sequence is a finite succession of symbols from a finite alphabet. In the case of deoxyribonucleic acid (DNA), sequences are composed from the set of nucleotide bases, denoted {A (adenine), C (cytosine), G (guanine), T (thymine)}. Although biologists are not in complete agreement over the mechanisms by which one DNA sequence *evolves* into another, it is generally assumed that the transformation consists of a series of the following types of changes:

1. Insertion of a character (nucleotide)
2. Deletion of a character
3. Substitution of one character for another

The similarity between two DNA sequences S and T can then be measured by assessing a cost for each of these three types of changes, and then finding the least expensive transformation of S into T. The cost corresponding to this transformation is called the *evolutionary distance* from DNA sequence S to DNA sequence T.

Transitions can be modeled by defining a node to represent a DNA sequence, and creating other neighboring nodes to represent all DNA sequences obtainable from the original one by making one of the three types of changes. Arcs are labeled with the cost of the change. Then a shortest path algorithm applied from the original node to any other desired node will yield the evolutionary distance between the two DNA sequences.

DNA sequences are quite long (millions of nucleotide bases), so for practical implementations, parallel computer hardware known as *systolic architectures* have been developed for research purposes. This approach involves a specialized spatial arrangement of processors and an appropriate flow or *pulsing* of data among the processors, in order to obtain the desired computational results much more quickly than could be achieved using general-purpose computing hardware. For further discussion of systolic architectures incorporating shortest path and other network based algorithms, refer to the work of (Lopresti 1987). A completely different but effective approach to this problem is based on dynamic programming methods; see Wagner and Fischer (1974) for a detailed description of this concise solution to the DNA sequencing problem.

3.9.2 Multiprocessor Network Traffic Scheduling (Bianchini and Shen 1987)

In the design of real-time signal processing computer systems, one of the most important issues is the efficient scheduling of data communication traffic among special-purpose processing elements. For example, certain types of digital filters can be implemented on a small set of specialized functional modules, and the determination of filter functionality lies in the specification of intermodule communication.

The process of mapping consists of first placing functional data operators onto processing elements. This is easily accomplished using well-known placement algorithms.

The second and more difficult phase of the problem is the design of the network data traffic. This requires routing each unit of traffic onto a path of network links between the source and destination processing elements, with the objective of maximizing the aggregate flow of network traffic that can be maintained in a system.

Traffic management is viewed as a multi-commodity fluid flow problem. The multi-commodity aspect arises because of the need to maintain the identity of data traffic between different source/destination pairs, although the traffic may simultaneously occupy the same data link. An optimal traffic pattern is obtained when a cut set of saturated links is formed.

The network formulation results in an extremely large linear program because of the exponential number of network paths that contribute explicitly to the size of the problem. An alternative is a *policy iteration* method that successively improves current traffic patterns by re-routing certain data units. To improve a traffic pattern, under-utilized paths are determined between each source/destination pair, and then it must be decided whether re-routing along the proposed new path is cost-effective. To do this, a minimum spanning tree for the network is found. It can be shown that the least cost path connecting any two nodes in a network lies on the minimum spanning tree. Therefore, if a minimum spanning tree is known, the traffic scheduler can examine each processing element adjacent to a saturated link, and if traffic can be re-routed away from the saturated link and onto a minimum spanning tree link, then the cost of the traffic pattern can be reduced, while at the same time smoothing congestion and perhaps creating capacity for flow of additional data.

3.9.3 Shipping Cotton from Farms to Gins (Glover et al. 1992; Klingman et al. 1976)

At a time when cotton production had decreased by 50% in the Upper Rio Grande River Valley of Texas and New Mexico, it was necessary to determine how best to utilize the processing capacity available in the area's 20 cotton gins. Analysts began by mapping the 150 farms producing cotton, and charting the distances to the gins that were scattered throughout the Valley.

The efficiency of the industry had been brought into question because of the excess ginning capacity that resulted from the decrease in cotton crop production. Local farmers and gin operators had resorted to individual, fragmented decisions and actions that did not contribute to overall prosperity or profitability in the region. A mathematical model was constructed that represented the entire system, with the hope that a comprehensive approach would encourage joint cooperation among all farmers and gin operators.

Because of the excess gin capacity, there were fears that some gins may have to close down and, indeed, such reductions were found to contribute favorably to profitability. Gin operation involves annual fixed charges to activate the gin, such as electrical connections, cleaning, and salaried personnel. Variable costs of operation then include regular time and overtime labor costs. If the regular shift capacity of a gin is consumed, any additional cotton must be processed at the overtime rate; but this use of the more expensive overtime capacity can be justified if it avoids the fixed activation costs of starting up an additional gin.

The problem was first viewed as a shipping cost problem, to identify the particular gin that should service each farmer's needs. But it was quickly discovered that the real issue was the need to quantify the utilization of the cotton gins. This information could provide justification for some tough decisions related to the closing of certain gins which simply could not operate economically. The model grew into a fixed-charge transshipment formulation that included:

Production levels at each farm

Shipping costs from each farm to each gin

Holding costs for storing cotton at each farm

Seasonal gin activation costs

Two levels of operating capacity at each gin

The network model initially involved around 5000 nodes and over 2 million arcs, but refinements reduced this to around 100,000 arcs. The solution indicated that substantial cost savings (a 20% reduction in ginning costs) could be achieved by closing some gins and working as a cooperative. Implementation was allowed to evolve over several seasons in order to obtain the full cooperation of all the farmers and gin operators in the region.

3.10 Summary

Network analysis is applicable to an enormous variety of problems that can be modeled as networks and optimized using network algorithms. Some of the systems represent physical networks for transportation or flow of commodities, while others are more abstract and can be used to model processes or plan and manage projects.

A maximum flow algorithm optimizes the total flow of an entity through a network in which links have capacities that limit the flow. This algorithm not only determines the greatest possible flow, but in so doing also locates and identifies the bottlenecks in the network.

Transportation models find the minimum cost flow from an origin, through a network, to a destination, subject to supply and demand requirements. The transportation Simplex algorithm is often used for this optimization problem. A slight refinement in the interpretation of the transportation model results in an assignment problem, which is used to model the matching or assignment of two sets of entities in the most advantageous way.

Maintaining network connectivity has important practical implications. Minimum cost spanning trees provide a simple and useful means of addressing the connectivity issue. When appropriate connections between nodes do exist in a network, it is often useful to find the shortest route between two specified nodes. Simple labeling algorithms provide solutions to this problem, and also inspire a broader approach known as dynamic programming. Dynamic programming has far-ranging applications, but generally can be viewed as a way to model decisions that take place in stages over a period of time.

Project activity networks are used to plan and coordinate large complex projects consisting of many tasks. Critical paths in networks determine the minimum project completion time, and identify those tasks or activities whose timely completion is critical to achieving this minimum project duration.

Key Terms

activity
acyclic graph
arcs
assignment problem
backward pass
bipartite graph
capacitated transshipment
chain
connected graph
critical activity
critical event
critical path
critical path method
crash completion time
crash cost
crash limit
crashing
critical event
cut
cut set
cycle
cyclic path
decision variable
degree of a node
demand
Dijkstra's algorithm
directed chain
directed graph
dynamic programming
dynamic programming principle of optimality
earliest time
events
expected project duration
expected time
flow
Ford-Fulkerson algorithm
forward pass
graph
Hungarian method
isolated node
latest time
longest path
maximum flow
minimum cost method

minimum row cost method
minimum spanning tree
multiple sinks
multiple sources
network
node-arc incidence matrix
nodes
northwest corner rule
path
PERT
predecessor
Prim's algorithm
project management
shortest network problem
shortest path
sink
slack time
source
spanning tree
stable matching
stages
states
Steiner tree
successor
supply
transportation problem
transportation Simplex
tree
undirected graph
unstable matching
variance

Exercises

3.1 Find the maximum flow through the networks shown in Figure 3.24. Identify the edges in the minimum cut set. In each case, assume node 0 is the source, and the highest-numbered node is the sink. Arc capacities are shown in boxes.

3.2 A data communications network can be described by the diagram in Figure 3.25. Every data link from node i to node j has a capacity which is denoted as a label on the data link in the diagram. Non-existent links have zero capacity. Data is being generated at node 1 and is to be routed through the network (not necessarily passing through all other nodes) to node 6 where the data will be used. The amount of data generated at node 1 is exactly the amount of data consumed at node 6. No data is generated or used at intermediate nodes, so all data that enters an intermediate node must leave it, and vice versa.

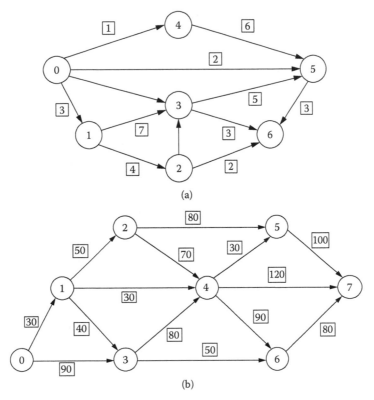

FIGURE 3.24
(a,b) Maximum flow in networks.

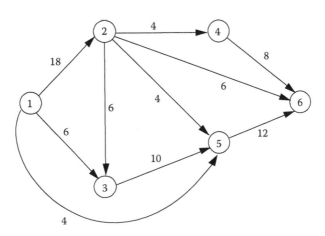

FIGURE 3.25
Communications network.

a. What is the maximum feasible amount of data that can flow through this network?

b. What is the flow on each of the data links in order to achieve this maximum?

c. Which links comprise the bottleneck in this network?

d. What is the complexity of the Ford-Fulkerson algorithm for maximum network flow?

3.3 Formulate and solve the following distribution problem to minimize transportation costs, subject to supply and demand constraints. Two electronic component fabrication plants, A and B, build radon-cloud memory shuttles that are to be distributed and used by three computer system development companies. Following are the various costs of shipping a memory shuttle from fabrication plants to the system development sites, the supply available from each fabrication plant, and the demand at each system development site.

Fabrication plant A is capable of creating a supply of 160 shuttles; and the cost to ship to site 1, 2, and 3 is $1000, $4000, and $2500, respectively. Fabrication plant B can produce 200 shuttles, and the shipping costs are $3500, $2000, and $4500 to the three sites. The demand at site 1 is 150, at site 2 is 120, and at site 3 is 90 memory shuttles.

a. Identify the decision variables, write the objective function, and give the constraints associated with this problem.

b. Solve this distribution problem.

3.4 Suppose that the countries of Agland, Bugland, and Chemland produce all the wheat, barley, and oats in the world. The world demand for wheat requires 125 million acres of land devoted to wheat production. Similarly, 60 million acres of land are required for barley, and 75 million acres of land are needed for oats. The total amount of land available for these purposes in Agland, Bugland, and Chemland is 70 million, 110 million, and 80 million acres of land, respectively. The number of hours of labor needed in the three countries to produce an acre of wheat is 18 hours, 13 hours, and 16 hours, respectively. The number of hours of labor needed to produce an acre of barley is 19 hours, 15 hours, and 10 hours in the three countries, respectively. And the labor requirements for an acre of oats are 12 hours, 10 hours, and 16 hours in the three countries, respectively. The hourly labor cost to produce wheat is $6.75 in each of the countries. The labor cost per hour in producing barley is $4.10, $6.25, and $8.50 in the three countries. To produce oats, the labor cost per hour is $8.25 in each country. The problem is to allocate land use in each country so as to meet the world food requirements and minimize the total labor cost. Formulate this problem as a transportation model, letting decision variable x_{ij} denote the number of acres of land allocated in country i for crop j.

3.5 Four workers are to be assigned to machines on the basis of the worker's relative skill levels on the various machines. Five machines are available, so one machine will have no worker assigned to it. In order to maximize profitability, we wish to minimize the total cost of the assignment. Use the cost matrix given in the following, and the Hungarian assignment algorithm, to determine the optimal assignment of workers to machines, and give the cost of the optimal assignment.

Workers	Machines				
	1	2	3	4	5
1	10	9	8	12	7
2	3	4	5	14	6
3	2	1	1	10	2
4	4	3	5	12	6

3.6   Four federally funded research projects are to be assigned to four existing research labs, with one project being allotted to each lab. The costs of each possible placement are given in the following table. Use the Hungarian method to determine the most economical allocation of projects.

Project	Sandy Lab	Furrmy Lab	Xenonne Lab	Liverly Lab
Cryogenic cache memory	12	15	10	14
Spotted owl habitat	8	10	6	9
Pentium oxide depletion	20	22	18	12
Galactic genome mapping	10	12	8	16

3.7   To solve a maximization assignment problem, first convert it to a minimization problem by multiplying each element in the cost matrix by −1, then adding sufficiently large constants to rows and column so that no element is negative. Then apply the Hungarian method to the new problem. Suppose the following matrix elements represent the *value* or productivity of associating certain workers with machines. Solve this assignment problem to maximize the productivity.

Workers	Machines			
	1	2	3	4
1	6	7	6	7
2	4	3	8	8
3	5	8	9	8
4	9	5	4	3

3.8   The following matrix contains the hazard insurance premiums that a company must pay in order for employee i to operate machine j. It is assumed that a low insurance premium implies that a worker can safely and proficiently operate a machine. Determine an assignment of workers to machines that will be the safest (least hazardous).

$$\begin{bmatrix} 36 & 24 & 36 & 12 \\ 14 & 28 & 40 & 26 \\ 12 & 22 & 28 & 38 \\ 28 & 22 & 38 & 38 \end{bmatrix}$$

What is the total insurance premium corresponding to the optimal assignment?

3.9   Prospective employees are to be assigned to jobs by the following mechanism: Each employee ranks his job preferences (rank 1 means highest preferences) and this information is contained in an array P where p_{ij} denotes employee i's ranking of job j. Similarly, each prospective employer ranks his preferences of employees, and matrix R is such that r_{ij} denotes employer i's ranking of employee j. Formulate this problem to determine an assignment of n jobs to n employees that optimizes the *mutual* satisfaction of employers and employees. (Assume that each employer corresponds to a different job.)

3.10 A group of m people, where m ≤ 40, is to be organized into teams of at most four people. Each team is associated with a workstation, of which ten are available. People may not express preferences for teammates; however, each ranks his workstation preference, and these preferences appear in a 40 × 10 matrix P where p_{ij} denotes the preference of person i for workstation j (low numbers in P indicate high preference). This is a variation of the classical assignment model. Formulate this problem to optimize the association of people to workstations.

3.11 Use Kruskal's algorithm to find the minimum cost spanning tree for the undirected graph in Figure 3.26. Identify the arcs in the tree, and state the cost of the minimum spanning tree.

3.12 Use Prim's algorithm to find the minimum cost spanning tree for the graph in Figure 3.26. Identify the arcs that comprise the minimum spanning tree, and state the cost of the minimum spanning tree for this graph.

3.13 Consider a graph in which the four nodes are located at the corners of a unit square, and the shortest possible arcs connect all pairs of nodes.

   a. Find the minimal spanning tree of this graph.

   b. Construct the Steiner tree obtained by placing a junction point in the center of the square. Is this an optimal Steiner tree?

   c. Determine the total length of the connections in this Steiner tree, and compare it with the length of the connections in the minimum spanning tree.

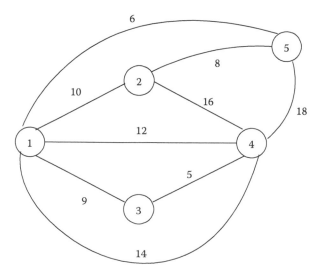

FIGURE 3.26
Minimum spanning tree.

3.14 How many different spanning trees are there in a fully connected undirected graph of five nodes?

3.15 How many arcs are there in a spanning tree of a fully connected undirected graph with 1000 nodes?

3.16 Use the backward labeling algorithm to find the shortest path from node 1 to node 9 in the graph in Figure 3.27. The labels shown on the arcs denote costs or distances between nodes.

    a. What are the arcs in the shortest path through this network?

    b. What is the length (cost) of the shortest path?

3.17 Following is the connectivity matrix of a graph. Use the shortest path labeling algorithm to find the shortest route from node 1 to node 6. The symbol ∞ denotes the absence of a path.

$$\begin{bmatrix} 0 & 5 & \infty & 6 & \infty & \infty \\ \infty & 0 & 3 & 1 & \infty & 8 \\ \infty & \infty & 0 & \infty & 2 & \infty \\ \infty & \infty & \infty & 0 & 3 & 6 \\ \infty & \infty & \infty & \infty & 0 & 2 \\ \infty & \infty & \infty & \infty & \infty & 0 \end{bmatrix}$$

3.18 Formulate the general problem of finding the minimum cost (shortest) path from node 1 to node n in a directed acyclic network of n nodes, where the distance from

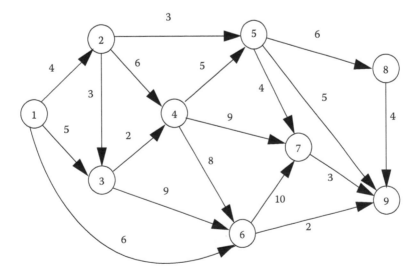

FIGURE 3.27
Shortest path.

node i to node j is denoted d_{ij}. *Hint:* Let the decision variables be restricted to have only the values zero or one, with the following interpretation:

$x_{ij} = 1$ means the arc from node i to node j is in the shortest path

$\phantom{x_{ij}} = 0$ otherwise

Give the objective function and the constraints, in terms of these decision variables.

3.19 Six thousand dollars is to be applied to a student's educational expenses in the following way:

Between $1000 and $3000 for books

Between $2000 and $4000 for tuition

Between $1000 and $2000 for tutors

The allocation is to be made in whole thousands of dollars. An analyst has quantified the anticipated payoffs (perhaps in terms of increased future earnings) as:

Books		Tuition		Tutors	
	Return		Return		Return
Invested					
$1K	$5K	$2K	$6K	$1K	$2K
$2K	$8K	$3K	$8K	$2K	$3K
$3K	$10K	$4K	$9K		

Use dynamic programming to determine the optimal allocation of the $6000. Show the tables you build as you solve this problem.

3.20 A student must select ten elective courses out of four different departments. From each department, at least one and no more than three courses must be chosen. The selection is to be made in such a way as to maximize the combined general *knowledge* from the four fields. The following chart indicates the knowledge acquired as a function of the number of courses taken from each field. Solve this as a dynamic programming problem. Show each of your tables in this staged decision-making process.

	Number of courses taken		
	1	2	3
Anthropology	25	50	60
Art	20	30	40
Economics	20	40	50
Physics	50	60	60

3.21 A space telescope being launched aboard a space shuttle is to be deployed and immediately will be transmitting data to earth-bound data processors at a prodigious rate. Suppose there are four teams of technical experts that can be allocated among two projects: one aimed at collecting and compressing data, and another whose responsibility is to catalog and store data. Because this data is extremely valuable and virtually irreplaceable, it is essential that you allocate the teams optimally to the two projects. Each of the two projects must have at least one team assigned to it. Use a dynamic programming table-oriented method to allocate the four teams. The following information is available:

Number of Teams Allocated	Payoff for Assigning Teams to Projects	
	Collecting and Compression Project	Cataloging and Storage Project
1	5	4
2	9	10
3	12	15

3.22 A small project consists of ten jobs whose durations in days are shown on the arcs in the activity diagram in Figure 3.28:

   a. Calculate early and late occurrence times for each event.

   b. What is the minimal project duration?

   c. Which activities are critical?

3.23 Suppose that for the aforementioned project, we have the following crash times and costs:

Task (i, j)	Minimum (Crash) Duration	Crash Cost ($/day)
(1, 2)	2	20
(2, 3)	3	15
(2, 4)	5	25
(3, 5)	2	20
(3, 6)	1	—
(4, 6)	3	20
(4, 7)	2	—
(5, 8)	5	15
(6, 8)	5	15
(7, 8)	3	20

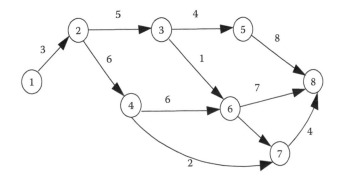

FIGURE 3.28
Activity diagram.

   a. What is the minimum (crashed) project duration?

   b. Determine the minimum crashing costs of schedules ranging from normal length down to the minimal length.

   c. If overhead costs amount to $75 per day, what is the optimal schedule length with respect to both crashing and overhead costs? Indicate the scheduled duration of each activity in this optimal schedule.

References and Suggested Readings

Aho, A. V., and J. E. Hopcroft. 1974. *The Design and Analysis of Computer Algorithms*. Delhi, India: Pearson Education India.

Ahuja, R. K., T. L. Magnanti, and J. B. Orlin. 1993. *Network Flows: Theory, Algorithms, and Applications*. Upper Saddle River, NJ: Prentice-Hall.

Battersby, A. 1967. *Network Analysis for Planning and Scheduling*. Basingstoke, UK: Macmillan.

Battersby, A. 1970. *Network Analysis for Planning and Scheduling*. New York: John Wiley & Sons.

Bazaraa, M. S, J. J. Jarvis, and H. D. Sherali. 2009. *Linear Programming and Network Flows*. New York: John Wiley & Sons.

Beightler, C. S., D. T. Phillips, and D. J. Wilde. 1976. *Foundations of Optimization*. Englewood Cliffs, NJ: Prentice-Hall.

Bellman, R. 1957. *Dynamic Programming*. Princeton, NJ: Princeton University Press.

Bellman, R. 1958. On a routing problem. *Quarterly of Applied Mathematics* 16 (1): 87–90.

Bern, M. W., and R. L. Graham. 1989. The shortest-network problem. *Scientific American* 260 (1): 84–89.

Bernot, M., V. Caselles, and J. Morel. 2009. *Optimal Transportation Networks: Models and Theory*. New York: Springer.

Bertsekas, D. P. 1991. *Linear Network Optimization: Algorithms and Codes*. Cambridge, MA: MIT Press.

Bertsekas, D. P. 1998. *Network Optimization: Continuous and Discrete Models*. Belmont, MA: Athena Scientific.

Bianchini, R. P., and J. P. Shen. 1987. Interprocessor traffic scheduling algorithm for multiple-processor networks. *IEEE Transactions on Computers* 36 (4): 396–409.

Bodin, L., B. Golden, and T. Goodwin. 1986. Vehicle routing software for microcomputers: A survey. *Proceedings of a Symposium on Impacts of Microcomputers on Operations Research*, University of Colorado, Denver, CO., pp. 65–72.

Bonald, T., and M. Feuillet. 2011. *Network Performance Analysis*. New York: John Wiley & Sons.

Bondy, J. A., and U. S. R. Murty. 2008. *Graph Theory*. New York: Springer.

Bradley, S. P., A. C. Hax, and T. L. Magnanti. *Applied Mathematical Programming*. Reading, MA: Addison-Wesley.

Christofides, N. 1975. *Graph Theory: An Algorithmic Approach*. New York: Academic Press.

Cooper, L., and M. W. Cooper. 1981. *Introduction to Dynamic Programming*. Elmsford, NY: Pergamon Press.

Cottle, R., and J. Krarup (Eds.). 1972. Optimization methods for resource allocation. *Proceedings NATO Conference*, Elsinor, Denmark.

Dekel, E., D. Nassimi, and S. Sahni. 1981. Parallel matrix and graph algorithms. *SIAM Journal on Computing* 10 (4): 657–675.

Denardo, E. V. 2003. *Dynamic Programming: Models and Applications*. North Chelmsford, MA: Courier Corporation.

Deo, N. 1974. *Graph Theory with Applications to Engineering and Computer Science*. Englewood Cliffs, NJ: Prentice-Hall.

Deo, N., and C. Y. Pang. 1984. Shortest-path algorithms: Taxonomy and annotation. *Networks* 14 (2): 275–323.

Dial, R., F. Glover, D. Karney, and D. Klingman. 1979. A computational analysis of alternative algorithms and labeling techniques for finding shortest path trees. *Networks* 9 (3): 215–248.

Dinic, E. A. 1970. Algorithm for solution of a problem of maximum flow in a network with power estimation. *Soviet Mathematics Doklady* 11: 1277–1280.

Dreyfus, S. E. 1969. An appraisal of some shortest-path algorithms. *Operations Research* 17 (3): 395–412.

Dreyfus, S. E., and A. M. Law. 1977. *Art and Theory of Dynamic Programming.* San Diego, CA: Academic Press.

Edmonds, J., and R. M Karp. 1972. Theoretical improvements in algorithmic efficiency for network flow problems. *Journal of the ACM (JACM)* 19 (2): 248–264.

Evans, J., and E. Minieka. 1992. *Optimization Algorithms for Networks and Graphs.* New York: Marcel Dekker.

Floyd, R. W. 1962. Algorithm 97: Shortest path. *Communications of the ACM* 5 (6): 345.

Ford Jr, L. R. 1956. *Network Flow Theory.* Santa Monica CA: RAND Corporation.

Francis, R. L., and J. A. White. 1976. *Facility Layout and Location.* Englewood Cliffs, NJ: Prentice-Hall.

Fulkerson, D. R. 1961. An out-of-kilter method for minimal-cost flow problems. *Journal of the Society for Industrial and Applied Mathematics* 9 (1): 18–27.

Glover, F., D. Klingman, and N. V. Phillips. 1992. *Network Models in Optimization and their Applications in Practice,* Vol. 36. New York: John Wiley & Sons.

Hall, R., and J. Partyka. 2016. Vehicle routing: Higher expectations drive transportation. *OR/MS Today* 43 (1): 40–47.

Hall, R. W. 2003. *Handbook of Transportation Science,* 2nd ed. Boston, MA: Kluwer Academic Publishers.

Harary, F. 1969. *Graph Theory.* Reading, MA: Addison-Wesley.

Hu, T. C. 1970. *Integer Programming and Network Flows.* Reading, MA: Addison-Wesley.

Hwang, F. K. 2017. The Steiner tree problem, 2012. In Y. Jiang, and Z.-P. Jiang (Eds.), *Robust Adaptive Dynamic Programming.* Hoboke, NJ: John Wiley & Sons.

Karney, D., and D. Klingman. 1976. Implementation and computational study on an in-core, out-of-core primal network code. *Operations Research* 24 (6): 1056–1077.

Karzanov, A. V. 1974. Determination of maximal flow in a network by method of preflows. *Soviet Mathematics Doklady* 15 (1): 434–437.

Kennington, J. L., and R. V. Helgason. 1980. *Algorithms for Network Programming.* New York: John Wiley & Sons.

Klingman, D. D., and R. F. Schneider. 1985. *Microcomputer-based Algorithms for Large Scale Shortest Path Problems.* Austin, TX: University of Texas.

Klingman, D., P. H. Randolph, and S. W. Fuller. 1976. A cotton ginning problem. *Operations Research* 24 (4): 700–717.

Klingman, D., and R. Russell. 1975. Solving constrained transportation problems. *Operations Research* 23 (1): 91–106.

Knuth, D. E. 1976. *Marriages Stables.* Montreal, Canada: Les Presses de l'Universite de Montreal.

Kuhn, H. W. 1955. The Hungarian method for the assignment problem. *Naval Research Logistics (NRL)* 2 (1–2): 83–97.

Lakhani, G., and R. Dorairaj. 1987. A VLSI implementation of all-pair shortest path problem. *ICPP*.

Lalande, J. F., M. Syska, and Y. Verhoeven. 2004. Mascopt-a network optimization library: Graph manipulation. Technical Report, RT-0293, INRIA, Sophia Antipolis Cedex (France), p. 25.

Lawrence, K. D., and S. H. Zanakis. 1984. *Production Planning and Scheduling: Mathematical Programming Applications.* Peachtree Corners, GA: Institute of Industrial Systems Engineers.

Lew, A., and H. Mauch. 2007. *Dynamic Programming: A Computational Tool,* Vol. 38. New York: Springer.

Lewis, T. G. 2009. *Network Science, Theory and Application.* Hoboken, NJ: John Wiley & Sons.

Lopresti, D. P. 1987. P-NAC: A systolic array for comparing nucleic acid sequences. *Computer* 20 (7): 98–99.

Marberg, J. M., and E. Gafni. 1987. An $O(n_2 m_{1/2})$ distributed max-flow algorithm. *Proceedings of the 1987 IEEE International Conference on Parallel Processing.* Los Angeles, CA: University of California.

Mathis, P. (Ed.). 2010. *Graphs and Networks,* 2nd ed. Hoboken, NJ: John Wiley & Sons.

Melnyk, S. A., and D. M. Stewart. 2002. Managing metrics. *APICS: The Performance Advantage,* 12 (2): 23–26.

Melzak, Z. A. 1961. On the problem of Steiner. *Canadian Mathematical Bulletin* 4 (2): 143–148.

Moder, J. J., and C. R. Phillips. 1970. *Project Management with CPM and PERT*, 2nd ed. New York: Van Nostrand Reinhold.

Moeller, G. L., and L. A. Digman. 1981. Operations planning with VERT. *Operations Research* 29 (4): 676–697.

More, J. J., and S. J. Wright. 1993. *Optimization Software Guide*. Philadelphia, PA: SIAM Publications.

Murty, K. 1992. *Network Programming*. Upper Saddle River, NJ: Prentice-Hall.

Näsberg, M. 1986. Two tools for marking for bucking analysis. In *OR Models on Microcomputers*. New York: Elsevier, pp. 23–33.

Nemhauser, G. L. 1966. *Introduction to Dynamic Programming*. New York: John Wiley & Sons.

Nilsson, N. J. 1971. *Problem-Solving Methods in Artificial Intelligence*. New York: McGraw-Hill.

Phillips, D. T., and A. Garcia-Diaz. 1981. *Fundamentals of Network Analysis*. Englewood Cliffs, NJ: Prentice-Hall.

Price, W. L. 1971. *Graphs and Networks: An Introduction*. New York: Auerbach Publishers.

Ravindran, A., D. T. Phillips, and J. J. Solberg. 1987. *Operations Research: Principles and Practice*. New York: John Wiley & Sons.

Reid, R. A., and W. A. Stark. 1986. Optimal replacement policy developed for items that fail. *Industrial Engineering* 18 (3): 23–27.

Salvendy, G. (Ed.). 1982. *Handbook of Industrial Engineering*. New York: John Wiley & Sons.

Sedgewick, R. 1990. *Algorithms*. Reading, MA: Addison-Wesley.

Sifaleras, A. 2015. Classification of network optimization software packages. In *Encyclopedia of Information Science and Technology*, 3rd ed. Hershey, PA: IGI Global, pp. 7054–7062.

Tarjan, R. E. 1982. Sensitivity analysis of minimum spanning trees and shortest path trees. *Information Processing Letters* 14 (1): 30–33.

Tarjan, R. E. 1983. *Data Structures and Network Algorithms*. Philadelphia, PA: SIAM.

Toint, P. L., and D. Tuyttens. 1992. LSNNO, a FORTRAN subroutine for solving large-scale nonlinear network optimization problems. *ACM Transactions on Mathematical Software (TOMS)* 18 (3): 308–328.

Wagner, R. A., and M. J. Fischer. 1974. The string-to-string correction problem. *Journal of the ACM (JACM)* 21 (1): 168–173.

Waterman, M. S. 1988. *Mathematical Methods for DNA Sequences*. Boca Raton, FL: CRC Press.

Weintraub, A. 1970. The shortest and the K-shortest routes as assignment problems. *Networks* 3 (1): 61–73.

Wheelwright, J. C. 1986. How to choose the project-management microcomputer software that's right for you. *Industrial Engineering* 18 (1): 46–50.

White, D. J. 1969. *Dynamic Programming*. San Francisco, CA: Holden-Day.

Wiest, J. D., and F. K. Levy. 1969. *A Management Guide to PERT/CPM*. Englewood Cliffs, NJ: Prentice-Hall.

Winter, P. 1987. Steiner problem in networks: A survey. *Networks* 17 (2): 129–167.

4

Integer Programming

4.1 Fundamental Concepts

Mathematical programming problems in which the variables are constrained to have integer values are called **integer programming** (IP) problems. Many engineering, industrial, and financial applications involve integer constraints. For example, in a manufacturing scenario, it would be difficult to implement a solution that specifies producing 10.4 cars or 7.2 tables. Fractional values are *infeasible*. For integer programming problems, the feasible region is neither continuous nor convex, as illustrated in Figure 4.1 for a simple two-dimensional integer problem. Observe that the feasible points for this problem do not lie at the extreme points of the region, or even on the boundaries; and in fact, the elegant solution techniques that have been developed for solving linear programming problems generally do not find solutions to integer problems. The Simplex method for linear programming converges to a solution at an extreme point which is typically a point with fractional variables.

Although the formulations of integer programming problems often look remarkably similar to those of continuous mathematical programming problems, the resemblance is in some ways deceptive. The algebraic expression of the objective function and the constraints in the two types of models may appear to have a similar form, but the additional constraint requiring that some or all of the variables have integer values generally makes solving the integer problem vastly more difficult, from a computational standpoint. Most integer programming problems are classified as *hard* optimization problems, and many integer programming problems belong to the class of NP-hard problems (described in Chapter 1). So, while a general linear programming problem may be solvable in polynomial time, finding an optimal *integer* solution to the same formulation usually requires an exponential amount of computation time.

Most integer programming problems are notoriously difficult, yet some integer problems are easy to solve. In particular, many linear network problem solutions, such as assignment and matching problems, transportation and transshipment problems, and network flow problems, always produce integer results, provided that the problem bounds are integers. In these problems, all of the extreme points of the feasible region represent integer solutions; therefore, if these problems are formulated and solved as linear programming problems, we find that the Simplex method yields integer solutions. Unfortunately, this occurs only for problems that have a network structure, and for the majority of integer problems, the linear programming formulation does not suggest an easy solution.

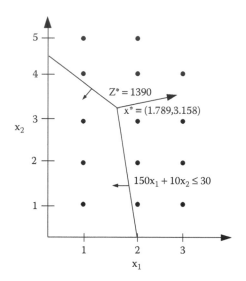

FIGURE 4.1
Graphical representation.

For integer programming problems with linear objective and constraints, one may wonder why we cannot simply solve the linear program (LP) and then round the answer to the nearest integer. The rounding approach turns out to be more difficult than it may seem. For example, if we have equality constraints, and we round down some variables, we will probably have to round up some others, and selecting which ones go up and which ones go down is itself an integer decision problem. Even when there are no equality constraints, it is easy to construct examples in which rounding up or down or to the nearest integer does not result in a feasible solution. Thus, in general, rounding does not yield satisfactory solutions.

That being said, there are some problems for which rounding can be effective. For example, in solving a problem for manufacturing tires, if the LP solution specifies making 1296.4 tires of a particular style, it is probably safe to round the answer down to 1,296 without drastically affecting feasibility or the objective function. In contrast, if the product being manufactured is a multi-million dollar aircraft, rounding is probably a poor solution. Rounding down a half a plane here or there could put a company right out of business. In some cases, a simple guideline for deciding whether rounding is an appropriate option might be to assess the damage (expressed as a percentage) to the objective function that results from rounding. In our examples, rounding down 1296.4 tires will almost certainly have a negligible impact on total profit, whereas rounding a small number of would probably have a significant effect.

An even more dramatic difficulty arises when using rounding for integer problems in which the variables are further constrained to have values of either zero or one. Consider a production planning problem for a large auto manufacturer such as General Motors, where it must be decided at which plants each car model should be built. A formulation for this problem might involve variables x_{ij}, each having a value of one or zero, depending on whether model i is produced at plant j, or not. Suppose there are ten plants, and each model can be assigned to only one location. An LP solution could easily recommend a small fraction of each model at each plant, yet rounding could produce a solution in which no models are produced anywhere. This situation is frequently encountered in integer programming; and in such cases, the LP solution gives virtually no insight into how to solve the integer problem.

4.2 Typical Integer Programming Problems

Mathematical programming problems in which all decision variables must have positive integer values are called **general integer programming** problems. If all the decision variables are restricted to have only the value zero or one, the problem is then called a **zero–one programming** (or **binary integer programming**) problem. In that case, the constraints on the variables are sometimes called *binary* or *Boolean* constraints, and the model is often referred to in abbreviated form as a **0–1 problem**. Variations on the aforementioned problems arise if some of the variables must be integer, others must be zero or one, while still others may have real values. Any problem involving such combinations is described as a **mixed integer programming** (MIP) problem. This section illustrates each of these types of integer problems with typical practical examples.

4.2.1 General Integer Problems

An illustration of general integer programming can be found in a simple version of the **portfolio selection problem**. An investor wishes to purchase a portfolio of financial instruments that will provide a maximum expected return. Many investment products, such as on the futures market for example, must be purchased in large lot sizes. We can define variables x_i to denote the number of units of security i in the portfolio. The objective function measures the expected return, and the problem will often have constraints limiting the amount of *risk* that the investor is willing to accept. In the simplest form of the problem, we could assume that the only constraint on the portfolio is a limit on the number of dollars that can be invested. Problems that have this basic underlying structure involve selecting as many investments as possible and figuratively *packing* them into a portfolio of limited size.

A three-dimensional view of this same idea is seen in a problem known as the **cargo loading problem**. Consider trying to pack boxes into trucks or shipping containers. The variables x_{ij} represent the number of boxes of type i to be loaded into container j. The constraints for this type of problem are complicated because they must define a spatially feasible packing.

The **employee scheduling problem** can also be formulated as a general integer problem, in which we define a number of shift patterns for workers. For example, a pattern could be to have a person work the day shift on Monday, Tuesday, and Wednesday, have two days off, and then work Saturday and Sunday evening. We then define variables x_i to specify the number of employees who are assigned to work using pattern i. The objective is to minimize total salary costs while ensuring that there are sufficient employees available in each shift.

4.2.2 Zero–One (0–1) Problems

Zero–one (0–1) problems are among the most common integer problems. *All* of the variables in the problem are required to take on a value of zero or one. Often, the variables have an abstract interpretation; they simply indicate whether or not some activity occurs, or whether or not some particular matching or assignment takes place.

One of the simplest 0–1 examples is the **capital budgeting problem**. Suppose we have a number of possible projects from which we must choose. Each project has a known value, and requires some level of resources such as funding, space, time, or services. We define the variables x_i to have a value 1 if project i is selected. The objective is to maximize total value subject to a constraint on total budget. (This problem at first appears to be another form of packing problem; but in this case, each project is to be chosen just once or not at all.)

Many scheduling problems can be formulated using 0–1 variables. For example, in a **production scheduling** environment, we could define variables x_{ik} to have a value 1 if job i is assigned to machine k, and zero otherwise. Or we might define variables $y_{ij} = 1$ if job i immediately precedes job j on an assembly line. We can then use these variables to develop constraints on the time available for resources, on due dates for individual jobs, and on total schedule costs.

A simple example of a scheduling problem is **examination timetabling**. Variable x_{ij} is given a value of 1 if examination i is assigned to period j. Conflicts are not allowed, so constraints are included to prevent two examinations from being assigned to the same period if any students need to be present at both exams. Additional constraints may reflect limits on the number of exams per period, or the total number of seats in an exam location. The objective function must in some way measure the quality of a given timetable.

Another popular variation is the **vehicle routing** problem. Suppose that a fleet of trucks on a given day must deliver goods from a central warehouse to a set of customers. The objective is to minimize the total cost of making all deliveries. The cost is normally approximated based on minimizing the number of trucks used and the total mileage and/or total hours of delivery time. One common formulation of this problem defines variables x_{ijk} to have a value 1 if customer i is assigned to truck j and is delivered immediately before customer k. Constraints are included to ensure that the assignment is feasible (perhaps based on the drivers' expertise, or on contractual agreements or regulations).

One of the most successful practical applications of integer programming has been in the **airline crew scheduling** problem. The airlines first design a flight schedule composed of a large number of **flight legs**. A flight leg is a specific flight on a specific piece of equipment, such as a 747 from New York to Chicago departing at 6:27 a.m. A **flight crew** is a complete set of people, including pilots, navigator, and flight attendants who are trained for a specific airplane. A **work schedule** or **rotation** is a collection of flight legs that are feasible for a flight crew, and that normally terminate at the point of origin. Variables x_{ij} have value 1 if flight leg i is assigned to crew j. The objective is to ensure that all flight legs are covered at minimum total cost. Most of the major world airlines now use integer programming to assign crews to flight legs, and many claim to be saving millions of dollars annually in operating costs.

A distributed computing problem arises in a multiprocessor computing environment where the programs and data files must be allocated to various machines in different locations. Variables x_{ij} have a value 1 if module i is assigned to processor j. The objective is to minimize the total execution costs (which may depend on the choice of processor) and communication costs (that are incurred when one processor needs to communicate with another).

4.2.3 Mixed Integer Problems

Section 4.1 introduced the problem of **production planning** at General Motors. In that problem, there are two sets of variables: it is necessary to decide which products are assigned to each plant, and then to determine production levels at each plant. We could define 0–1 variables $x_{ij} = 1$ if product i is assigned to plant j. We might then define variables y_{ij} to represent the number of units of product i to produce at plant j. If production levels are fairly high, we might treat the y_{ij} variables as real valued, and round them to integers in the end. Additional constraints must prevent a product from being produced if it is not assigned to the plant. The problem can be modeled as a large mixed integer problem with both 0–1 and real-valued variables.

A related problem involves **warehouse location**: given a set of potential locations for warehouses for a distributor, select the locations that will minimize total delivery costs.

We can define 0–1 variables x_j to have a value 1 if location j is selected. Once it is decided which locations are going to be used, then we must solve some kind of a transportation problem to get the products from the producers to the warehouses, and from the warehouses to the customers. Real-valued variables y_{ij} are defined to represent the amount of product transported from supplier i to warehouse j, and real-valued variables z_{jk} denote the amount of product distributed from warehouse j to customer k. The total cost is a function of the distances that the products must travel.

A further variation, which can be considered as a general version of warehouse location, is called the **fixed charge problem**. Suppose there is a fixed cost (with a 0–1 variable) for opening a warehouse. Once the warehouse is open, the remaining costs are essentially continuous. There are a number of practical problems that lend themselves to this type of formulation. For example, when a telecommunications company installs fiber optic cable, there is a fixed cost for actually laying the cable, but then there is a real-valued cost corresponding to the capacity of the cable. This leads to a related problem called **capacity planning**.

4.3 Zero–One (0–1) Model Formulations

This section presents a few examples of mathematical formulations of some classical 0–1 programming problems. These basic formulations frequently occur in actual practice, often in the form of subproblems within larger practical applications. We emphasize these models because many of the most practical advances in integer programming in recent years have been in the area of 0–1 models.

4.3.1 Traveling Salesman Model

Suppose you want to visit a number of cities and then come back to your point of origin. This is one of the most challenging and most extensively studied problems in the field of combinatorics. The formulation is deceptively simple, and yet it has proven to be notoriously difficult to solve. Define 0–1 variables $x_{ij} = 1$ if city i is visited immediately prior to city j. Let d_{ij} represent the distance between cities i and j. Suppose that there are n cities that must be visited. Then the **traveling salesman problem** (TSP) can be expressed as:

$$\text{minimize} \quad \sum_{i=1}^{n} \sum_{j=1}^{n} d_{ij} x_{ij}$$

$$\text{subject to} \quad \sum_{i=1}^{n} x_{ij} = 1 \quad \text{for all cities j}$$

$$\sum_{j=1}^{n} x_{ij} = 1 \quad \text{for all cities i}$$

$$\sum_{i \in S} \sum_{j \in S} x_{ij} \leq |S| - 1 \quad \text{for all } |S| < n$$

The first constraint says that you must go *in* to city j exactly once, and the second constraint says that you must *leave* every city i exactly once. These constraints ensure that there are two edges adjacent to each city, one in and one out, as we would expect. However, this does not prevent so-called *sub-tours*. A **sub-tour** occurs when there is a loop containing a subset of the cities. Instead of having one tour of all of the cities, the solution can be composed of two or more sub-tours. The third constraint eliminates sub-tours; it states that no proper subset of cities, S, can have a total of $|S|$ edges.

The TSP has many practical industrial applications. Consider the problem of placing components on a circuit board. To minimize the time required to produce a board, one of the primary considerations is often the distance that a placement head must travel between components. Another example occurs in routing trucks or ships delivering products to customers. (When we allow multiple trucks, this problem becomes the vehicle routing problem described earlier.) Another application occurs in a production environment when it is desired to minimize sequence-dependent setup times. When multiple jobs are to be processed on a machine, the total setup time for each job frequently depends on which job preceded it. This situation can be modeled as a TSP, where we sequence jobs rather than sequencing the order in which cities are visited.

4.3.2 Knapsack Model

Two versions of the knapsack problem have been discussed in Section 4.2 when portfolio selection and the capital budgeting problem were reviewed. Assume that we have a number of items, and we must choose some subset of the items to fill our *knapsack*, which has limited space. Each item, i, has a value v_i and takes up w_i units of space in the knapsack. Let the 0–1 variables $x_i = 1$ if item i is selected, and let b represent the total space in the knapsack. Then we can formulate the knapsack problem as follows:

$$\text{maximize} \quad \sum_{i=1}^{n} v_i x_i$$

$$\text{subject to} \quad \sum_{i=1}^{n} w_i x_i \leq b$$

The 0–1 version of the knapsack problem states that every item is unique, and that each can either be selected or not (as in the capital budgeting problem). A slight generalization of the knapsack problem states that you can choose more than one copy of each item, so that the variables can take on general integer values (probably with upper bounds on each variable), as with the portfolio selection problem.

4.3.3 Bin Packing Model

Bin packing is a generalization of the knapsack problem. Suppose that we are given a set of m bins of equal size, b; and a set of n items that must be placed in the bins. Let w_i be the size of item i. We define the 0–1 variable $x_{ij} = 1$ if item i is placed in bin j. Bin packing is usually expressed as a problem of minimizing the number of bins required to pack all of the items. We can let $y_j = 1$ if we need to use bin j. (Note that if $y_j = 0$, then the corresponding bin has no capacity.) The objective function minimizes the number of bins required

$$\text{minimize} \qquad \sum_{j=1}^{m} y_j$$

$$\text{subject to} \qquad \sum_{i=1}^{n} w_i x_{ij} \leq y_j b, \text{ for all } j$$

$$\sum_{j=1}^{n} x_{ij} = 1 \text{ for all } i$$

Bin packing has applications in industry where, for example, there is a limited amount of work that can be assigned to each person working at stations on an assembly line. This model may also be applicable when deciding which products should be produced at each of several possible manufacturing plants, or which customer should be assigned to each delivery truck. Of course, each of these problems involves additional criteria and constraints.

4.3.4 Set Partitioning/Covering/Packing Models

Many problems in combinatorial optimization include (as subproblems) partitioning a group of items into *optimal* subsets. For example, vehicle routing requires that we allocate customers to vehicles. Airline crew scheduling requires that we allocate flight legs to a crew. Municipal garbage pickup requires that we allocate specific street blocks to trucks. Each of these subproblems can be modeled in the following form as a **set partitioning problem**:

$$\text{minimize} \qquad \sum_{j} c_j x_j$$

$$\text{subject to:} \qquad \sum_{j} a_{ij} x_j = 1 \text{ for all } i = 1, \ldots, m$$

$$x_j = 0 \text{ or } 1 \text{ for all } j$$

where $a_{ij} = 1$ if item i is included in (potential) subset j. Each column of the m × n constraint matrix A represents a feasible combination of items. For example, each column might represent the items that could feasibly be loaded into a truck for delivery to customers; or the items could be road segments that require garbage collection, and a column would represent a feasible route for a truck to pick up garbage. The cost c_j represents the cost of delivering (or traveling, or producing) that subset of items. A variable $x_j = 1$ if we decide to include that particular subset in our solution.

In the set partitioning problem, all of the items must be included exactly once. In vehicle routing, for example, we might typically require that exactly one truck travel to each customer. In a slightly different problem, the **set covering problem**, we require that each item be selected *at least* once. For example, in the garbage collection problem, and in the crew scheduling problem, every street (every flight leg) must be covered at least once; but it is also feasible to cover the same street (flight leg) twice, if this turned out to be the most efficient solution. (The second truck would not pick up any garbage, and the second flight crew would ride as passengers.) Set covering differs from set partitioning in that the constraints are "≥" inequalities instead of equalities.

The **set packing problem** describes another similar situation. In some production scheduling problems, we are given a list of orders, and we have possible subsets of orders that can be combined on different machines. In some cases, there may not be sufficient resources to satisfy all of the demand. The problem is to select the optimal subset of orders to maximize some profit function, p_j. This problem can be formulated as:

$$\text{maximize} \quad \sum_j p_j x_j$$

$$\text{subject to} \quad \sum_j a_{ij} x_j \leq 1 \text{ for all } i = 1, \ldots, m$$

$$x_{ij} = 0 \text{ or } 1 \text{ for all } j$$

We select as many items as possible, but we are not allowed to process any items more than once. We will revisit this type of problem in greater detail in Section 4.8, where we discuss column generation.

4.3.5 Generalized Assignment Model

Section 3.3 described the assignment problem, which is considered to be one of the easiest combinatorial problems to solve. The **assignment problem** can be formulated as follows:

$$\text{minimize} \quad \sum_i \sum_j c_{ij} x_{ij}$$

$$\text{subject to} \quad \sum_j x_{ij} = 1 \text{ for all } i = 1, \ldots, n$$

$$\sum_i x_{ij} = 1 \text{ for all } j = 1, \ldots, n$$

$$x_{ij} = 0 \text{ or } 1 \text{ for all } i, j$$

This classical representation can be illustrated by a set of jobs that must be allocated to a group of workers. The term c_{ij} represents the *cost* of assigning job i to employee j. The first constraint requires every job to be assigned to exactly one employee; and the second constraint states that every employee must do exactly one job.

The **generalized assignment problem** is a simple extension in which every job must be assigned to one employee, but each employee has the capacity to perform more than one job. In particular, suppose that each employee, j, has a limited amount of time, (b_j hours) available, and that job i will take employee j a total of a_{ij} hours. Then, the generalized assignment problem can be formulated as:

$$\text{minimize} \quad \sum_i \sum_j c_{ij} x_{ij}$$

$$\text{subject to} \quad \sum_j x_{ij} = 1 \text{ for all } i = 1, \ldots, m$$

$$\sum_i a_{ij} x_{ij} \leq b_j \text{ for all } j = 1, \ldots, n$$

$$x_{ij} = 0 \text{ or } 1 \text{ for all } i, j$$

As discussed earlier, the generalized assignment problem has applications in the vehicle routing problem, where every customer order must be assigned to one truck, but a single truck can hold more than one customer order, subject to capacity constraints.

4.4 Branch-and-Bound

4.4.1 A Simple Example

Branch-and-bound algorithms are widely considered to be the most effective methods for solving medium-sized general integer programming problems. These algorithms make no assumptions about the structure of a problem except that the objective function and the constraints must be linear. Even these restrictions can be relaxed without changing the basic framework of the technique.

In its simplest form, **branch-and-bound** is just an organized way of taking a hard problem and splitting it into two or more smaller (and hence easier) subproblems. If these subproblems are still too hard, we *branch* again and further subdivide the problems. The process is repeated until each of the subproblems can be easily solved. Branching is done in such a way that solving each of the subproblems (and selecting the best answer found) is equivalent to solving the original problem.

Consider the following simple example in two variables. A manufacturer has 300 person-hours available this week and 1,800 units of raw material. These resources can be used to build two products, A and B. The requirements and the profit for each item are given as follows:

Product	Person-Hours	Raw Material	Profit ($)
A	150	300	600
B	10	400	100

Let x_1 and x_2 represent the *integer* number of units of products A and B, respectively. We can formulate this problem as an integer linear programming problem:

$$
\begin{aligned}
\text{maximize} \quad & z = 600x_1 + 100x_2 \\
\text{subject to} \quad & 150x_1 + 10x_2 \le 300 \\
& 300x_1 + 400x_2 \le 1800 \\
& x_1, x_2 \ge 0 \text{ and integer}
\end{aligned}
$$

This problem is illustrated in Figure 4.1. The feasible region is given by the discrete set of integer points within the constraint region. The optimal LP solution occurs at $x_1 = 1.789$ and $x_2 = 3.158$ with a profit of $z = 1{,}389.47$. Unfortunately, we cannot sell a fractional number of items. One obvious alternative is to round down both values to $x_1 = 1$ and $x_2 = 3$, for a profit of $900. We will call the feasible integer solution $x^I = (1, 3)$ the **current incumbent** solution, which is the best answer found thus far. When we find a better integer solution, we will update the current incumbent. Before reading any further, try to locate the optimal integer solution to the problem in Figure 4.1, and consider how integer solutions might be found in general.

The basic branch-and-bound algorithm results from the following observations:

- The feasible integer solution $x = (1, 3)$ with $z = 900$ was fairly easy to find. The optimal integer solution cannot have a lower value of z than \$900. Thus, we write $z^I = 900$ and call this a *lower bound* on the optimal solution. Each time we find a higher valued integer solution, we replace the lower bound z^I. This is the *bound* part of branch-and-bound methods.

- Over the whole feasible region, the largest possible value of $z = 1389.47$, which is the real valued solution obtained from the LP. We call this an upper bound on the optimal integer function value.

- The graphical solution shows that $x_2 = 3.158$. This is infeasible because it is a fractional solution. Since x_2 must be an integer, then clearly either $x_2 \leq 3$ or $x_2 \geq 4$. This is equivalent to saying that x_2 cannot lie part way between 3 and 4.

Consider the following two subproblems:

$$[A] \quad \text{maximize} \quad z = 600x_1 + 100x_2$$
$$\text{subject to} \quad 150x_1 + 10x_2 \leq 300$$
$$300x_1 + 400x_2 \leq 1800$$
$$x_1 \geq 0 \text{ and integer}$$
$$x_2 \geq 4 \text{ and integer}$$

$$[B] \quad \text{maximize} \quad z = 600x_1 + 100x_2$$
$$\text{subject to} \quad 150x_1 + 10x_2 \leq 300$$
$$300x_1 + 400x_2 \leq 1800$$
$$x_1, x_2 \geq 0 \text{ and integer}$$
$$x_2 \leq 3$$

Observe that if we find the best integer solution of both of these subproblems, then one of them must be the optimal solution to the original problem. These subproblems are represented graphically in Figure 4.2, where the diagram is identical to Figure 4.1 except that the range of values for x_2 between 3 and 4 is now infeasible. We say that we have **separated** on variable x_2.

Consider problem [A] first. The LP solution occurs at $x = (0.667, 4)$ with an objective function value of $z = \$800$. Notice that x_2 is now integer valued. We will see that each time we separate, the chosen variable will always be integer, although it does not necessarily stay integer on subsequent iterations.

By definition, the linear programming solution is the largest value possible for the problem. Therefore, the value $z = 800$ is an upper bound on all possible solutions in the feasible region for problem [A]. Any integer solution to [A] must be ≤ 800. However, we already have a feasible integer solution with $z^I = 900$. Therefore, problem [A] can be ignored as it cannot contain any answer better than 900. In branch-and-bound terminology, we say that problem [A] has been *fathomed*.

In general, a subproblem is called **fathomed** whenever it is no longer necessary to branch any further. A subproblem is fathomed when the LP solution is less than the current lower bound for a maximization problem, when the LP solution is infeasible, or when the LP produces an integer solution.

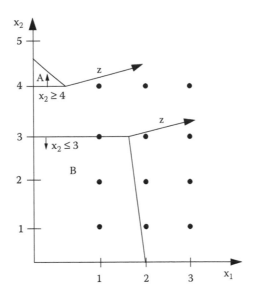

FIGURE 4.2
Separate into two subproblems.

Problem [B] has its optimal LP solution at x = (1.8, 3) with a function value of z = 1,380. This value gives us a new upper bound on the optimal integer solution. At each iteration of the branch-and-bound process, the upper and lower bounds can be revised until they eventually converge to the optimal solution. We now know that the optimal value lies between 900 and 1,380. Variable x_2 is integer valued, but x_1 is still fractional. We can now further divide problem [B] into two subproblems based on the fact that $x_1 \leq 1$ or $x_1 \geq 2$ as follows:

$$[B1] \quad \text{maximize} \quad z = 600x_1 + 100x_2$$
$$\text{subject to} \quad 150x_1 + 10x_2 \leq 300$$
$$300x_1 + 400x_2 \leq 1800$$
$$x_1, x_2 \geq 0 \text{ and integer}$$
$$x_1 \leq 1$$
$$x_2 \leq 3$$

$$[B2] \quad \text{maximize} \quad z = 600x_1 + 100x_2$$
$$\text{subject to} \quad 150x_1 + 10x_2 \leq 300$$
$$300x_1 + 400x_2 \leq 1800$$
$$x_2 \geq 0 \text{ and integer}$$
$$x_1 \geq 2 \text{ and integer}$$
$$x_2 \leq 3$$

For problem [B1], it is easy to see that the optimal LP solution occurs at point x = (1, 3) with a function value z = 900. Since x is now integer valued, it must be optimal for this subproblem. This subproblem is considered to be fathomed because it gives us an integer solution: there is no need for further branching as the solution cannot get any better below this node. It is also considered fathomed because the solution of 900 is no better than the one we already obtained earlier. In either case, problem [B1] is finished.

Problem [B2] consists of the single point $x = (2, 0)$ with a function value of $z = 1,200$. This solution is both integer, and better than the previous lower bound. Since x is integer, subproblem [B2] is fathomed and no further branching is required. Our new lower bound increases to $z^I = 1,200$ and $x^I = (2, 0)$ becomes the new current incumbent.

At this point, we observe that all of our subproblems have been fathomed. Therefore, $x^I = (2, 0)$ is the optimal integer solution, and $z^I = \$1,200$ is the optimal function value.

It is often convenient to display this procedure in the form of a **branch-and-bound tree**. The tree corresponding to the previous example is illustrated in Figure 4.3. Each subproblem is represented by a node in the tree. Each node must either be fathomed or split into subproblems, which are shown by lower level nodes.

In Figure 4.3a, node 0 represents the original problem. We construct nodes 1 and 2 (for subproblems [A] and [B], respectively) by constraining x_2 in Figure 4.3b. Node 1 is fathomed and node 2 is further subdivided into nodes 3 and 4 in Figure 4.3c, corresponding to problems [B1] and [B2].

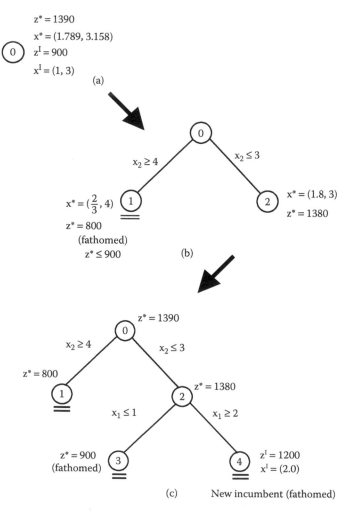

FIGURE 4.3
Branch-and-bound example: (a) node 0: original problem, (b) subproblems [A] and [B], and (c) subproblems [B1] and [B2].

4.4.2 A Basic Branch-and-Bound Algorithm

We will now give a more precise description of the previous procedure. The problem is expressed with a maximization objective, and a similar framework can be followed with minimization problems. A node in the tree is called an **active node** if it has not been fathomed and we have not separated on it yet.

Step 0: Initialize

Let the set A denote the list of currently active nodes. A node in the tree is active if we have not either solved it or subdivided it yet. Initially, the set A = {the original problem}, node 0, and $z^I = -\infty$.

Step 1: Done?

If the set A is empty, then stop. The current incumbent, x^I is optimal.

Step 2: **Branching**

Select a node, j, from the active list A (and remove it from A) according to some *Branching Rule*.

Step 3: **Solve**

Solve the LP relaxation of node j. (That is, relax/ignore the integer restrictions.)

Let z^* denote the optimal LP solution at point x^*.

Step 4: **Fathoming Criterion 1**

If the LP has no feasible solution, then node j is fathomed; go to Step 1.

Step 5: **Fathoming Criterion 2**

If $z^* \leq z^I$, then this subproblem cannot contain any integer solution better than the current incumbent: node j is fathomed; go to Step 1.

Step 6: **Fathoming Criterion 3**

If x^* is integer, then it becomes the new incumbent. Set $x^I = x^*$ and $z^I = z^*$. Node j is fathomed; go to Step 1.

Step 7: **Separation**

Otherwise, we must separate node j into two or more subproblems (according to some *Separation Rule*.) Select some fractional variable in x^* and construct two new subproblems. Add these new nodes to the set A and go to Step 1.

4.4.3 Knapsack Example

The manager of an Operations Research department in a large company has a list of projects that she would like to initiate. Each project has an expected payback expressed (in thousands of dollars) as the net present value over a 10-year period. Although all of the projects would be beneficial, there are simply not enough resources (in person days) available this month to do all of them. The estimates of resources and return are:

Project	1	2	3	4	5	6	7	8
Estimated value	15	20	5	25	22	17	30	4
Days	51	60	40	62	63	50	70	10

There are 250 person-days available this month. Which projects should be selected? At the end of this month, the manager must write a report summarizing the results from completed projects; any projects that are not completed cannot be included among the successful projects in the report.

Define $x_j = 1$ if project j is selected, and 0 otherwise. The "node 0" problem can be modeled as:

maximize $15x_1 + 20x_2 + 5x_3 + 25x_4 + 22x_5 + 17x_6 + 30x_7 + 4x_8$

subject to $51x_1 + 60x_2 + 40x_3 + 62x_4 + 63x_5 + 50x_6 + 70x_7 + 10x_8 \leq 250$

                 $x_j = 0$ or 1

When the 0–1 constraints are relaxed to solve the LP, we replace them with the linear constraints: $0 \leq x_j \leq 1$

Step 0: $A = \{0\}$, and $z^I = -\infty$.

Step 1: A is not empty.

Step 2: Select node 0 from A. (A is now empty.)

Step 3: $z^* = 96.3$ at the optimal LP solution at point $x^* = \{0, 0, 0, 1, 1, 0.9, 1, 1\}$.

Step 4: The solution is feasible.

Step 5: $z^* > z^I$.

Step 6: x^* is not an integer.

Step 7: Separate node 0 on a fractional variable (x_6 is the only fractional value). Construct node 1, the same problem as node 0 with the additional constraint that $x_6 = 0$. Similarly, construct node 2, the same problem as node 0 with the constraint that $x_6 = 1$. Let $A = \{1, 2\}$.

Step 1: A is not empty.

Step 2: Select a node from A. Suppose we choose node 2; $A = \{1\}$. Add constraint $x_6 = 1$.

Step 3: $z^* = 96.25$ at the optimal LP solution at point $x^* = \{0, 0, 0, 1, 0.92, 1, 1, 1\}$.

Step 4: The solution is feasible.

Step 5: $z^* > z^I$.

Step 6: x^* is not an integer.

Step 7: Separate node 2 on a fractional variable. (x_5 is the only fractional value). Construct node 3, the same problem as node 2 with the additional constraint that $x_5 = 0$. Similarly, construct node 4, the same problem as node 2 with the constraint that $x_5 = 1$. Let $A = \{1, 3, 4\}$.

Step 1: A is not empty.

Step 2: Select a node from A. If we choose node 4, then $A = \{1, 3\}$. Add constraint $x_5 = 1$.

Step 3: $z^* = 96$ at the optimal LP solution at point $x^* = \{0, 0, 0, 1, 1, 1, 1, 0.5\}$.

Step 4: The solution is feasible.

Step 5: $z^* > z^I$.

Step 6: x^* is not an integer.

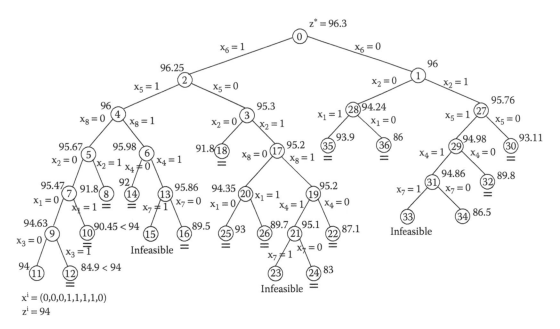

FIGURE 4.4
Branch-and-bound tree for the knapsack example.

Step 7: Separate node 4 on a fractional variable. (x_8 is the only fractional value). Construct node 5, the same problem as node 4 with the additional constraint that $x_8 = 0$. Similarly, construct node 6, the same problem as node 4 with the constraint that $x_8 = 1$. Let $A = \{1, 3, 5, 6\}$.

The algorithm continues until the set A is empty. The complete branch-and-bound tree for this problem is illustrated in Figure 4.4.

4.4.4 From Basic Method to Commercial Code

It is possible to construct examples in which the basic algorithm explicitly enumerates all possible integer solutions. If we assume, for simplicity, that there are n variables, and that each variable has m possible integer values, then our branch-and-bound tree could have as many as m^n nodes at the lowest level of the tree. The amount of computation required increases exponentially and the problem would become computationally intractable for even moderate values of m and n. For example, when m = 3 and n = 20, the number of potential integer solutions is over 3 billion. Of course, we hope that the vast majority of potential nodes will be implicitly eliminated using the various fathoming criteria. A good branch-and-bound algorithm will try to find the optimal solution as quickly as possible; but if we hope to solve problems of any practical size, the algorithms must be designed very carefully. In particular, the three components of the algorithm that are most critical to the performance of various branch-and-bound implementations are:

1. **Branching strategy**: Selection of the next node (in the active list) to branch on in Step 2.

2. *Bounding strategy*: Many techniques have been suggested for improving the LP bounds (in Step 5) on the solution of each subproblem.

3. *Separation rule*: The selection of which variable to separate on in Step 7.

4.4.4.1 Branching Strategies

To control the selection of the next node for branching, it is typical to restrict the choice of nodes from the list of currently active nodes in one of the following ways.

The Backtracking or LIFO (Last In, First Out) Strategy

Always select a node that was most recently added to the tree. Evaluate all nodes in one branch of the tree completely to the bottom, and then work back up to the top following all indicated side branches. A typical order of evaluating nodes is illustrated in Figure 4.5a. The numbers inside each node represent the order in which they are selected.

The Jumptracking (Unrestricted) Strategy

As the name implies, each time the algorithm selects a node, it can choose any active node anywhere in the tree. For example, it might always choose the active node corresponding to the highest LP solution, z^*. A possible order of solving subproblems under jumptracking is illustrated in Figure 4.5b.

At first glance, the **backtracking** procedure appears to be unnecessarily restrictive. The major advantages are conservation of storage required and a reduction in the amount of computation required to solve the corresponding LP at each node. Observe that the number of active subproblems in the list at any time is equal to the number of levels in the current branch of the tree. Using jumptracking, the size of the active list can grow exponentially. Each node in the active list corresponds to a linear programming problem with its own set of constraints. Consequently, storage space for subproblems is an important consideration.

Computation time is an even more serious issue with jumptracking. Observe that each time we solve a subproblem, we solve an LP complete with a full Simplex tableau. When we move down the tree, we add one new constraint to the LP. This can be done relatively efficiently if the old tableau is still available.

To do this using the jumptracking strategy, we would have to save the Simplex tableau for each node (or at least enough information to generate the tableau easily). Hence, backtracking can save a large amount of LP computation time at each node. The efficiency of solving subproblems is crucial to the success of a branch-and-bound method because practical problems will typically generate trees with literally thousands of nodes.

The major advantage of jumptracking is that, by judicious selection of the next active node, we can usually solve the problem by examining far fewer nodes. Observe that when we find the optimal integer solution, many of the nodes can be eliminated by the bounding test. Jumptracking will normally find the optimal solution sooner than backtracking. To illustrate this, suppose that the integer solution is represented by a node at the bottom of the branching tree. With backtracking, each time we choose a branch, one is *correct* and the other is *wrong*. If we choose the wrong branch, we must evaluate all nodes in that branch before we can get back on the correct branch. Using jumptracking, we can return to the correct branch as soon as we realize that we may have made a mistake. When we find the optimal solution, many of the nodes in the *wrong* branch will be fathomed at a higher level of the tree by the bounding test.

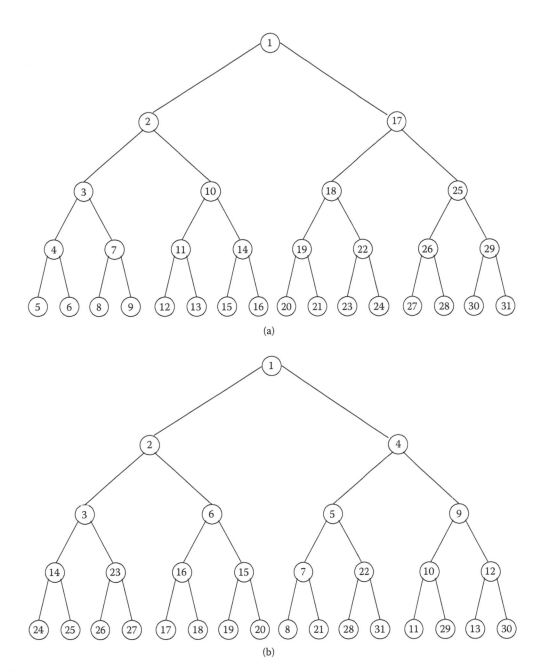

FIGURE 4.5
Branching strategies: (a) back tracking and (b) jump tracking.

In short, there is a trade-off between backtracking and jumptracking, and many commercial algorithms use a mixed strategy. Backtracking is used until there is a strong indication of being in the wrong branch; then there is a jump to a more promising node in the tree and a resumption of a backtracking strategy from that point. The amount of jumptracking is determined by the definition of *wrong*.

4.4.4.2 Bounding Strategies

In the branch-and-bound algorithm, suppose we have selected a node (subproblem) to branch on. We must now choose a fractional basic variable to separate on. Whether we round the variable up or down, the objective function value will normally decrease. The up and down penalties for each basic variable give us an estimate (lower bound) on the reduction in the value of z that would occur if we were to add the integer constraint. We can then use this information to pick the most promising basic variable.

Consider the example in Section 4.4.1. The optimal LP tableau is:

Basis	x_1	x_2	x_3	x_4	Solution
z	0	0	$3\tfrac{13}{19}$	$\tfrac{3}{9}$	$1389\tfrac{9}{19}$
x_1	1	0	$\tfrac{2}{285}$	$-\tfrac{1}{5700}$	$1\tfrac{15}{19}$
x_2	0	1	$-\tfrac{1}{190}$	$\tfrac{1}{380}$	$3\tfrac{3}{19}$

Define f_i to be the fractional part of each basic variable. In the example, $f_1 = 15/19$ and $f_2 = 3/19$, are the fractional parts of x_1 and x_2, respectively. Define $\bar{a}_{ij}$ to be the element of the optimal LP tableau; and define $\bar{c}_j$ to be the j-th reduced cost from the tableau. We define the **down penalty** D_i to be the decrease in the objective function that would result from decreasing the variable to the next lower integer value. The down penalty for branching down on the basic variable in the i-th row is:

$$D_i = \min_{j \in N} \left\{ \frac{\bar{c}_j f_j}{\bar{a}_{ij}}, \text{ where } \bar{a}_{ij} > 0 \right\}$$

Similarly, we can derive a formula for the **up penalty** for variable x_i, which will indicate the amount by which the objective function would decrease if we increased the basic variable in the i-th row to the next highest integer. The up penalty, U_i, is given by:

$$U_i = \min_{j \in N} \left\{ \frac{\bar{c}_j (f_j - 1)}{\bar{a}_{ij}}, \text{ where } \bar{a}_{ij} < 0 \right\}$$

In the example, the down penalty corresponding to branching down on basic variable x_1 is given by:

$$D_1 = \min_{j \in N} \left\{ \frac{\bar{c}_j f_j}{\bar{a}_{1j}}, \text{where } \bar{a}_{1j} > 0 \right\}$$

$$= \left\{ \frac{\left(3\frac{13}{19}\right)\left(\frac{5}{19}\right)}{\left(\frac{2}{285}\right)} \right\}$$

$$= 414\frac{9}{19}$$

Either: Separate on X_1 or: Separate on X_2

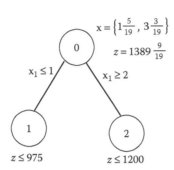
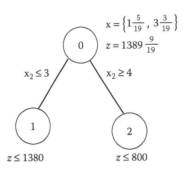

FIGURE 4.6
Up and down penalties for fractional basic variables and the corresponding potential branches.

Consider the row of the tableau corresponding to x_1. We can show that decreasing x_1 by $f_1 = 15/19$ implies we must increase x_3 by $\left[\left(\frac{15}{19}\right)/\left(\frac{2}{285}\right)\right]$ to maintain the equation. This, in turn, would produce the given decrease in the objective function row. (See Salkin and Mathur [1989] for a detailed proof.) Similarly, $D_2 = 9\frac{9}{19}$, $U_1 = 189\frac{9}{19}$, and $U_2 = 589\frac{9}{19}$. The potential effect on the new branch-and-bound tree is shown in Figure 4.6.

4.4.4.3 Separation Rules

We can think of up and down penalties as a kind of *look-ahead* feature, in that they give us an estimate of the LP objective function value for separating on each basic variable without actually solving all of the possible LP problems. We could, of course, improve these estimates by actually solving the corresponding LP tableaus, but this would be far more expensive. With branch-and-bound algorithms, we will always be faced with the trade-off between better (more accurate) bounds and computational cost.

Consider the two potential branch-and-bound trees in Figure 4.6. Which tree allows a more efficient solution? One simple general rule is to construct, at each node, a *good* branch and a *bad* one. Hopefully, we can follow the good branch, find the optimal integer solution, and then fathom the bad branch without having to separate further.

Thus, an effective separation rule is to separate on the variable that has the largest up or down penalty; then branch to the active node with the *highest* lower bound on the new function value; that is, the one most likely to lead to an optimal integer solution.

In the example, we would separate on variable x_2 and then branch to subproblem [B] with $x_2 \leq 3$. When we solve [B], we will find the optimal integer with a function value of 1,200. Because problem [A] has an upper bound of $z \leq 800$, it will be fathomed without solving the corresponding LP.

4.4.4.4 The Impact of Model Formulation

For linear programming models, it does not make much difference how the original problem is formulated, provided that the objective function and the constraints are correct. In integer programming, however, the formulation itself can have a dramatic effect on algorithm performance. As an example, consider the original problem formulation:

$$\text{maximize} \qquad z = 600x_1 + 100x_2$$

$$\text{subject to:} \qquad 150x_1 + 10x_2 + x_3 = 300$$

$$300x_1 + 400x_2 + x_4 = 1800$$

$$x_1, x_2 \geq 0 \text{ and integer}$$

Observe that, for any feasible integer solution to this problem, x_3 must be a multiple of 10, and x_4 must be a multiple of 100. Suppose we first reduced the original problem to lowest common terms (before adding the slack variables):

$$\text{maximize} \qquad z = 6x_1 + x_2$$

$$\text{subject to:} \qquad 15x_1 + x_2 + x_3 = 30$$

$$3x_1 + 4x_2 + x_4 = 18$$

$$x_1, x_2 \geq 0 \text{ and integer}$$

This *new* problem is identical to the original as far as the LP is concerned, but it is *not* the same integer problem! The new optimal Simplex tableau is:

Basis	x_1	x_2	x_3	x_4	Solution
z	0	0	$\frac{7}{19}$	$\frac{3}{19}$	$13\frac{17}{19}$
x_1	1	0	$\frac{4}{57}$	$-\frac{1}{57}$	$1\frac{15}{19}$
x_2	0	1	$-\frac{1}{19}$	$\frac{5}{19}$	$3\frac{3}{19}$

The most obvious immediate consequence of this new formulation is simply that z must be an integer multiple of 100. The upper bound on z denoted $\bar{z}$ is now 1,300. The reduction has no effect on the up and down penalties except that we get the decrease in the reduced units of z. Because the optimal value of z must be integer, the up and down penalties can be strengthened. For example, D_1 in the old version reduced z to 975. Using the new tableau, z will become 9.75, which can be replaced by 9 as an upper bound on the down problem. Since the initial rounded solution is z = 9, the corresponding branch is fathomed, that is, we can branch up on x_1 for free. The complete revised branch-and-bound tree with up and down penalties is illustrated in Figure 4.7.

Notice also that the slack variables, x_3 and x_4 are integer valued in both problems and they will be candidates for branching. The slack and surplus variables (and the objective function variable z) will always be integer valued (in a pure integer problem) if all of the problem coefficients are integer. Thus, for example, if one of the constraint coefficients is 0.5, it would be advantageous to multiply the corresponding constraint by 2 to produce all integer coefficients. In general, any rational fractions can be removed by multiplying by the denominator. Refer to Johnson et al. (2000) for additional formulations.

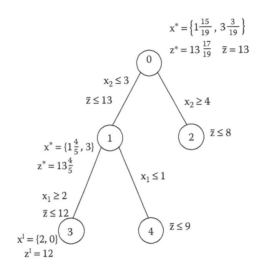

FIGURE 4.7
A complete branch-and-bound tree for the example problem using all penalty information.

4.4.4.5 Representation of Real Numbers

In Chapter 2 on linear programming, we mentioned some of the problems associated with round-off error and numerical stability. The Simplex tableau will normally contain imperfect machine-representations of real numbers that have a limited number of significant digits. As the algorithm proceeds, this inaccuracy in the representation of problem parameters will be compounded during each iteration so that the results can eventually become very inaccurate. This problem becomes much more critical in the context of integer programming because we often solve the LP several thousand times. Most commercial LP codes include a *re-inversion* feature that computes a new basis inverse matrix after a specified number of iterations.

We have the additional problem that it is difficult even to recognize when we have found an integer solution. The values of x_i will not yield exact integer answers. We must assume that they are actually integers when they get *close enough* to an integer value within some prespecified tolerance.

In the example earlier, we expressed all of our calculations in the form of precise rational fractions to avoid any rounding error. Unfortunately, this is not a very practical approach in large-scale problems.

4.5 Cutting Planes and Facets

There is an extensive literature concerning the use of cutting planes to solve integer programming problems. Early algorithms were theoretically intriguing, but not very effective in practice. However, some recent developments in the application of special cutting planes for problems with specific structure have produced some rather surprising results. One example is presented in Section 4.6 for the pure 0–1 problem. This section briefly discusses the general concepts and provides some background.

Given any integer programming problem, consider the set of feasible integer points. If the extreme points of the LP are all integers, then the problem is easy; the LP solution will be an integer solution. If the extreme points are not integer, then we can always *tighten up* the constraints (and possibly add new ones) in such a way that the new reduced LP does have integer extreme points.

For an intuitive motivation of this statement, suppose that the LP has an optimal extreme point solution that is not an integer. Then, it should be possible to add a new constraint that makes that extreme point infeasible (by at least a small amount) without excluding any feasible integer solutions. (We will illustrate shortly that this is always possible.) We can repeat this process until all extreme points are integers.

The general idea is illustrated in Figure 4.8. Given a feasible region defined by the constraints of a linear programming formulation, we are interested in only the integer points inside the region. In the figure, the outside polygon defines the LP feasible region: the inside polygon defines a unique tightened region that does not exclude any integer solutions. We call the reduced region the **convex hull** of the set of feasible integers. It is also referred to as the **integer polytope** of the problem. (A **polytope** is an n-dimensional polygon.)

A constraint is called a **face** or **facet** of the integer polytope if it defines an $(n - 1)$-dimensional set of points on the surface of the convex hull. In the two-dimensional example, a facet is a line of feasible points between two integer extreme solutions. In a three-dimensional cube, for example, the facets are simply the two-dimensional faces of the cube. A constraint that meets the cube along only an edge (one dimension) is not a facet. Clearly (at least in three dimensions), there must be one facet constraint for each face, and no others are needed to define the integer polytope.

If we could find all the facets of an integer problem, then all of the extreme points would be integers and the LP solution method would easily find the optimal integer solution. Unfortunately, for general problems, it is extremely difficult to find the facets of the convex hull. Much of the current research in integer programming is devoted to finding some facet-defining constraints for very specific problems.

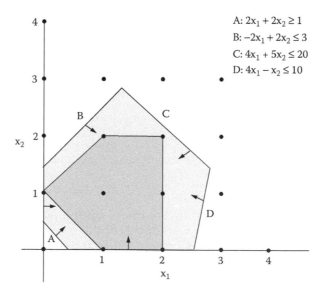

A: $2x_1 + 2x_2 \geq 1$
B: $-2x_1 + 2x_2 \leq 3$
C: $4x_1 + 5x_2 \leq 20$
D: $4x_1 - x_2 \leq 10$

FIGURE 4.8
The convex hull of the set of integer solutions.

The preceding observations have led many researchers to try to develop algorithms that would try to approximate the convex hull of the integer polytope. In particular, it is not necessary to find all of the facets—only the ones that define the integer optimum. Consider the following general algorithm:

1. Solve the LP.
2. If the solution is integer, then it must be optimal.
3. Otherwise, generate a **cutting plane** that excludes the current LP solution, but does not exclude any integer points, and then return to Step 1.

By our definition, a cutting plane is not necessarily a facet. A cutting plane is only guaranteed to take a slice of non-integer solutions out of the feasible region. In general, facets are hard to find, while cutting planes are easy; but, of course, the best cutting plane would be a facet.

Consider the example problem from Section 4.4.4.4, the branch-and-bound example after the coefficients have been reduced. The optimal Simplex tableau is:

Basis	x_1	x_2	x_3	x_4	Solution
z	0	0	$\frac{7}{19}$	$\frac{3}{19}$	$13\frac{17}{19}$
x_1	1	0	$\frac{4}{57}$	$-\frac{1}{57}$	$1\frac{15}{19}$
x_2	0	1	$-\frac{1}{19}$	$\frac{5}{19}$	$3\frac{3}{19}$

As a simple example of a cutting plane, observe that one row of the tableau can be written as:

$$x_1 + \frac{4}{57}x_3 - \frac{1}{57}x_4 = 1\frac{15}{19}$$

Every feasible solution to this problem must satisfy this constraint, which is derived by elementary row operations on the original constraints. To obtain an integer solution for x_1, at least one of the non-basic variables will have to increase, and these must also be integer. This leads to the simple cutting plane:

$$x_3 + x_4 \geq 1$$

At the current LP optimum, x_3 and x_4 are both equal to zero. Therefore, this constraint must make the current point infeasible. Furthermore, every feasible integer solution must satisfy this constraint, so no integers have been excluded. That is, this constraint satisfies the criteria for a cutting plane.

Notice that there is no branching involved here; at each iteration, we define a smaller feasible region, solve the new LP, and repeat the process, continuing until all of the basic variables are integers.

This procedure looks intuitively appealing because the cuts are easy to find and there are none of the complicated storage and bound problems associated with branch-and-bound methods. However, it is not a very efficient or effective technique. As an exercise, the reader should try a few iterations on the example problem. Convergence is generally very slow, which means that we have to generate a large number of new constraints. In fact, for this particular cut, we cannot even prove that the procedure is always finite.

A wide variety of better cutting planes have been proposed, of which the best known is called a **Gomory fractional cut**. This method is based on the premise that, in any integer solution, all of the fractional parts (in the tableau) must cancel one another. Consider the previous example for x_1. From the tableau:

$$x_1 + \frac{4}{57}x_3 - \frac{1}{57}x_4 = 1\frac{15}{19}$$

We first separate each coefficient into two parts: an integer component and a *positive* fractional part:

$$x_1 + 0x_3 + \frac{4}{57}x_3 - x_4 + \frac{56}{57}x_4 = 1 + \frac{15}{19}$$

Grouping all of the *integer* parts together on the right-hand side, we obtain:

$$\frac{4}{57}x_3 + \frac{56}{57}x_4 = \left[-x_1 + x_4 + 1\right] + \frac{15}{19}$$

Observe that, for any integer solution, the part in square brackets must also be integer. Moreover, because the variables must be non-negative, the left-hand side has to be positive. In fact, the left-hand side must be equal to: $^{15}/_{19}$ or $1^{15}/_{19}$ or $2^{15}/_{19}$ or $3^{15}/_{19}$, and so on. In other words:

$$\frac{4}{57}x_3 + \frac{56}{57}x_4 \geq \frac{15}{19}$$

This is the Gomory fractional cut. Because the non-basic variables, x_3 and x_4 are equal to zero at the current LP solution, the Gomory cut always cuts off the corner of the feasible region containing the optimal solution. If any variable has a fractional solution, it is always possible to construct a Gomory cut. This method has the property that it will converge in a finite number of iterations.

The main disadvantages associated with the Gomory fractional cut method are: (1) the method can converge slowly; and (2) unlike branch-and-bound methods, integer solutions are not obtained until the very end. Pure cutting plane methods are therefore not considered to be very practical for large problems.

4.6 Cover Inequalities

One of the most successful approaches to 0–1 problems has been the introduction of cover inequalities (Crowder et al. 1983). A **cover inequality** is a specialized type of cutting plane. It defines a constraint that is added to the original problem in the hope that the extreme point solutions will occur at 0–1 points. After generating as many cover inequality constraints as possible, the reduced problem is solved using a standard branch-and-bound algorithm. This technique was able to dramatically decrease computation time on large, sparse 0–1 programming problems, and practical problems with over 10,000 0–1 variables were solved to optimality. Prior to the introduction of this method, problems with 500 0–1 variables were considered very difficult.

As before, the problem is formulated as a standard linear program with the additional restriction that all variables must be either 0 or 1. The constraints are partitioned into two types. Type I constraints are called Special Ordered Set (SOS) constraints. Type II constraints are simply all of the non-SOS inequalities. The simplest form of SOS constraint is as follows:

$$\sum_{j \in L} x_j \leq 1 \quad \text{for some subset L of the variables}$$

In practical problems, we will often find that the vast majority of constraints are SOS. For example, if the variables x_{ij} are equal to 1 if resource i is assigned to location j, then we will have a number of SOS constraints which state that each resource can be assigned to at most one location. We may also get SOS equality constraints if resource i must be assigned to exactly one location.

SOS constraints have a very useful property with respect to 0–1 integer programming. Observe that, when we consider only one constraint (plus the non-negativity constraints on the variables), *every* extreme point solution occurs at a 0–1 point. For example, consider a simple system: $x_1 + x_2 + x_3 = 1$; $x_1, x_2, x_3 \geq 0$. The extreme points occur at $(1,0,0)$, $(0,1,0)$, $(0,0,1)$, and $(0,0,0)$. Unfortunately, when several SOS constraints intersect, fractional LP solutions are introduced, but the property of having many 0–1 extreme points is still very attractive.

In a sense, SOS constraints produce *easy* problems, while the remaining inequalities are *difficult*. In general, the vast majority of extreme points using non-SOS constraints will lead to fractional solutions. Cover inequalities can be considered a simple technique for converting an individual non-SOS constraint into a set of equivalent SOS inequalities.

Before we present a precise definition, consider the following simple constraint as an example:

$$3x_1 + 4x_2 + 5x_3 \leq 6$$

Observe that if we consider only 0–1 solutions, no two of these x_j's are allowed to have a value equal to 1. In particular, we can express this as:

$$x_1 + x_2 \leq 1$$

$$x_1 + x_3 \leq 1$$

$$x_2 + x_3 \leq 1$$

All of these constraints are **cover inequalities**; if any two variables are equal to 1, then the left-hand side will be greater than (or *cover*) the right-hand side. As an example, if x_1 and $x_2 = 1$, then $3x_1 + 4x_2 = 7 > 6$. In fact, we can represent all three of these constraints in one by observing that only one of these x_j's can equal 1 in any feasible 0–1 solution:

$$x_1 + x_2 + x_3 \leq 1$$

Here, we can replace the original non-SOS inequality with its cover. As far as any 0–1 solutions are concerned, the two constraints are equivalent. With respect to the LP solution, however, the cover inequality is much more restrictive. For example, the point (1, 0.75, 0) is feasible for the LP but infeasible under the cover.

As a more general illustration, consider the inequality:

$$3x_1 + 4x_2 + 5x_3 + 6x_4 + 7x_5 + 9x_6 \leq 12$$

Any subset of x_j's that results in a sum greater than 12 can be eliminated by a cover inequality such as $x_2 + x_3 + x_4 \leq 2$ because we cannot have all three of these variables equal to one. (The sum would be at least 15.)

A **cover** for a single inequality is a subset of variables, the sum of whose (positive) coefficients is greater than (or *covers*) the right-hand side value, b. A cover assumes that the inequality is in less than or equal ($\leq$) form, and that all of the coefficients are positive (or zero). We can convert a greater than or equal to constraint into less than or equal to form by multiplying through by −1. We can also represent an equality constraint by two inequalities (one $\geq$ and a $\leq$) and then multiply the $\geq$ by −1. Each of these would be considered separately. If the constraint has a negative coefficient for variable x_j, we can perform a temporary variable substitution of $x_j = 1 - x_j'$ to make all coefficients positive.

Suppose, for example, that a problem contains the constraint:

$$4x_1 - 5x_2 + 3x_3 - 4x_4 - 7x_5 + 5x_6 = 1$$

We can *replace* this constraint with two inequalities:

$$4x_1 - 5x_2 + 3x_3 - 4x_4 - 7x_5 + 5x_6 \leq 1$$
$$4x_1 - 5x_2 + 3x_3 - 4x_4 - 7x_5 + 5x_6 \geq 1$$

(We do not really replace the constraint. We simply transform it for the purpose of finding cover inequalities.) The second ($\geq$) constraint can be written as:

$$-4x_1 + 5x_2 - 3x_3 + 4x_4 + 7x_5 - 5x_6 \leq -1$$

Substitute $x_1 = 1 - x_1'$, $x_3 = 1 - x_3'$, and $x_6 = 1 - x_6'$ to get

$$4x_1' + 5x_2 + 3x_3' + 4x_4 + 7x_5 + 5x_6' \leq 11$$

Similarly, for the first inequality, we get:

$$4x_1 + 5x_2' + 3x_3 + 4x_4' + 7x_5' + 5x_6 \leq 17$$

We can then use each of these independently to construct cover inequalities. The preceding constraint implies (among others) that:

$$x_1 + x_3 + x_4' + x_5' \leq 3 \text{ (that is, the variables cannot all have the value 1)}$$

Converting back to original variables, we get:

$$x_1 + x_3 - x_4 - x_5 \leq 1$$

and we could add this new SOS constraint to the original LP, and resolve it.

In general, any non-SOS constraint can be written in the form:

$$\sum_{j \in K} a_j x_j \leq b$$

where K refers to the subset of non-zero coefficients and we can assume that $a_j > 0$. We have deleted the subscript i for the row to simplify the notation.

Let S be any subset of K such that:

$$\sum_{j \in S} a_j > b$$

The set S defines a **cover**. S is called a **minimal cover** if:

$$\sum_{j \in S} a_j - a_k < b \text{ for all } k \in S$$

that is, every element of S must *cover* b. In our example, for

$$4x_1 + 5x_2' + 3x_3 + 4x_4' + 7x_5' + 5x_6 \leq 17$$

we could say that the set $S = \{1, 2, 3, 4, 5, 6\}$ is a cover. The sum of the coefficients is greater than 17. However, there are a number of *smaller* covers. If we remove x_2', the set is still a cover. If we also remove x_3, the result, $S = \{1, 4, 5, 6\}$ is still a cover. However, if we remove any other element, S is no longer a cover; the sum will not be greater than 17. This set is called a **minimal cover**, and the cover inequality is:

$$x_1 + x_4' + x_5' + x_6 \leq 3$$

or, equivalently,

$$x_1 - x_4 - x_5 + x_6 \leq 1$$

If the set S is a cover, then every 0–1 solution must satisfy the **cover inequality**:

$$\sum_{j \in S} x_j \leq |S| - 1$$

There is a simple procedure for finding a minimal cover. Begin with $S = K$. Pick any index to delete from S such that the remaining indices still form a cover. Repeat until no index can be deleted without making the coefficient sum less than or equal to b. By repeating this process several times in a systematic way, we could generate all possible minimal cover inequalities. However, for large practical problems, the number of cover inequalities can be exponential. Therefore, we need a method for *efficiently* finding a *good* cover.

Unfortunately, the approach described earlier is not very practical for large problems. Suppose that one of the non-SOS constraints contains 50 variables, and each cover inequality has approximately 25 variables; then the constraint allows only half of the variables to be used in any 0–1 solution. The number of potential minimal cover inequalities is $\binom{50}{25} \approx 1.26 \times 10^{14}$. Generating all possible covers is not a very practical strategy, for even if we could generate all covers, we would discover that most of them were unnecessary in the following sense. The original purpose behind constructing these constraints was to force the LP into a 0–1 extreme point. Most of the covers, although perfectly valid, will have no effect on the current optimal solution to the LP. The preferred approach would be to solve the LP, and then, if the solution contains fractional values, to look for a *single* cover inequality that makes the current LP solution infeasible.

To illustrate this process, consider the following simple problem:

$$\text{maximize} \quad z = 12x_1 + 13x_2 + 11x_3 + 10x_4$$

$$\text{subject to:} \quad 12x_1 + 13x_2 + 12x_3 + 11x_4 \leq 29$$

$$x_j = 0 \text{ or } 1$$

Solving this problem as an LP (with constraints $0 \leq x_j \leq 1$), we find that $x^* = (1, 1, 0.333, 0)$, with $z^* = 28.667$. We want to find a set S such that:

1. The set S forms a cover of the constraint:

$$\sum_{j \in S} a_j > b$$

therefore,

$$\sum_{j \in S} x_j \leq |S| - 1$$

2. The current LP solution violates the cover inequality:

$$\sum_{j \in S} x_j^* > |S| - 1$$

It is fairly easy to show that, if $x_j^* = 0$, then j will never occur in the set S. Because every $x_j^* \leq 1$, if any of them are zero, the constraint will never violate the cover inequality. It is also easy to prove that, if $x_j^* = 1$, then we can always include it in the set S. If the corresponding j is not in a set S that satisfies the aforementioned criteria, then adding j to S will still be a cover. Therefore, in our example, we will include x_1 and x_2 and ignore x_4. The only question is whether to include x_3. Observe that when we do not include it, we do not get a cover; but, when we do add it to S, we get a cover and the current solution violates the cover inequality, as required:

$$x_1 + x_2 + x_3 \leq 2$$

We now add this constraint to the original problem and solve the LP again. If the new solution is fractional, we look for another cover inequality.

We now present a simple algorithm for finding effective cover inequalities. Let x^* be the optimal solution to the linear programming problem with $0 \leq x_j \leq 1$, and suppose that we want to find a valid cover inequality for one of the non-SOS constraints that will cut off the current LP solution. Consider any non-SOS constraint of the form:

$$\sum_{j \in K} a_j x_j \geq b + 1$$

(We will repeat this procedure for each of the non-SOS constraints separately.)
Define the elements of S using the 0–1 variables s_j, where:

$$s_j \begin{cases} 1 & \text{if } j \in s \\ 0 & \text{Otherwise} \end{cases}$$

We claim that this problem is equivalent to solving the following 0–1 knapsack problem:

$$\text{minimize} \qquad z = \sum_{j \in K} \left(1 - x_j^*\right) s_j$$

$$\text{subject to} \qquad \sum_{j \in S} a_j x_j \leq b$$

The constraint ensures that the solution will be a cover. If the optimal value of z in this problem is less than 1, then the corresponding cover inequality will make the current LP solution infeasible. For a proof of this claim, refer to the work of Crowder et al. (1983).

In this subproblem, we do not actually require the optimal value of z. It is only necessary to find a z value less than 1, so we can use a variation of the *biggest bang for your buck* heuristic, which will be described in the following, to find an approximate solution efficiently. This method may miss a valid cover; but if it does find one, it will be acceptable.

We present a method for finding an approximate solution to the following 0–1 knapsack problem:

$$\text{maximize} \qquad z = \sum_{j \in S} t_j x_j$$

$$\text{subject to} \qquad \sum_{j \in S} a_j x_j \leq b$$

The LP version of the knapsack problem is very easy to solve optimally. The algorithm sorts all of the variables in decreasing order of *bang for buck*. The cost coefficient t_j represents the value (bang) that we get from each x_j, while a_j represents the cost (buck) or weight associated with the limited resource b. Process the variables in decreasing order of $\{t_j / a_j\}$, and set $x_j = 1$ as long as the constraint is still satisfied. Let k be the index of the first variable that will not fit in the knapsack. Define the amount of space left in the knapsack (the residual) as:

$$r = b - \sum_{j < k} a_j$$

and set x_k equal to the fraction just large enough to use all remaining capacity:

$$x_k = \frac{r}{a_k}$$

The rest of the x_j's for $j > k$ are set to 0.

This simple one-pass assignment gives the optimal objective function for the LP and has only one possible fractional variable. Let z^* be the objective function value. The optimal value z for the 0–1 knapsack problem will be less than or equal to z^*. If z^* is not integer valued, we can round it down, and use it to approximate the 0–1 knapsack solution. Thus, we do not actually solve the 0–1 knapsack problem.

The *bang for buck* heuristic also gives us a *lower bound* on the 0–1 knapsack problem. If we ignore the fractional variable x_k, we have a feasible 0–1 solution and, therefore, the optimal 0–1 solution is bounded *below* by $z^* - t_k x_k$ and *above* by z^*. In particular, if the LP has no fractional variable, the solution z^l must be optimal.

Our situation presents a type of *reverse* knapsack problem: minimize a cost function and have at least $(b + 1)$ selected for inclusion in the knapsack. We can apply the same *bang for buck* heuristic; only we select the variable with the *smallest* ratio first, and keep selecting until the solution is feasible.

Consider the previous example: $12x_1 + 13x_2 + 12x_3 + 11x_4 \leq 29$ and $x^* = (1, 1, 0.333, 0)$. The knapsack problem becomes:

$$\text{minimize} \qquad z = 0s_1 + 0s_2 + 0.667s_3 + 1s_4$$

$$\text{subject to} \qquad 12s_1 + 13s_2 + 12s_3 + 11s_4 \geq 30$$

The heuristic solution is: $s_1 = s_2 = s_3 = 1$ or $S = \{1, 2, 3\}$ with the value of $z = 0.667$, which is less than 1. Therefore, the corresponding cover inequality, $x_1 + x_2 + x_3 \leq 2$ cuts off the current LP solution, as required.

Cover inequalities are included, as an option, in most of the higher quality commercial packages. These implementations usually develop as many cover inequalities as possible in a preprocessor, and then solve the reduced problem using branch-and-bound or other techniques. Some implementations may use the technique repeatedly, after each iteration of branch-and-bound.

In large practical test problems, Crowder et al. (1983) have discovered that the main advantage of cover inequalities does not rely on getting 0–1 extreme points. However, the objective function value for the resulting LP is much closer to the final integer optimum. In other words, the cover inequalities appear to be defining very strong cuts into the feasible region. This has a dramatic effect on the branch-and-bound routine because tree nodes will now be fathomed much earlier, and the bounds will tend to be considerably stronger. As mentioned at the outset, it is possible to solve pure 0–1 problems with up to 10,000 0–1 variables to optimality in a reasonable amount of computer time.

Since then, many other inequalities have been developed and incorporated into commercial software. We are now solving problems with millions of 0–1 variables routinely. See Johnson et al. (2000) for several additional examples.

4.7 Lagrangian Relaxation

4.7.1 Relaxing Integer Programming Constraints

At each node of the branch-and-bound algorithm, we solved a **relaxation** of the corresponding integer programming problem, relaxing the *hard* constraints to produce an *easy* subproblem. Namely, we *relaxed* the integer constraints, and solved the resulting LP. The solution to the easier problem is an upper bound on the original (maximization) problem because we have ignored some of the original restrictions.

With **Lagrangian relaxation**, we find that it is not always necessary to relax the integer constraints. In some special problem instances, we could relax other constraints and leave the integer restrictions in the problem, and still produce an easy *integer* problem. Recall from Chapter 3 that some integer problems, such as network problems, can be easy to solve.

Consider the following general definition of an integer programming problem:

$$\text{maximize} \quad z = c^T x$$

$$\text{subject to} \quad Ax \le b$$

$$Dx \le e$$

$$x \text{ integer}$$

This formulation is the same as before except that we have divided the set of constraints into two groups. Assume that the constraints of the form $Ax \le b$ are relatively easy, while the constraints $Dx \le e$ are *hard*. If we could ignore the second set of constraints, then the integer problem would be easy to solve.

Unlike the LP relaxation, we will not ignore the hard constraints completely. Instead, we will add a *penalty* term to the objective function that adds a *cost* for violating these restrictions. This penalized function is called the **Lagrangian** and is written in the form:

$$\text{maximize} \quad L(x,u) = c^T x - u^T (Dx - e)$$

$$\text{subject to} \quad Ax \le b$$

$$x \text{ integer}$$

$$u \ge 0$$

The vector u contains one entry for each of the constraints in the set $Dx \le e$. The variable u_i represents the *penalty* associated with violating constraint i in this group. Observe that, if we choose any fixed values for these penalties, then the resulting function becomes a linear function of x, and because the remaining constraints are easy, we can maximize this function with respect to x.

To simplify the discussion, suppose that there is only one hard constraint:

$$\sum_{j=1}^{n} d_j x_j - e \le 0$$

and therefore, the penalty u is a single scalar term. Initially, set u = 0 and solve the easy integer problem ignoring the hard constraint. Having done this, we are likely to discover that the solution violates the hard constraint, which means that:

$$\sum_{j=1}^{n} d_j x_j - e > 0$$

If we now keep x fixed and increase u, we will *decrease* or *penalize* the Lagrangian function.

Suppose we now choose some *fixed* positive penalty value for u, and rewrite the Lagrangian as a function of x:

maximize 　　$L(x,u) = \sum_{j=1}^{n} \left(c_j - u d_j \right) x_j + u e$

subject to 　　$Ax \leq b$

　　　　　　　x integer

This problem is, once again, an easy integer problem for any fixed value of u. The penalty on the hard constraint will eventually force the LP to move to an integer solution that is feasible when u is large enough.

If we make u too large, the term (dx − e) becomes negative. That is, if we put too much emphasis on satisfying the constraint, it will be over-satisfied, and we will have gone too far. The value of u is no longer penalizing the objective function. Larger values of u will now *increase* L(x, u). At this point, we can penalize the objective function by using a *smaller* value of u.

The optimal value of the Lagrangian function is expressed as a *min-max* problem:

minimize 　　maximum 　　$L(x,u)$
$u \geq 0$ 　　　　x integer

which means that we want to find the value of u that has the greatest penalty effect on L(x, u). This problem in itself is rather difficult; however, we can take advantage of the fact that, when we fix u and maximize over x, the problem is easy. Similarly, when we fix x, and minimize over u, the problem becomes an unconstrained linear function of u, and is also easy to solve. More accurately, it is easy to decide whether u should increase or decrease (if possible) to minimize L(x, u).

4.7.2 A Simple Example

Consider the following example problem, which is illustrated in Figure 4.9:

maximize 　　$z = x_1 + 2x_2$

subject to 　　$2x_1 + x_2 \leq 2$

　　　　　　　$x_1, x_2 = 0 \text{ or } 1$

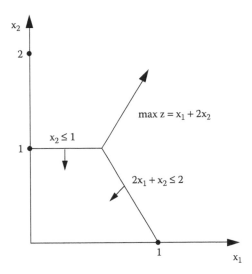

FIGURE 4.9
Simple Lagrangian problem.

Observe that, if a 0–1 problem does not have any constraints, it is trivial to maximize. That is, if the objective function coefficient c_j is positive, then set $x_j = 1$; otherwise, set $x_j = 0$. We can express the problem in Lagrangian form as:

$$\underset{u \geq 0}{\text{minimum}} \quad \underset{x \in (0,1)}{\text{maximum}} \quad x_1 + 2x_2 - u\left(2x_1 + x_2 - 2\right)$$

We begin with $u = 0$, and note the maximum of the problem is $L(x, 0) = 3$ with $x_1, x_2 = 1$. However, this point violates the constraint, so we substitute these values of x into the Lagrangian, and consider the result as a function only of u.

$$\underset{u \geq 0}{\text{minimum}} \ 3 - u$$

This function can be minimized by choosing u as large as possible. We could try $u = 5$, for example; and when we substitute this value into the original Lagrangian, we get:

$$\underset{x=(0,1)}{\text{maximum}} \ x_1 + 2x_2 - 5\left(2x_1 + x_2 - 2\right)$$

$$= \underset{x=(0,1)}{\text{maximum}} \ -9x_1 - 3x_2 + 10$$

The optimal solution to this problem is to set both decision variables to zero. The corresponding function value is $L(x, 5) = 10$. This time, when we substitute x into the Lagrangian in terms of u, we find:

$$\underset{u \geq 0}{\text{minimum}} \ 0 - u(0 - 2) = 2u$$

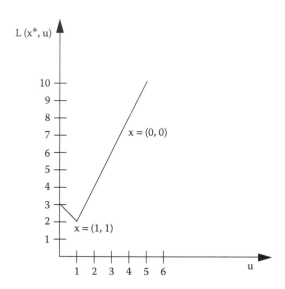

FIGURE 4.10
L(x*, u) as a function of u.

This subproblem tells us to *decrease* u as far as possible. We already know however that when u = 0, it will tell us to increase u. So, the correct value of u must lie somewhere between these two extremes.

Observe that, for any value of u, we can solve for x and find the value of L(x, u). Figure 4.10 illustrates what we have learned so far about L(x, u) as a function of u.

Recall that we want to minimize L(x, u) as a function of u. When u = 0, we found x = (1, 1) and the function was defined by the decreasing line as u increases. This expression is valid as long as x = (1, 1); but then at some point, the optimal solution for x changes, and we get a new linear function describing L(x, u). We now know what that linear function is when u = 0 and u = 5, yet we do not know how it behaves in between these two points. The two line segments in Figure 4.10 represent our best guess at the moment. In particular, it looks as if the minimum value of the Lagrangian will be found when u = 1, so we try that next.

Substituting u = 1 into the Lagrangian gives:

$$\text{maximum } x_1 + 2x_2 - 1\left(2x_1 + x_2 - 2\right)$$
$$\text{x=(0,1)}$$

$$= \text{maximum} - x_1 + x_2 + 2$$
$$\text{x=(0,1)}$$

The maximum of this function is L(x, 1) = 3 when x = (0, 1). If we substitute x = (0, 1) into the original function, we get:

$$L(x, u) = 2 - u(-1) = 2 + u$$

This new section of the Lagrangian is added to the previous approximation to get the function illustrated in Figure 4.11.

From this function, we obtain a new estimate of the minimum value of u = 0.5. Once again, we substitute this value into the Lagrangian and solve for x.

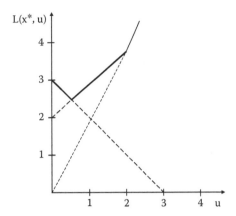

FIGURE 4.11
L(x*, u) as a function of u.

$$\text{maximum } x_1 + 2x_2 - 0.5\left(2x_1 + x_2 - 2\right)$$

$$x = (0,1)$$

$$= \text{maximum } 1.5x_2 + 1$$

$$x = (0,1)$$

The maximum of $L(x, 0.5) = 2.5$ occurs at $x = (0, 1)$ or $x = (1, 1)$. It is easy to verify that this is the true minimum of the Lagrangian. That is, we will not find any new solutions that we have not already described in Figure 4.11.

Let us summarize a number of very useful properties of the Lagrangian, and indicate how we can make use of these properties.

- The Lagrangian method always finds an integer solution, although the solution found is not necessarily feasible.
- If the solution, x^I is feasible, and if the original function, z^I at x^I is equal to the value of the Lagrangian, then x^I is optimal for the original integer problem.
- Most important, if z^* is the solution of the relaxed LP, $L(x, u)$ is the optimal solution to the Lagrangian, and z^I is the (unknown) optimal integer function value, then

$$z^I \leq L(x, u) \leq z^*$$

A proof of these relationships can be found in Fisher (1985).

In other words, the value of the Lagrangian always gives a bound on the optimal integer solution that is at least as good as the LP. Therefore, if we use the Lagrangian instead of the LP in any branch-and-bound algorithm, we may get better results. The LP bound is *never* better than the bound from the Lagrangian. In our simple example problem, the optimal integer function value $z^I = 2$ when $z^I = (0, 1)$. The LP solution occurs at $x = (0.5, 1)$ with $z^* = 2.5$. The LP and the Lagrangian both give the same upper bound.

4.7.3 The Integrality Gap

Let L^I be the optimal solution to the Lagrangian when the integer variables are forced to take integer values, and let L^* be the optimal solution to the Lagrangian when the integer variables are allowed to take on real values (i.e., when we drop the integer constraint on the Lagrangian). It can be proved that the optimal solution for L^* is precisely equal to the optimal solution z^* to the LP. (In fact, the penalty terms u in the Lagrangian will be identical to the corresponding dual variables in the constraints.) Therefore, we can expand the preceding inequality to be:

$$z^I \le L^I \le L^* = z^*$$

We use the term **integrality gap** to describe the difference between L^I and L^* which is the amount by which the Lagrangian decreases when we add the integer constraints. In the example problem, when we solved the Lagrangian without integer restrictions, we obtained integer solutions anyway. Adding the integer constraints does not change the function value. There is no integrality gap. Because the optimal solution to the Lagrangian is equal to the LP solution in this example, the Lagrangian will never give a better bound. Indeed, we saw that $z^* = 2.5$, and $L^I = 2.5$. When we construct the Lagrangian to get an *easy* integer problem, we actually do not want it to be *too* easy; we want an integrality gap so that the Lagrangian bound is better than the LP bound. We provide an example of this type of Lagrangian function in the next section.

4.7.4 The Generalized Assignment Problem

Consider the generalized assignment problem which was introduced and formulated in Section 4.3.5. This problem is similar to the standard assignment problem, where we want to assign jobs to machines for example, except that, in this case, we can assign more than one job to the same machine subject to some capacity limitations.

The generalized assignment problem has a wide variety of practical applications. We could, for example, be assigning computer programs to a set of processors, or customer orders to a set of delivery vehicles, or university students to sections of a course. The capacity might represent a space restriction (in the truck) or a limit on total available time. The problem formulation can be written as follows:

$$\text{maximize} \quad z = \sum_{i=1}^{n} \sum_{j=1}^{m} c_{ij} x_{ij}$$

$$\text{subject to} \quad \sum_{j=1}^{m} x_{ij} = 1 \text{ for } i = 1, 2, \ldots, n$$

$$\sum_{i=1}^{n} a_i x_{ij} \le b_j \text{ for } j = 1, 2, \ldots, m$$

$$x_{ij} = 0 \text{ or } 1 \text{ for all } i, j$$

where there are n jobs and m machines. The variables $x_{ij} = 1$ if job i is assigned to machine j. The payoff for assigning job i to machine j is c_{ij}. Each machine has a capacity b_j, and each job requires a_i units of machine capacity. The first set of constraints forces each job to be assigned to exactly one machine, while the second set of constraints ensures that no machine has its capacity exceeded.

In the standard assignment problem, the size of each job and the capacity of each machine are equal to one. We have already seen in Chapter 3 that the basic assignment problem is relatively easy to solve. Surprisingly, when we generalize the machine capacity constraint, we create an integer programming problem that is difficult to solve. The LP solution is not necessarily an integer solution.

A straightforward formulation of the Lagrangian is to move the capacity constraints into the objective function:

$$L_1(x, u) = \min_{u \geq 0} \max_{x=(0,1)} \sum_{i=1}^{n} \sum_{j=1}^{m} c_{ij}x_{ij} - \sum_{j=1}^{m} u_j \left(\sum_{i=1}^{n} a_i x_{ij} - b_j \right)$$

$$\text{subject to} \sum_{j=1}^{m} x_{ij} = 1 \text{ for } i = 1, 2, \dots, n$$

When $u_j = 0$, this problem is trivial to solve. We can consider each job independently, and simply put it on the best machine (with the highest c_{ij}). This solution will generally violate some of the capacity constraints, so we can increase the corresponding penalty terms, u_j, and construct a new simple problem with:

$$\overline{c}_{ij} = c_{ij} - u_j a_i$$

This penalizes placing all jobs on the machines whose capacities are exceeded. Now, we solve this new problem where we again place each job on the best possible machine using the values c_{ij}. Unfortunately, this formulation is a little too easy. The solution of the Lagrangian (in terms of x) would give 0–1 answers even if we solved it as an LP. Therefore, there is no *integrality gap* and the optimal Lagrangian function value will be the same as the LP function value for the original problem. The corresponding Lagrangian will not produce better bounds than the LP.

The same problem could also be formulated in the following way as a Lagrangian:

$$L_1(x, u) = \min_{u} \max_{x=(0,1)} \sum_{i=1}^{n} \sum_{j=1}^{m} c_{ij}x_{ij} - \sum_{j=1}^{m} u_i \left(\sum_{i=1}^{n} x_{ij} - 1 \right)$$

$$\text{Subject to} \sum_{i=1}^{n} a_i x_{ij} \leq b_j \text{ for } j = 1, 2, \dots, m$$

This formulation can be interpreted as considering each machine separately. Initially, we start with $u = 0$ and assign the best possible jobs to each machine without violating the capacity restrictions. Each machine can be thought of as defining an independent knapsack problem. Although the knapsack problem is not as easy as the simple assignment solution that we used in the previous formulation, it is still a relatively easy problem in many practical situations.

The solution obtained will generally assign some jobs to more than one machine and other jobs will be unassigned, which are both infeasible because every job must be assigned to exactly one machine in any feasible solution. When a job i is assigned to more than one machine, the corresponding penalty term will be positive and we can use a positive value of u_i to penalize the Lagrangian. However, when a job i is unassigned, the term will be equal to -1, and we use a negative value of u_i to penalize the infeasibility. Thus, we do not restrict u_i to have a non-negative value.

In this formulation, if we solve the Lagrangian as an LP, we will get a fractional solution. In particular, each knapsack (machine) may have one fractional part of a job assigned to it. By solving the Lagrangian as a sequence of knapsack problems, we get an integer solution, and therefore, the problem will, in general, have an integrality gap. The integer restriction on the Lagrangian will decrease the objective function value. Hence, the Lagrangian will give a better upper bound than the standard LP bound.

This approach has been used successfully by Fisher et al. (1986) to obtain practical solutions to the vehicle routing problem, in which a given set of customer orders must be assigned to delivery trucks. Each order takes a fixed amount of space in the truck, and there is a capacity restriction on the size of each vehicle.

4.7.5 A Basic Lagrangian Relaxation Algorithm

A succinct general description of a Lagrangian relaxation algorithm is given in the following. We omit implementation details because specific implementations vary considerably, depending on the application.

1. Select an initial value for u^0 (say $u^0 = 0$), and find the maximum of the Lagrangian with respect to x with u fixed. Suppose the solution is L^0 at x^0. Define $k = 0$ to be the current iteration.
2. Substitute the current solution x^k into the Lagrangian objective function to get a linear function of u. If the i-th coefficient of u is negative, then the Lagrangian can be reduced by increasing the i-th component of u^k. If it is positive, then we can decrease the Lagrangian by decreasing the i-th component of u^k provided it is feasible to do so.
3. Determine a value of u^{k+1} such that the Lagrangian $L^{k+1} < L^k$. (There are many methods for doing this, some of which rely on trial and error.)
4. If no decrease can be found, stop. Otherwise, set $k = k + 1$, and go back to step 2.

4.7.6 A Customer Allocation Problem

We will illustrate the basic method of Lagrangian relaxation by solving a distribution problem. Many companies operate multiple distribution warehouses to supply products to their customers. One of the common problems facing such companies is to determine which set of customers should be assigned to each warehouse. Because of the additional delivery costs, it usually does not make economic sense to have a customer's demand satisfied by more than one warehouse. This is referred to as a **single sourcing** constraint.

Consider a delivery problem in which four customers must be served from three warehouses. The cost of satisfying each customer from each warehouse is illustrated in the following table. Each customer has a demand that must be met, and each warehouse has a maximum capacity.

Customers	Warehouses			Demand d_i
	1	2	3	
1	475	95	665	19
2	375	150	375	15
3	360	180	180	12
4	360	180	360	18
Capacity b_j	18	27	20	

The problem can be formulated as a generalized assignment problem where $x_{ij} = 1$ if customer i is served by warehouse j. Every customer must be served by exactly one warehouse, and every warehouse has a capacity constraint on the set of customers it can service.

$$\text{minimize} \qquad \sum_i \sum_j c_{ij} x_{ij}$$

$$\text{subject to:} \qquad \sum_j x_{ij} = 1 \qquad \text{for all customers i}$$

$$\sum_i d_i x_{ij} \le b_j \qquad \text{for all warehouses j}$$

$$x_{ij} = 0 \text{ or } 1$$

If we solve the problem as an LP with ($0 \le x_i \le 1$), we get a total cost of 890, but two of the customers are served from two warehouses. This violates the 0–1 constraint on the variables.

We construct a Lagrangian function by penalizing the customer constraints:

$$L(x, u) = \text{maximum}_u \ \text{minimum}_{x=\{0,1\}} \sum_i \sum_j c_{ij} x_{ij} + \sum_i u_i \left[\sum_j x_{ij} - 1 \right]$$

$$\text{subject to:} \qquad \sum_i d_i x_{ij} \le b_j \qquad \text{for all warehouses j}$$

or, equivalently:

$$L(x, u) = \text{maximum}_u \ \text{minimum}_{x=\{0,1\}} \sum_i \sum_j \left[c_{ij} + u_i \right] x_{ij} - \sum_i u_i$$

$$\text{subject to:} \qquad \sum_i d_i x_{ij} \le b_j \qquad \text{for all warehouses j}$$

Observe that because this problem is a *minimization* in x, we construct the Lagrangian as a *maximization* in u. When we substitute any fixed value of u into the Lagrangian, this problem becomes a simple multiple knapsack problem. We can treat each warehouse as an independent knapsack problem, and find the least expensive customers for that warehouse. However, because we have dropped the customer constraint, there is no reason why a customer cannot be assigned to more than one warehouse, or in fact, to no warehouse. In particular, if we set u = 0 initially, we discover that the optimal solution is x = 0. (No customers are assigned to any warehouse!) To make customers *attractive* to the warehouses, at least some of the costs must be negative; that is, we must choose initial values for the u vector to be negative enough to make some of the Lagrangian costs negative. We will choose u = (−475, −375, −360, −360), a somewhat arbitrary choice, but one in which the new Lagrangian costs have at least one negative cost for every customer. We can subtract the second smallest cost in each row to ensure that the smallest cost will be negative. (In the first row, we can subtract 475 from each element.) Now, every customer is *desired* by at least one warehouse. The new Lagrangian costs are:

	Warehouses			
Customers	1	2	3	Demand d_i
1	0	−380	190	19
2	0	−225	0	15
3	0	−180	−180	12
4	0	−180	0	18
Capacity b_j	18	27	20	

When we solve the knapsack problem for each warehouse we find:

Warehouse 1: Does not take any customers (all costs are zero).

Warehouse 2: Would like to take all of them, but can take only customers 2 and 3 due to capacity constraints for a *cost* of −405.

Warehouse 3: Takes customer 3 for a cost of −180.

The value of the Lagrangian function is the sum of these costs minus the sum of the penalties, u_i: $0 − 405 − 180 − (−1570) = 985$. This first approximation is already a better bound on the solution than the LP solution, which has a value of 890.

When we now examine the customer constraints, we see that no warehouse took customer 1 or 4, and two warehouses took customer 3. To encourage at least one warehouse to take customers 1 and 4, we want to decrease the cost for those customers (that is, decrease u_1 and u_4).

In order to decrease the number of warehouses that want customer 3, we increase the cost slightly. There are many popular methods for doing this, but they all essentially involve trial and error. We can change all three u_i values at once, or we can change them one at a time. We can take small steps, and keep increasing them until the Lagrangian stops increasing, or we can take large steps (too far) and then back up. Without elaborating on details, we will briefly illustrate the first couple of steps.

Suppose we decide to change the three u_i values by 200. (A small change by 1 or 2 does in fact increase the Lagrangian.) Then, the new u values are (−675, −375, −160, −560) and the costs will be:

	Warehouses			
Customers	1	2	3	Demand d_i
1	−200	−580	−10	19
2	0	−225	0	15
3	200	20	20	12
4	−200	−380	−200	18
Capacity b_j	18	27	20	

The three knapsack problem solutions are:

Warehouse 1: Takes customer 4 (customer 1 will not fit) for a cost of −200.

Warehouse 2: Takes customer 1 for a cost of −580.

Warehouse 3: Takes customer 4 for a cost of −200.

The value of the Lagrangian is: −200 − 580 − 200 − (−1770) = 790. We thought that we were moving in a direction that increased the Lagrangian; and, in fact, the Lagrangian will increase for the fixed previous value of x. Unfortunately, as we continue to increase the change in u, we eventually get a new minimum solution x, and the Lagrangian starts to decrease. Apparently, we have gone too far; so let us try again, using a smaller change for the values of u by 10. The new u vector is: (−485, −375, −350, −370), and the resulting cost matrix is:

Customers	Warehouses			Demand d_i
	1	2	3	
1	−10	−390	180	19
2	0	−225	0	15
3	10	−170	−170	12
4	−10	−190	−10	18
Capacity b_j	18	27	20	

The knapsack solutions are:

Warehouse 1: Takes customer 4 for a cost of −10.

Warehouse 2: Takes customers 2 and 3 for a cost of −395.

Warehouse 3: Takes customer 3 for a cost of −170.

The Lagrangian function is: −10 − 395 − 170 − (−1580) = 1005. At this stage, customer 1 is still unserved, and customer 3 is still served by two warehouses. Decreasing u_1 further and increasing u_3 should lead to a further increase in the Lagrangian.

In fact, the value of the optimal solution to the Lagrangian for this problem is 1,355, which also happens to be the optimal integer function value (with customer 4 assigned to warehouse 1; customers 2 and 3 to warehouse 2, and customer 1 to warehouse 3). Thus, for this particular example problem, the Lagrangian bound is tight.

4.8 Column Generation

Many integer programming problems can be stated as a problem of determining what patterns or combinations of items should be assigned to each of a set of *orders*. Problems of this type arise frequently in some of the most important industrial and organizational applications, and are typified by the following examples.

In problems involving **vehicle routing**, customer orders are to be assigned to trucks and routes. A pattern might be a set of customers that could feasibly fit on one truck load (and be delivered by a driver without violating any workday or time delivery constraints).

In **airline crew scheduling**, work pieces (flight legs) must be assigned to airline crews (teams including pilots, navigators, flight attendants, etc.). A pattern might be one (or several) day(s) of work for one crew consisting of several feasible flight legs (with constraints for required rest time between flights, layovers, constraints on legal flying hours per day, etc.).

Various **cutting stock problems** involve choosing which orders should be cut from each piece of stock material. In this context, a pattern would include a set of *orders* that could be cut from one piece of material. The orders might be pieces of fabric cut out for dresses, or large rectangular sheets of paper cut from a large roll.

An example of a **shift scheduling problem** is determining how to assign hospital work shifts to nurses or doctors. In shift scheduling, a pattern might consist of a feasible set of shifts that a nurse could work over a two week rotation.

Each of these problems could be solved in the following way:

1. Construct *all possible* feasible assignment patterns.
2. Define $x_i = 1$ if we decide to use pattern i.
3. Define c_i to be the total cost of using pattern i.
4. Define $a_{ij} = 1$ if customer/order/leg/shift j is included in pattern/route/work-stretch i.

To simplify the discussion, we will use the example of vehicle routing. Given a set of customer orders that will be assigned to one truck, we can calculate the (minimum) cost of paying a driver to visit all of the locations and return to the warehouse. We could then solve the following 0–1 integer programming problem:

$$\text{minimize} \qquad \sum_{i=1}^{n} c_i x_i$$

$$\text{subject to} \qquad \sum_{i=1}^{n} a_{ij} x_i = 1 \text{ (for each customer j)}$$

Customer j may be included in many different possible routes. We want to find a minimum cost set of routes such that every customer is covered exactly once. This type of problem, called a **set partitioning problem**, has very special structure; and there are a number of specialized software codes for this problem that can solve extremely large problem instances (with millions of variables) optimally (Barnhart et al. 1998).

For small enough problem instances, the exhaustive enumeration or construction procedure suggested earlier might be a reasonable way to find optimal solutions. Unfortunately, the number of possible routes is an exponential function of the number of customers. Count the number of ways you can feasibly select a subset of customers, and you will discover that this approach is not at all practical.

Instead, we are going to begin by constructing a small subset of potential routes. It is important here that the number of routes be greater than the number of customers, but not exponential. These routes should include each customer at least a couple of times; but the routes do not have to be particularly good ones. The usual procedure is to use a simple heuristic to construct reasonable routes.

We now solve this problem as a linear programming problem (with $0 \leq x_i \leq 1$), and then use the *dual* values to help us find a new route (column). We add this new column to the problem and solve the linear program again. We continue this process until no new column can be added, and we then solve the 0–1 integer problem optimally. This final problem does not give the optimal solution to the original problem because it typically accounts for only a small fraction of the possible feasible routes. However, the solution to the linear program *is* optimal in the sense that there is no new column that can be added that could reduce the cost of the linear programming problem. Because the LP is a lower bound on the IP, the true integer solution is bounded by the optimal LP and the IP solution that we obtain.

Consider the following simple vehicle routing example. Suppose that a fleet of trucks must travel on a square grid road network, and each edge in the road network takes one hour of travel time. Each driver can travel at most 10 hours. Each truck must begin at the depot (marked with the letter "D"), visit one or more customers (marked with numbers from "1" through "6"), and then return to the depot. The network is illustrated in Figure 4.12. In this network, for example, the route from "D" to "1" to "D" will take six hours; the route from "D" to "4" to "5" to "D" will take eight hours; and the route from "D" to "1" to "2" to "D" will take 10 hours.

To initiate the procedure, select at least six feasible routes, and compute the cost of each route. These routes form the initial set of columns. We have chosen the following set of columns, where each customer is in two routes. (We intentionally chose poor routes to illustrate that the quality of these routes does not matter at this point, although normally, reasonable routes should be selected.)

	Cost						
	10	8	10	10	10	10	
Customer	x_1	x_2	x_3	x_4	x_5	x_6	RHS
1	1					1	1
2					1	1	1
3				1	1		1
4	1	1					1
5			1	1			1
6		1	1				1

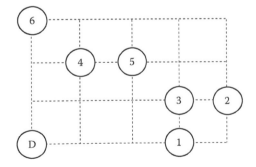

FIGURE 4.12
Vehicle routing example.

Solving this as a linear program gives: $z = 28$; $x_2, x_4, x_6 = 1$. The solution is integral by coincidence (with $0 \le x_i \le 1$). The dual variables are $(-2, -10, 0, -8, -10, 0)$. First, we will illustrate how we solve this problem. Later, we will illustrate *why* it works.

The duals (one for each row implies one for each customer) represent the sensitivity of the objective function to changes in the RHS. (Increasing the RHS by 1 would result in an increase in the objective function of the corresponding dual variable). In our case, it would decrease the objective function because the duals are negative.

Consider the following special problem of finding a single route that starts at the depot, visits some customers, and returns to the depot in at most 10 hours. The *cost* of the route is the total time; however, for every customer that is visited, increase the cost by the corresponding dual variable. For example, a route that goes from "D" to "4" to "5" to "D" will cost $8 - 8 - 10 = -10$. We claim that if we had initially added a column with customers 4 and 5, and then computed the Simplex tableau for the current basic feasible solution, the new reduced cost would be precisely -10. Since it is negative, it can immediately enter the basis.

	Cost							
	10	8	10	10	10	10	8	
Customer	x_1	x_2	x_3	x_4	x_5	x_6	x_7	RHS
1	1					1		1
2					1	1		1
3				1	1			1
4	1	1					1	1
5			1	1			1	1
6		1	1					1

Solving this as a linear program again gives: $z = 28$; $x_2, x_4, x_6 = 1$. However, the dual variables are now $(-10, 0, -10, 0, 0, -8)$. Again, by inspection, we find a route with a negative reduced cost. The new column corresponds to "D" to "1" to "3" to "D" for a cost of $8 - 10 - 10 = -12$.

Solving this as a linear program again gives: $z = 27$; $x_2, x_3, x_5, x_6, x_7, x_8 = 0.5$. The new dual variables are now $(-4, -6, -4, -3, -5, -5)$. By inspection again, we find a route with a negative reduced cost. The best new column corresponds to "D" to "5" to "4" to "6" to "D" for a cost of $10 - 5 - 3 - 5 = -3$.

After a few more iterations, we find a solution with $z = 20$ and $x_9, x_{10} = 1$ corresponding to two routes: customers $\{1, 2, 3\}$ and $\{4, 5, 6\}$. The final dual variables are: $(0, -2, -8, 0, -2, -8)$. The reader should verify that there are no feasible routes with a negative reduced cost, and therefore, this is the optimal solution to the LP. In fact, because this is by chance an integer solution, it is also the optimal integer solution.

	Cost									
	10	8	10	10	10	10	8	8	10	10
Customer	x_1	x_2	x_3	x_4	x_5	x_6	x_7	x_8	x_9	x_{10}	RHS
1	1					1		1	1		1
2					1	1			1		1
3				1	1			1	1		1
4	1	1					1		1		1
5			1	1			1		1		1
6		1	1						1		1

Normally, column generation produces a fractional LP solution, and no new column can be created with a negative reduced cost. This means that no such column exists. Column generation is an optimal procedure for the linear programming problem. Moreover, the optimal solution to the LP is a lower bound on the optimal solution to the corresponding integer programming problem.

Current software packages with column generation use specialized software to solve the resulting partitioning problem optimally. Methods have been designed that use the special structure of the problem to solve very large problems to optimality. The solution is optimal for the given set of columns. There is no guarantee that there is no new column that could be added to produce a lower integer answer. However, the integer function value is often quite close to the linear function value. In general practice, column generation tends to produce very good solutions.

4.9 Software for Integer Programming

An essential component in any solver for integer programs or mixed integer programs is the underlying linear programming solver used for generating lower bounds, separating and selecting subproblems. Dramatic improvements in LP solvers, coupled with faster, more powerful hardware, have led to a wide range of software for integer programs, incorporating a variety of the techniques discussed earlier in this chapter. Performance enhancements have been remarkable, but software solvers for integer programming problems still are often characterized by their requirements for significant amounts of memory and computation time.

No one strategy works for all integer programming models. The cutting plane methods that successfully exploit the special structure of the traveling salesman problem are not the same techniques that would be effective on an integer problem having a different structure. Commercial codes have the advantage of ease of use, but for many practical (large-scale) integer problems, successful solution may require a skillful analyst to develop a customized solver, based on just the right branching strategy, bounding strategy, or tailored cutting plane mechanism.

AIMMS software offers an integrated modeling system for solving pure integer, mixed integer, and 0–1 programming problems. Input and output are supported by the algebraic modeling language and graphical user interface for which AIMMS is well-known. AIMMS modeling and optimization platform supports building and solving problems for applications such as workforce and financial portfolio optimization, production planning and scheduling, logistics, and transportation. AIMMS allows analysts to evaluate multiple action plans and assess the impact of different and continuously changing scenarios. **LINDO Integrated Modeling Language** is a comprehensive tool for expressing integer optimization models. **LINDO API** includes an integer solver that works together with linear, nonlinear, and quadratic solvers.

IBM ILOG CPLEX Optimizer (commonly referred to as CPLEX) is a powerful suite of solvers including solvers for integer and mixed programming that can run on different platforms. The CPLEX solvers have been used to solve large real-life optimization problems with millions of variables and constraints. They are often integrated with convenient and powerful modeling languages, such as GAMS and AMPL modeling systems for large-scale optimization of linear and nonlinear mixed integer programs, a combination that offers advanced features for solving difficult integer programming problems for which other software systems may be inadequate.

SAS/OR systems, described in previous chapters, also have capabilities for solving pure, mixed, and 0–1 integer programming problems. **SAS OPTMODEL** provides an integrated modeling environment, with special features for solving mixed integer problems by using parallel branch-and-bound techniques with cutting planes and heuristics, and decomposition algorithms for mixed integer programming problems.

MIPIII Mixed-Integer Optimizer from Ketron Management Science allows the user to match the problem structure to an appropriate solver, and exploits the user's knowledge of the model through the use of pre-solvers, special approaches for knapsack problems, the use of branching priority lists, and a choice of stopping criteria.

Gurobi Optimizer provides solvers for mixed integer solutions of linear, and quadratic programs. It uses advanced implementations of new MIP algorithms using parallel non-traditional search techniques and cutting planes.

Google OR Tools offers an interface to several MIP solvers. By default, it uses COIN-OR branch and cut implementation, an open source solver from the Computational Infrastructure for Operations Research project (COIN-OR). However, one can also use other MIP solvers (such as Gurobi) with Google OR Tools wrapper. Google's OR Tools are offered for various platforms (Windows, Mac OS and Linux) and languages (C++, Java, and Python).

Software for specialized applications often provides unique and convenient user interfaces as well as efficient solution techniques directed specifically to the type of application. For example, software for scheduling systems may yield a competitive edge in manufacturing and production environments. The underlying scheduling methodology forming the backbone of such software systems may be based on a classical 0–1 programming model (matching activities to time slots subject to constraints) or may perform priority or rule based scheduling with release dates, deadlines, or due dates. Other considerations include ease of modeling the processes and operations in a production scheduling system, and the ability to incorporate materials handling, quality assurance, shop floor data, and production activity control subsystems (Seyed 1995).

Many integer programming problems can be viewed as routing problems, and numerous software packages are available to solve problems cast into this framework (Hall and Partyaka 2016, Horner 2018). For an overview of approaches, see Bosch and Trick (2014).

4.10 Illustrative Applications

4.10.1 Solid Waste Management (Antunes 1999)

Along with extensive political, social, and economic changes in Portugal during the past several decades, urban population growth has increased dramatically. Authorities are faced with the resulting problem of disposing of significant amounts of municipal solid waste generated in population centers such as Lisbon, Coimbra, and Oporto. By the 1990s, the Centro Region Coordination Agency was looking at growth rate projections that indicated that the waste management problem would rapidly extend beyond the major urban centers to affect smaller municipalities and rural areas as well.

The *collection* of waste was already being handled effectively; in fact, by 1991, approximately 90% of households were served by modern fleets of vehicles collecting garbage. But *disposal* of the collected waste was being dealt with using the worst possible techniques: huge open air dumps that periodically burned the waste. And whereas hazardous, dangerous, and toxic waste was being managed and monitored by the national government under a separate initiative, the massive amounts of ordinary solid waste were the responsibility of regional authorities. The Centro Region needed to develop a clear view of how solid waste would be managed, from the time it is generated, through the phases of collection and reduction, until it is finally disposed of in a sanitary landfill that is built and operated according to strict and appropriate regulations.

Storage space for solid waste is a major consideration. Volume reduction based on composting is effective only on the organic portion of the waste, which is a small and decreasing proportion of total waste in the Centro Region. Subsequent separation from glass, metal, and plastics represents an added expense in the composting regimen. Incineration is the most effective way of reducing waste volume, but set-up costs are extremely high, and the environmental concerns over fly-ash and hazardous bottom-ash combine to argue against the use of incineration on a large scale.

Compaction is less effective than incineration, but it is cheaper and has the additional advantage that it can be applied early in the process, during either the generation or the collection phases. Thus, compaction can substantially decrease transportation costs between collection points, transfer stations, and landfills.

With these issues in mind, an analyst developed a mixed integer programming model having nearly 10,000 variables, about 100 of which were 0–1 variables, and about 10,000 constraints. The model combines elements of a p-median problem with a capacitated facility location problem. The model included 18 possible sites for sanitary landfill locations, and 86 possible sites for transfer station locations. Problem parameters were based on projections for the year 2014 in order to accommodate anticipated population growth rates.

Multiple objectives were considered during the development of this solid waste management model. On the one hand, it is aesthetically desirable to locate sanitary landfills as far as possible from urban centers (subject to the very legitimate *not-in-my-backyard* reaction of rural residents). But it is also expeditious to keep the landfills as close as possible to waste producers, to minimize costs of transportation.

The minimum cost objective was ultimately given greater weight. In achieving this objective, a number of constraints were imposed. Landfills and transfer stations have a minimum capacity (in tons per day) to take advantage of economies of scale. There is a maximum distance to be traveled by the trucks during their daily collection and transfer trips. Landfills are placed in the municipalities with the largest populations. Finally, collection and transfer trucks are routed to avoid mountainous regions with narrow winding roads, both for economic reasons and out of respect for the fragility of natural resources in the national parks.

Because of the complexity of the model, the analyst initially assumed that it would not be possible to solve the mixed integer problem using a general exact method on the PC equipment available for this study. A greedy heuristic based on capacitated transshipment methods was developed, and the results obtained in this way were included in the initial reports presented to the Centro Region Coordination Agency for consideration. However, a new version of XPRESS-MP software running on slightly faster processors allowed the model to be solved exactly with reasonable computational effort.

The final solution developed in conjunction with the Agency called for eight land-fills, each with a specified capacity, and eight transfer stations, also each having a specified capacity. It was possible to delay the capital investment needed for three of the transfer stations (without violating the maximum truck trip distance constraints) so that initial expenditures could be concentrated on the more urgently needed sanitary landfills.

The results of this study brought some credible and rational order to a previously chaotic situation. The solution developed during this study led the representatives from throughout the Centro Region to adopt a Strategic Municipal Solid Waste Plan that serves as a guide during the present process of implementing the proposed waste management system.

4.10.2 Timber Harvest Planning (Epstein et al. 1999)

The Chilean forestry industry consists primarily of large private firms that own pine and eucalyptus plantations and are vertically integrated, comprising pulp plants, sawmills, and paper market operations. Short-term harvest scheduling (over a three-month period) amounts to matching stands of timber, of a given age and quality, to market demands that are typically defined by the length and diameter of each piece of timber. The process of cutting harvested trees into products having required lengths and diameters is called *bucking*. Bucking sequences are expressed in terms of lengths to be cut from timbers of decreasing diameters.

Different types of harvesting equipment are used in different terrains. Steep slopes require towers or cables, while flat areas can be harvested using tractors or skidders. In either case, bucking can be done on the ground and the resulting pieces transported to their respective destinations, or entire logs can be delivered to a central location for bucking. Transportation costs (which can include road building costs) play a significant role in the decisions that select timber from a certain origin and assign it to a destination.

Determining an optimal harvest plan is a difficult combinatorial problem that involves selecting mature timber stands available at specified locations, and assigning them according to product demand; obtaining optimal bucking patterns to utilize the timber itself in the most valuable way; and minimizing transportation costs, subject to the firm's harvesting equipment limitations and trucking capacities.

A principal component of the harvest plan is the specification of optimal bucking patterns, from among exponentially-many possible patterns. The solution is based on an LP model, and incorporates a branch-and-bound approach using column generation to create the bucking sequences. In the branch-and-bound tree for generating bucking patterns, a path from the root node to the bottom of the tree represents a bucking sequence; the terminal node in the tree represents the product (a piece of timber having a certain diameter cut to required length); and the terminal node's level in the tree denotes the product's position in the bucking process.

The **column-generation** technique improved the harvest value by 3% to 6% over the fixed bucking patterns that had been in use previously when harvest planning was done manually by experienced human planners. Furthermore, transportation costs were cut substantially when the model solution revealed the savings that could be obtained by bucking and shipping directly to market destinations rather than transshipping through intermediate central bucking locations.

Other applications of operations research in the Chilean forestry industry include systems for:

- Scheduling trucks among timber stands, mills, and destination ports.
- Selecting stands for harvest, and partitioning the timber for logs, sawtimber, and pulpwood, using mixed integer LP models.
- Determining the optimal placement of harvesting equipment and the optimal locations of access roads within the forest.
- Long-term planning over a 50-year horizon to maintain steady and consistent supplies of timber, which involves the purchase, sale, and rental of timber lands; choosing appropriate silviculture regimes for different plantations; and planning for mills and other industrial processing plants.

4.10.3 Propane Bottling Plants (Sankaran and Raghavan 1997)

During recent years, the importation, bottling, and distribution of liquefied petroleum gas (LPG) in India has transitioned from a government-controlled operation into a private-sector enterprise. Two major import and storage facilities (ports), already in place, provide supplies of LPG. Industrial customers acquire LPG in bulk directly from these locations, but the needs of other domestic residential and commercial establishments are supplied through a network of dealer agencies. Customers use LPG contained in cylinders, and when empty, these cylinders are picked up by dealers and replaced by filled cylinders. Each dealer town must have a bottling plant where empty cylinders can be replenished for future distribution to customers.

Because the sources of LPG and the customer market are already established, the problem was to determine the pattern and mechanisms for distributing LPG from the two storage facilities to the customers. Tanker trucks can transport LPG from the source to dealer locations for bottling, but it is also feasible to operate mobile bottling plants. Considerations for mobile operations include not only capital investment and operating and distribution costs, but also public safety and firefighting capabilities at all intermediate storage points.

Strategic decisions for dealer and bottling facility location are complicated by the fact that any necessary future increases in capacity at a given location can be undertaken only if such increases are provided for in the original layout. Thus, a significant portion of expansion costs are incurred during original construction, although the payoff from such expansion will not be realized until the projected market growth actually takes place.

The problem facing the Shri Shakti company is optimally locating the bottling plants, determining the long-run size of each facility, and projecting the target date at which each facility will commence operating at full (expanded) capacity. The integer programming model used for this problem involves about 400 dealer towns and 2,500 constraints, and seeks to minimize total cost of operations in the target year. Costs include:

- Fixed annual costs that are independent of volume throughput at the plants
- Costs of transporting LPG from the two ports to the plants
- Cost of bottling
- Costs of transporting bulk and cylinder LPG and empty cylinders among bottlers, dealers, and customers

Determining the amounts of LPG to be distributed through the network dictates the location and size (capacity) of each proposed facility. Complicating the problem were uncertainties about competition, corporate takeovers, market growth, and initially some inaccuracies in data defining the distances between sites.

A solution to this problem was developed using a linear programming-based branch-and-bound method. Subsets of the problem were originally solved in which the subproblems were defined by geographical or political boundaries. Combining these separate solutions, however, often resulted in certain customers being served by distant in-area suppliers instead of by closer plants just across a boundary. In order to remedy this inefficiency, a novel and indirect method was designed for solving the full-scale problem. Specially tailored software routines in Fortran were linked to extended versions of LINDO software for mathematical programming.

Analysts working on this application created a well formulated model, developed a comprehensive and accurate database, and engaged in illuminating discussions with Shri Shakti's board of directors, government advisors, and financial experts during development of these solutions. The credibility of the resulting model and the proposed solutions provided a much-needed foundation for successful planning, negotiating, and funding for this newly privatized industry in India.

4.11 Summary

Many important engineering, industrial, organizational, and financial systems can be modeled as mathematical programming problems in which the variables are restricted to integer values, 0–1 values, or a mixture of integer and real values. Solving integer problems usually requires significantly more computational effort than is needed for solving continuous (real) linear programming problems.

Certain 0–1 models have become rather famous because their structures seem to arise in so many different kinds of practical applications. Specialized methods for solving such problems have been devised that take advantage of the mathematical structure inherent in the problems. These classical models include the traveling salesman problem, knapsack and bin packing problems, set partitioning, and generalized assignment problem. Many complex problems can be solved by identifying subproblems that have the characteristics of these well-known models, and creating a solution to the large and difficult problem by solving some simple subproblems.

Among the most effective methods for solving general integer programming problems are branch-and-bound algorithms. These methods repeatedly break large problems, which are not yet solved, into easier subproblems, imposing integer constraints along the way, until a solution to the original problem is finally found. Solutions to real-valued LP problems are used to guide the process, so that the computation does not escalate into an enumeration of exponentially many possible solutions.

A number of other approaches have been developed and refined over the years. Cutting plane and cover inequality methods repeatedly introduce new constraints into integer problems in order to exclude non-integer extreme points from the feasible region, and then use simple LP solutions to locate the optimum, which then occurs at an integer point. Lagrangian relaxation incorporates constraints into the objective function by placing a penalty on any violated constraint. Any solution that violates a constraint has a lower value than a solution with no constraint violation. The penalties must be chosen appropriately for the given problem. The technique of column-generation is applicable to problems such as vehicle routing and workforce scheduling, in which customers or workers must be assigned to trucks or work patterns. Incomplete initial solutions are iteratively built up into complete optimal solutions.

Most methods for solving integer programming problems rely on solving linear sub-problems using a standard technique such as the Simplex method. Thus, the performance of many integer solution methods depends greatly on the efficiency of the underlying LP methods. Recent improvements in LP solvers have contributed substantially to our present capabilities for solving large practical integer problems efficiently.

Key Terms

active node
airline crew scheduling
assignment problem
backtracking
bin packing problem
binary integer programming
branch-and-bound
branch-and-bound tree
branching strategy
bounding strategy
capacity planning
capital budgeting problem
cargo loading problem
column-generation
convex hull
cover
cover inequality
current incumbent
cutting plane
cutting stock problem
employee scheduling problem
examination timetabling
facet
fathomed
fixed charge problem
flight crew
flight legs
general integer programming
generalized assignment problem
Gomory fractional cut
integer polytope
integer programming
integrality gap
jumptracking

Exercises

4.1 A certain single-processor computer is to be used to execute five user programs. These programs may be run in any order; however, each requires a specific set of files to be resident in main memory during its execution. Furthermore, a certain amount of time is required for each file to be brought into main memory prior to use by a user program. The facts are summarized as follows:

User Program	Files Needed for Its Execution
1	B, C, E
2	A, B, C
3	A, B, D
4	A, D, E
5	B, C

File Name	Amount of Time Required to Bring It into Memory
A	30
B	20
C	25
D	35
E	50

Initially, no files are in memory. The five user programs are to run in sequence, but any order is feasible. At most, three files will fit in memory at one time. Clearly, because some of the files are used by multiple programs, it would be wise to try to schedule the programs to take advantage of files already in memory, so as to minimize the *change-over* (setup) times between programs. Define decision variables and formulate this problem to sequence the five user programs to minimize total change-over times. Note the similarity of this problem to one of the classical integer programming models discussed in this chapter.

4.2 Suppose you have a directed acyclic graph having n nodes, in which node 1 is designated as an origin and node n is designated as a destination. In Chapter 3, we described the problem of finding the shortest path from the origin to the destination. Formulate this problem as a 0–1 integer programming problem. (*Hint:* Let decision variable $x_{ij} = 1$ if the arc from node i to node j is in the shortest path.)

4.3 A small university computer laboratory has a budget of $10,000 that can be used to purchase any or all of the four items described in the following. Each item's value has been assessed by the lab director, and is based on the projected utilization of the item. Use a branch-and-bound technique to determine the optimal selection of items to purchase to enhance the computing laboratory facilities. Show your branch-and-bound tree, and give the total cost and total value of the items chosen for purchase.

Item	Cost	Value
NanoRobot	$4,000	8
WinDoze simulator	$2,500	5
Network pods	$3,000	12
BioPrinter	$4,500	9

4.4 Bruno the Beach Bum wishes to maximize his enjoyment of the seashore by taking along an assortment of items chosen from the following list. Help Bruno pack his beach bag with the most valuable set of items by using a branch-and-bound technique. Bruno's beach bag is rated for a 20-pound load.

Item	Weight	Value
Coconut oil	4	16
Sun shades	2	10
Snorkel and fins	8	16
Folding chair	10	30
Bummer magazine	5	30

Enumerate the number of packings (sets of items) for this problem, and draw a complete tree of possibilities. How many of these sets are feasible packings? How many subproblems are actually solved by your branch-and-bound procedure? What is the optimal feasible set of items?

4.5 If a problem formulation has n variables and each variable has m possible integer values, then a branch-and-bound tree could have as many as m^n terminal nodes. Verify this for the case $m = 4$ and $n = 3$.

4.6   Consider the following 0–1 integer programming problem:

$$\text{maximize} \quad 5x_1 - 7x_2 - 10x_3 + 3x_4 - 4x_5$$

$$\text{subject to} \quad x_1 + 3x_2 - 5x_3 + x_4 + x_5 \le 3$$

$$2x_1 - 3x_2 + 3x_3 - 2x_4 - 2x_5 < -3$$

$$2x_2 - 2x_3 + 2x_4 + x_5 \le 3$$

$$x_i = 0 \text{ or } 1 \text{ for all } i$$

Solve this problem completely, using a branch-and-bound algorithm.

4.7   Suppose you wish to solve the following general integer programming problem using branch-and-bound techniques.

$$\text{maximize} \quad 3x_1 + 5x_2 + 2x_3$$

$$\text{subject to} \quad x_1 + 5x_2 + 3x_3 \le 8$$

$$2x_1 + x_2 + 5x_3 \le 7$$

$$4x_1 + 2x_2 + 3x_3 \le 8$$

$$x_1 + 3x_2 + 3x_3 \ge 6$$

$$x_1, x_2, x_3 \ge 0 \text{ and integer}$$

Use up and down penalties to determine which variable would be branched on first. (*Note:* There is no *correct* answer, but you should be able to justify your choice.)

4.8   Consider the following integer programming problem:

$$\text{maximize} \quad -4x_1 - 5x_2$$

$$\text{subject to} \quad x_1 + 4x_2 \ge 5$$

$$3x_1 + 2x_2 \ge 7$$

$$x_1, x_2 \ge 0 \text{ and integer}$$

Calculate the penalties for branching up and down on variables x_1 and x_2.

4.9   Solve the problem given in Exercise 4.8 using Gomory fractional cuts.

4.10 Consider the following integer programming problem:

$$\text{maximize} \quad -3x_1 - 4x_2$$

$$\text{subject to} \quad 2x_1 + x_2 \ge 1$$

$$x_1 + 3x_2 \ge 4$$

$$x_1, x_2 \ge 0 \text{ and integer}$$

    a. Compute the up and down penalties for branching on variable x_2. Which way would you branch first? Explain why.

    b. What can you say about variable x_3, the surplus variable on constraint 1, with respect to up and down penalties? Explain.

4.11 Suppose that you are solving a large 0–1 linear programming problem, and the LP solution has

$$x^* = (0.3,\ 0.9,\ 0.1,\ 0.9,\ 0.9,\ 0.8,\ 0.9,\ 0.9,\ 0.7,\ 0)$$

One of the constraints in the problem is:

$$10x_1 - 2x_2 - 4x_3 + 7x_4 + -6x_5 - 11x_6 + 9x_7 - 3x_8 + x_9 + 12x_{10} \le -1$$

In Section 4.6, we used a knapsack model to find a cover inequality that cuts off the current LP solution. Describe the knapsack for this particular problem.

4.12 Suppose we are given a 0–1 linear programming problem in which one of the constraints is

$$3x_1 + 4x_2 - 7x_3 - 3x_4 + 5x_5 - 6x_6 + 3x_7 \ge 0$$

Find a cover inequality that cuts off the current LP solution $x^* = (0,\ \frac{1}{2},\ 0,\ 1,\ 1,\ \frac{2}{3},\ 0)$

4.13 A certain 0–1 linear programming problem involves the constraint

$$x_1 + 3x_2 + 4x_3 + 5x_4 \le 6$$

and the current LP optimum occurs at $x^* = (0.3, 0.3, 0.2, 0.8)$.

    Find a minimal cover inequality that cuts off the point x^*.

4.14 Solve the problem in Exercise 4.6 again by first constructing cover inequalities, and then using branch-and-bound if necessary.

4.15 We wish to assign three customers to two warehouses having limited capacity. Each customer must be assigned to precisely one warehouse. The assignment costs and the capacities are given in the following table. Solve this problem using Lagrangian relaxation.

	Warehouse 1	Warehouse 2	Demand
Customer 1	2	8	18
2	5	3	15
3	7	3	14
Capacity	30	18	

4.16 Suppose that you are the manager of a small store that is open seven days per week. You require the following minimum number of staff to work each day:

Sunday	5
Monday	3
Tuesday	4
Wednesday	4
Thursday	5
Friday	7
Saturday	7

Each employee can work only five days per week, and must have the weekend off (Saturday and Sunday) once every two weeks. The objective is to meet the demand using the minimum number of employees. Describe a formulation of this problem using column-generation. (*Hint:* Try to construct a work pattern for a two week period.) Describe the meaning of the rows and columns in the master problem. Provide an initial formulation of the LP; that is, pick a starting set of columns, and write out the LP. Perform a few iterations of column generation. Describe how you would formulate the subproblem.

4.17 In Section 4.8, it was suggested that column generation can be used to solve the cutting stock problem. The simplest (one-dimensional) cutting stock problem can be illustrated by the following example. Suppose we have a large supply of steel reinforcing bars to be used in the construction of concrete pillars. The bars are all 50 feet long. We have a set of orders for re-bars of the following lengths:

Length	Quantity
15 feet	3
10 feet	2
13 feet	5
18 feet	4
19 feet	5
23 feet	1

These orders are to be cut from some of the 50 feet long pieces. It is not economical to keep an inventory of the leftover pieces, so we sell them as scrap. We want to minimize the total cost of scrap for cutting this set of orders. Suppose that it costs (net) 0.50 per inch to throw away a scrap piece of re-bar. Formulate this as a column-generation problem. Generate the initial solution, and perform one iteration of column generation. Explain your algorithm for solving the subproblem.

4.18 Big City Wheel Trans (for disabled public transit users) has a large list of clients who must be picked up and delivered to locations around the city. Each client has a specific required pick-up time, and we assume that each customer travels alone. Describe how to formulate this problem using column-generation. Suppose that the primary objective is to minimize the number of vehicles required to satisfy all demand. Describe what the subproblem would look like and what algorithm you could use to solve it. Recall that the subproblem we solved in Section 4.8 was solved *by inspection*, but in this exercise, you should define an algorithm to solve the subproblem.

4.19 Formulate the examination timetabling problem as a 0–1 programming problem. Let c_{ik} be the number of students who must take both exams i and k. Define a penalty of $(100 \times c_{ik})$ for having examinations i and k in the same time period, and a penalty of $(5 \times c_{ik})$ for having examinations i and k in adjacent time periods. The objective is to minimize the total penalty costs. Let n denote the number of examinations to be scheduled, and m denote the number of time periods available for exams.

References and Suggested Readings

Adams, W. P., and H. D. Sherali. 1990. Linearization strategies for a class of zero-one mixed integer programing problems. *Operations Research* 38 (2): 217–226.

Antunes, A. P. 1999. Location analysis helps manage solid waste in Central Portugal. *Interfaces* 29 (4): 32–43.

Barnhart, C., E. L. Johnson, G. L. Nemhauser, M. W.P. Savelsbergh, and P. H. Vance. 1998. Branch-and-price: Column generation for solving huge integer programs. *Operations Research* 46 (3): 316–329.

Beale, E. M. L. 1979. Branch-and-bound methods for mathematical programming systems. *Annals of Discrete Mathematics* 5: 201–219.

Beasley, J. E. 1993. *Lagrangian Relaxation, in Modern Heuristic Techniques for Combinatorial Problems.* Oxford, UK: Blackwell Scientific.

Benders, J. F. 1962. Partitioning procedures for solving mixed variables programming problems. *Numerische Mathematik* 4: 238–252.

Bosch, R., and M. Trick. 2014. Integer programming. In G. Kendall, and E. Burke (Eds.), *Search Methodologies.* Boston, MA: Springer.

Chen, D., R. G. Batson, and Y. Dang. 2010. *Applied Integer Programming: Modeling and Solution.* Hoboken, NJ: John Wiley & Sons.

Cooper, M. W. 1981. A survey of methods for pure nonlinear integer programming. *Management Science* 27 (3): 353–361.

Crainic, T., and J. Rousseau. 1987. The column generation principle and the airline crew scheduling problem. *INFOR: Information Systems and Operational Research* 25: 136–151.

Crowder, H., E. L. Johnson, and M. Padberg. 1983. Solving large-scale zero-one linear programming problems. *Operations Research* 31: 803–834.

Desrosiers, J., and F. Soumis. 1989. A column generation approach for the urban transit crew scheduling problem. *Transportation Science* 23: 1–13.

Desrosiers, J., F. Soumis, and M. Desrochers. 1984. Routing with time windows by column generation. *Networks* 14: 545–565.

Epstein, R., R. Morales, J. Serón, and A. Weintraub. 1999. Use of OR systems in the Chilean forest industries. *Interfaces* 29 (1): 7–29.

Erlenkotter, D. 1978. A dual-based procedure for uncapacitated facility location. *Operations Research* 26: 992–1009.

Fisher, M. 1981. The Lagrangian relaxation method for solving integer programming problems. *Management Science* 34: 1–18.

Fisher, M. 1985. An applications-oriented guide to Lagrangian relaxation. *Interfaces* 15 (2): 10–21.

Fisher, M.L., R. Jaikumar, and L. N. Van Wassenhove. 1986. A multiplier adjustment method for the generalized assignment problem. *Management Science* 32: 1095–1103.

Francis, R. L., L. F. McGinnis, Jr., and J. A. White. 1992. *Facility Layout and Location: An Analytical Approach*. Englewood Cliffs, NJ: Prentice-Hall.

Garfinkel, R. S., and G. L. Nemhauser. 1972. *Integer Programming*. New York: John Wiley & Sons.

Geoffrion, A. M., and R. E. Marsten. 1972. Integer programming algorithms: A framework and state-of-the-art survey. *Management Science* 18: 465–491.

Glover, F. 1975. Improved linear integer programming formulations of nonlinear integer problems. *Management Science* 22: 455–460.

Gomory, R. E. 1963. *An Algorithm for Integer Solutions to Linear Programming, in Recent Advances in Mathematical Programming*. New York: McGraw-Hill.

Hall, R. W., and J. G. Partyka. Vehicle Routing Software Survey: Higher expectations drive transformation. *OR/MS Today* 43 (1): 40–49.

Hoffman, K. L., and M. Padberg. 1991. Improving LP-representations of zero-one linear programs for branch-and-cut. *ORSA Journal on Computing* 3: 121–134.

Horner, P. 2018. Innovation powers dynamic VR sector. *OR/MS Today* 45 (1): 43–45.

Hu, Te C. 1970. *Integer Programming and Network Flows*. Reading, MA: Addison-Wesley.

Johnson, E. L., G. L. Nemhauser, and M. W. P. Savelsbergh. 2000. Progress in linear programming-based algorithms for integer programming: An exposition. *INFORMS Journal on Computing* 12 (1): 2–23.

Jünger, M., T. M. Liebling, D. Naddef, G. L. Nemhauser, W. R. Pulleyblank, G. Reinelt, G. Rinaldi, and L. A. Wolsey. 2009. *50 years of Integer Programming 1958–2008: From the Early Years to the State-of-the-art*. Heidelberg, Germany: Springer Science & Business Media.

Kolesar, P. J. 1967. A branch-and-bound algorithm for the knapsack problem. *Management Science* 13: 723–735.

Lawler, E. L. (Ed.). 1985. *The Traveling Salesman Problem: A Guided Tour of Combinatorial Optimization*. Chichester, UK: John Wiley & Sons.

Martello, S., and P. Toth. 1990. *Knapsack Problems: Algorithms and Computer Implementations*. Chichester, UK: John Wiley & Sons.

Martin, R. K. 1999. *Large Scale Linear and Integer Programming*. Boston, MA: Kluwer Academic.

Nemhauser, G. L., and L. A. Wolsey. 1988. *Integer and Combinatorial Optimization*. New York: John Wiley & Sons.

Papadimitriou, C. H., and K. Steiglitz. 1982. *Combinatorial Optimization*. Englewood Cliffs, NJ: Prentice-Hall.

Parker, R. G., and R. L. Rardin. 1988. *Discrete Optimization*. Orlando, FL: Academic Press.

Piñedo, M. 1995. *Scheduling: Theory, Algorithms and Systems*. Englewood Cliffs, NJ: Prentice-Hall.

Plane, D., and C. McMillan. 1971. *Discrete Optimization: Integer Programming and Network Analysis for Management Decisions*. Englewood Cliffs, NJ: Prentice-Hall.

Reiter, S., and D. B. Rice. 1966. Discrete optimizing solution procedures for linear and nonlinear integer programming problems. *Management Science* 12 (11): 829–850.

Salkin, H., and K. Mathur. 1989. *Foundations of Integer Programming*. New York: North-Holland.

Sankaran, J. K., and N. R. Raghavan. 1997. Locating and sizing plants for bottling propane in south India. *Interfaces* 27 (6): 1–15.

Schrijver, A. 1986. *Theory of Linear and Integer Programming*. New York: John Wiley & Sons.

Seyed, J. 1995. Right on schedule. *OR/MS Today* 22: 42–44.

Sharda, R. 1993. *Linear and Discrete Optimization and Modeling Software: A Resource Handbook*. Atlanta, GA: Lionheart Publishing.

Syslo, M. M., N. Deo, and J. S. Kowalik. 1983. *Discrete Optimization Algorithms*. Englewood Cliffs, NJ: Prentice-Hall.

Taha, H. A. 1975. *Integer Programming: Theory, Applications, and Computations*. Orlando, FL: Academic Press.

Vielma, J. P. 2015. Mixed integer linear programming formulation techniques. *SIAM Review* 57 (1): 3–57.

Williams, H. P. 1990. *Model Building in Mathematical Programming*, 3rd ed. New York: John Wiley & Sons.

Winston, W. L. 1987. *Operations Research: Applications and Algorithms*. Boston, MA: Duxbury Press.

Wolsey, L. A. 1998. *Integer Programming*. New York: John Wiley & Sons.

5

Nonlinear Optimization

Nonlinear optimization involves finding the best solution to a mathematical programming problem in which the objective function and constraints are not necessarily linear. Because nonlinear models include literally all kinds of models *except linear* ones, it is not surprising that this category is a very broad one, and nonlinear optimization must incorporate a wide variety of approaches to solving problems.

The world is full of systems that do not behave linearly. For example, allowing a tree to grow twice as long does not necessarily double the resulting timber harvest; and tripling the amount of fertilizer applied to a wheat field does not necessarily triple the yield (and might even kill the crop!). In a distributed computing system networked for interprocessor communication, doubling the speed of the processors does not mean that all distributed computations will be completed in half the time, because interactions among processors now could foil the anticipated speedup in throughput.

This chapter examines optimization from a very general point of view. We will consider both unconstrained and constrained models. **Unconstrained optimization** is often dealt with through the use of differential calculus to determine maximum or minimum points of an objective function. Constrained models may present us with systems of equations to be solved. In either case, the classical underlying theories that describe the *characteristics* of an optimum do not necessarily provide the practical *methods* that are suitable for efficient numerical computation of the desired solutions. Nevertheless, a thorough grasp of the subject of nonlinear optimization requires an understanding of both the mathematical foundations of optimization as well as the algorithms that have been developed for obtaining solutions. This chapter is intended to provide insights from both of these perspectives. We will first look at an example of a nonlinear programming problem formulation.

Example 5.1

Suppose we want to determine a production schedule over several time periods, where the demand in each period can be met with either products in inventory at the end of the previous period or production during the current period. Let the T time periods be indexed by $i = 1, 2, \ldots, T$, and let D_i be the known demand at time period i. Equipment capacities and material limitations restrict production to at most E_i units during period i. The labor force L_i during period i can be adjusted according to demand, but hiring and firing is costly, so a cost C_L is applied to the square of the net change in labor force size from one period to the next. The productivity (number of units produced) of each worker during any period i is given as P_i. The number of units of inventory at the end of period i is I_i, and the cost of carrying a unit of inventory into the next period is C_I. The production scheduling problem is then to determine feasible labor force and inventory levels in order to meet demand at minimum total cost. The decision variables are the L_i and I_i for $i = 1, \ldots, T$. The initial labor force and inventory levels are given as L_0 and I_0, respectively. Therefore, we wish to

$$\text{minimize} \quad \sum_{i=1}^{T} C_L \left(L_i - L_{i-1} \right)^2 + C_I I_i$$

subject to

$$L_i \cdot P_i \le E_i \qquad \text{equipment capacities for } i = 1, \ldots, T$$

$$I_{i-1} + L_i \cdot P_i \ge D_i \qquad \text{demand for } i = 1, \ldots, T$$

$$I_i = I_{i-1} + L_i \cdot P_i - D_i \qquad \text{inventory for } i = 1, \ldots, T$$

$$L_i, I_i \ge 0 \qquad \text{for } i = 1, \ldots, T$$

This nonlinear model has a quadratic objective function, but linear constraints, and it happens to involve discrete decision variables. Other nonlinear models may involve continuous processes that are represented by time-integrated functions or flow problems described by differential equations.

5.1 Preliminary Notation and Concepts

A nonlinear function is one whose terms involve transcendental functions of the decision variables or in which there are multiplicative interactions among the variables, or in which there are other operations such as differentiation, integration, vector operations, or more general transformations applied to the decision variables. Examples include $\sin(x)$, $\tan(y)$, e^x, $\ln(x + z)$, x^2, xy, xe^y, and x^y. When an objective function or problem constraint involves such nonlinearities, we lose the guarantee that permitted us so conveniently to solve *linear* programming problems: namely that we could operate on a system of linear equations and if a solution existed, it could be found at one of the (finite number of) extreme points or vertices of the feasible region. In dealing with nonlinear programming models, we will see that points of optimality can occur anywhere interior to the feasible region or on the boundary.

We will also see that there are no general methods suitable for application to all the different types of nonlinear programming problems. Indeed, many of the diverse types of problems already presented in this book can be cast as nonlinear optimization problems: integer programming problems can be expressed as nonlinear models; systems of differential equations (as might be needed in continuous simulation models) can be viewed as nonlinear programming problems; and interior point methods for solving linear programming problems have a nonlinear aspect. So, it comes as no surprise that no single algorithm can be expected to cover the entire class of nonlinear optimization. Instead, special forms of nonlinear models have been identified, and algorithms have been developed that can be used on certain ones of these special cases. We will begin by describing and discussing the most significant properties of nonlinear models that will lead us to an understanding of some of these methods.

A nonlinear function may have a single maximum point, as seen at the point $x = a$ in Figure 5.1, or multiple maximum or minimum points, as seen in Figure 5.2. If we suppose the region of interest to be the interval $[a, f]$, then there is a **global maximum** at the point $x = f$, but also local maxima at the points $x = a$ and $x = c$. A local minimum occurs at $x = b$ and a **global minimum** occurs at $x = d$. Notice in the figure that local optima may occur where the slope of the function is zero or at a boundary of the region.

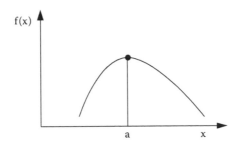

FIGURE 5.1
Single maximum point.

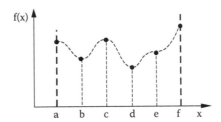

FIGURE 5.2
Multiple optima.

More formally, a **local maximum** of the function f(x) occurs at a point x* in a feasible region R if there is a small positive number ε such that

$$f(x^*) > f(x) \text{ for all } x \in R \text{ for which } |x - x^*| < \varepsilon$$

A **global maximum** of the function f(x) occurs at a point x* if

$$f(x^*) > f(x) \text{ for all } x \in R$$

Corresponding definitions for **local** and **global minima** can be given.

Clearly, the shape or curve of the function will play an important role in optimization. Two useful characteristics of the shape are convexity and concavity. For a **convex function**, given any two points x_1 and x_2 in a region of interest, it will always be true that

$$f(\alpha x_1 + (1-\alpha)x_2) \le \alpha f(x_1) + (1-\alpha) f(x_2) \text{ for } 0 \le \alpha \le 1$$

In Figure 5.3, let x = b be a linear combination of points x_1 and x_2, corresponding to $\alpha x_1 + (1 - \alpha)x_2$ in the definition earlier. Notice that any function value f(b) is always less than (or below) any point on the straight line connecting $f(x_1)$ with $f(x_2)$. This is precisely the characteristic of a convex function that will be useful in mathematical optimization. An alternate description of convexity is that the first derivative is non-decreasing at all points. As x increases, the slope of the function is increasing or *curving upward*.

For a **concave function**, the inequality is reversed, and any point on the function is always greater than (or above) the point on the straight line connecting $f(x_1)$ and $f(x_2)$. In Figure 5.4, the derivative of the function is always non-increasing or *curving downward*.

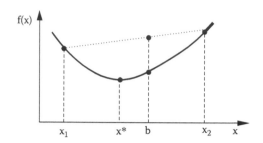

FIGURE 5.3
Convex function.

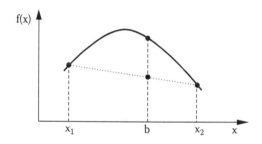

FIGURE 5.4
Concave function.

Notice that a straight-line function is both convex and concave. Definitions of convexity and concavity can be extended mathematically to include functions of multiple variables, but the concepts of *curving upward* and *curving downward*, respectively, are still preserved.

If a convex nonlinear function is to be optimized and there are no constraints, then a global minimum (if one exists) is guaranteed to occur at the point x* where the first derivative f'(x) of f(x) is zero. Figure 5.3 illustrates an unconstrained convex function with a minimum at x*. Figure 5.5 presents an example of a convex function (e^{-x}) with no minimum. Similarly, for a concave function, a global maximum is guaranteed to occur where f'(x*) = 0.

If there are constraints, then the shape of the feasible region is also important. Recall that a **convex region** or **convex set** is one in which the line segment joining any two points

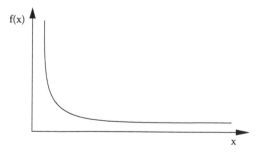

FIGURE 5.5
Convex function with no minimum.

in the set is contained completely within the set. If the feasible region forms a convex set, then the guarantees for global maxima and minima remain in effect, as described earlier.

More generally, the feasible region for a nonlinear programming problem is convex whenever each of the constraint functions is convex, and the constraints are of the form $g_i(x) \leq b_i$. For example, the reader should verify that the function $f(x) = x^2$ is convex; but the *regions* defined by $x^2 = 4$, or by $x^2 \geq 9$, are *not* convex. A **local minimum** is guaranteed to be a global minimum for a convex objective function in a convex feasible region, and a local maximum is guaranteed to be a global maximum for a concave objective function in a convex feasible region.

Many functions that arise in nonlinear programming models are neither convex nor concave. The function pictured in Figure 5.2 is a good example of a function that is convex in one region and concave in another region, but neither convex nor concave over the entire region of interest. Local optima are *not* necessarily global optima. Furthermore, a point x for which $f'(x) = 0$ may be *neither* a maximum nor a minimum. In Figure 5.2, the function $f(x)$ at the point $x = e$ has a zero slope. When viewed from the direction of $x = d$, it appears that $x = e$ may be a maximum; whereas when viewed from the direction of $x = f$, the function appears to be decreasing to a minimum. In fact, $x = e$ is an **inflection point**. For example, the function $f(x) = x^3$ has an inflection point at $x = 0$.

For unconstrained problems with just one variable x, necessary and sufficient conditions for local optima of a twice differentiable function $f(x)$ at $x = x^*$ can be summarized as follows:

Necessary conditions:

$$\frac{df}{dx} = 0 \text{ at } x = x^*$$

$$\frac{d^2f}{dx^2} \geq 0 \text{ for a local minimum at } x = x^*$$

$$\frac{d^2f}{dx^2} \leq 0 \text{ for a local maximum at } x = x^*$$

Sufficient conditions:

$$\frac{df}{dx} = 0 \text{ at } x = x^*$$

$$\frac{d^2f}{dx^2} > 0 \text{ for a local minimum at } x = x^*$$

$$\frac{d^2f}{dx^2} < 0 \text{ for a local maximum at } x = x^*$$

When the second derivative is equal to zero, the existence of a local optimum is not certain.

For unconstrained problems involving multiple variables $x = (x_1, x_2, ..., x_n)$, the necessary condition for a point $x = x^*$ to be optimal is for the partial derivative of the objective function $f(x)$, with respect to each variable x_i, to be zero at $x = x^*$; that is,

$$\frac{\delta f}{\delta x_i} = 0 \text{ for } i = 1, 2, \ldots, n$$

We define the gradient of a function $f(x_1, x_2, \ldots, x_n)$ to be the vector of first partial derivatives, and denote it as

$$\nabla f(x_1, x_2, \ldots, x_n) = \left[\frac{\delta f}{\delta x_1}, \frac{\delta f}{\delta x_2}, \ldots, \frac{\delta f}{\delta x_n} \right]$$

Then the necessary conditions can be stated more succinctly as

$$\nabla f(x) = 0 \text{ at } x = x^*$$

This condition is also sufficient for a minimization problem if $f(x)$ is convex (and for a maximization problem if $f(x)$ is concave). In fact, for a convex (concave) function, x^* is also a *global* optimum.

To determine whether a function $f(x_1, x_2, \ldots, x_n)$ is convex or concave, it is useful to examine the **Hessian matrix** H_f corresponding to f. The Hessian matrix H_f is an $n \times n$ symmetric matrix in which the (i, j)-th element is the second partial derivative of f with respect to x_i and x_j. That is,

$$H_f(x_1, x_2, \ldots, x_n) = \left[\frac{\delta^2 f}{\delta x_i \delta x_j} \right]$$

The function f is a convex function if H_f is positive definite or positive semidefinite for all x; and f is concave if H_f is negative definite or negative semidefinite for all x.

If the convexity (or concavity) criterion is met, then optimization may be as simple as setting the n partial derivatives equal to zero and solving the resulting system of n equations in n unknowns. However, since these are generally nonlinear equations, this system may not be at all simple to solve. And if the objective function is not convex (or concave), we lose the sufficiency condition, and $x = x^*$ could be a local minimum, a local maximum, or a *stationary point* instead of an optimum.

The search for an optimal solution to a general nonlinear programming problem must find and examine many candidate solutions to rule out local optima and inflection points. And it is not sufficient to examine just those points at which first derivatives are zero, for an optimum could occur at a point where there is a discontinuity and the derivatives do not exist. For example, in Figure 5.6, the function $|x|$ has a minimum at 0,

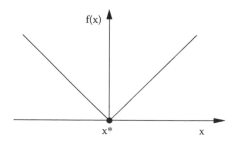

FIGURE 5.6
$f(x) = |x|$.

a non-differentiable point. Or the optimum could occur anywhere on the boundary of the feasible region. For constrained problems, the shape of the feasible region (which certainly may be non-convex) merely contributes further to the difficulty of the search.

Clearly, a single algorithm capable of making all of these considerations could not operate efficiently. Therefore, the remaining sections of this chapter present a number of different algorithms that have been developed for solving special classes of nonlinear programming problems.

5.2 Unconstrained Optimization

The simplest unconstrained optimization problem occurs when the objective function f involves just a single variable, is differentiable, and is concave for a maximization problem or convex for a minimization problem. In that case, the equation $df/dx = 0$ can be solved analytically to obtain the optimum value x^* because the necessary and sufficient conditions for optimality are met. If, however, this equation cannot be solved easily, it may be reasonable to resort to an iterative search procedure. Because there is only one variable, a **one-dimensional search** suffices.

5.2.1 One-Dimensional Search

The process begins by establishing an upper limit x_u and a lower limit x_l, within which an optimum is known to exist, and choosing an initial trial solution x to be halfway between the bounds:

$$x = \frac{(x_u + x_l)}{2}$$

Suppose a function $f(x)$ is to be maximized and that $f(x)$ is concave between x_u and x_l. Then the general idea is to examine the slope of $f(x)$ at the current trial solution x. If the slope is positive, then $f(x)$ is increasing and the optimum x^* is greater than x, so x is a new lower bound on the set of trial solutions to be examined. If the slope is negative, then $f(x)$ is decreasing and the optimum x^* is less than x, so x is a new upper bound. Each time a new bound is established, a new trial solution is computed (and choosing the midpoint is but one of several sensible rules). The sequence of trial solutions thus generated converges to the maximum at x^*. In practice, the process terminates when the bounds x_u and x_l enclose an interval of some predetermined size ε, denoting an error tolerance. The algorithm can be stated succinctly as follows.

5.2.1.1 One-Dimensional Search Algorithm

1. Establish an error tolerance ε. Determine an x_u such that $df(x_u)/dx \leq 0$ and an x_l such that $df(x_l)/dx \geq 0$.
2. Compute a new trial solution $x = (x_u + x_l)/2$.
3. If $x_u - x_l \leq \varepsilon$, then terminate. The current approximation is within the established error tolerance of x^*.

4. If $df(x)/dx \geq 0$, set $x_l = x$.

5. If $df(x)/dx \leq 0$, set $x_u = x$.

6. Go to Step 2.

Example 5.2

The algorithm can be illustrated by the problem of maximizing

$$f(x) = x^4 - 16x^3 + 91x^2 - 216x + 180$$

over the range $3.2 \leq x \leq 5.0$, which is shown in Figure 5.7. The function is certainly concave in the range $3.2 \leq x \leq 5.0$, so we will apply the search to that range.

The derivative $df(x)/dx = 4x^3 - 48x^2 + 182x - 216$ will be used during the procedure.

1. $x_u = 5.0$, $x_l = 3.2$, and let $\varepsilon = 0.15$
2. $x = (5.0 + 3.2)/2 = 4.1$
3. $5.0 - 3.2 = 1.8 > \varepsilon$
4. $df(x)/dx$ at $x = 4.1$ is equal to $-0.996 < 0$, so set $x_u = 4.1$ and leave $x_l = 3.2$
2. $x = 3.65$
3. $4.1 - 3.2 = 0.9 > \varepsilon$
4. $df(x)/dx$ at $x = 3.65$ is equal to $3.328 > 0$, so set $x_l = 3.65$ and leave $x_u = 4.1$
2. $x = 3.875$
3. $4.1 - 3.65 = 0.45 > \varepsilon$
4. $df(x)/dx$ at $x = 3.875$ is equal to $1.242 > 0$, so set $x_l = 3.875$ and leave $x_u = 4.1$
2. $x = 3.988$
3. $4.1 - 3.875 = 0.225 > \varepsilon$

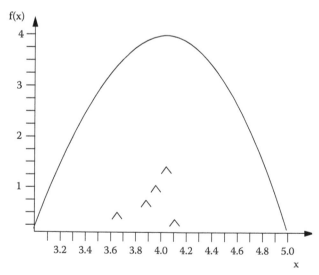

FIGURE 5.7
$f(x) = x^4 - 16x^3 + 91x^2 - 216x + 180$.

4. df(x)/dx at x = 3.988 is equal to 0.12 > 0, so set x_l = 3.988 and leave x_u = 4.1
2. x = 4.044
3. 4.1 − 3.988 = 0.112 < ε, so the process terminates with the current trial solution x = 4.044 and a function value of 3.99

Notice that at the point x = 4.044, the derivative of f(x) is −0.44 (close to zero); and at the true optimum of x = 4, where f(x) = 4, the derivative is exactly zero, a necessary condition for optimality.

Other methods for a one-dimensional search include the **Fibonacci method** and a related technique called the **golden section method**. These methods are discussed and compared in Wilde (1964). The golden section search is based strictly on the use of function evaluations, and is particularly useful when first derivatives are not available. (See Exercise 5.6.)

While a single-variable search method may seem too simplistic for practical nonlinear optimization problems, such methods are often incorporated into more elaborate multivariate search procedures, and therefore warrant our awareness and understanding.

5.2.2 Multivariable Search: Gradient Method

If our objective is to maximize a function f(x) where x = $(x_1, x_2, ..., x_n)$, then the previous single-variable search is not applicable. Recalling the necessary and sufficient conditions for the optimality of a solution x*, the necessary condition is that

$$\frac{\delta f}{\delta x_i} = 0 \text{ at } x = x^* \text{ for all } i = 1, ..., n$$

and this is *sufficient* if f(x) is also concave. So, it is tempting simply to approach the problem as being that of solving a system of n equations, setting all the partial derivatives equal to zero. This would allow us to find the stationary points by solving the equations $\nabla f(x) = 0$. However, f(x) and its partial derivatives are general nonlinear functions, and unless this system of equations has some special structure, this system cannot be solved analytically. So again, we turn to the use of iterative methods. And while the one dimensional search technique does not apply directly, it does provide a framework for how to proceed.

In a one dimensional search, at each iteration we examined the derivative of the function in order to decide whether to increase or decrease the current approximation to x*. There were only the two choices along one dimension. Now, in an n-dimensional search space, at each iteration there are infinitely many directions to change the current $(x_1, x_2, ..., x_n)$, and we can examine the partial derivatives to choose to move in that direction that yields the fastest possible improvement in f(x). Whereas in a one dimensional search, we tried to reach a point x at which df(x)/dx = 0, now our aim is ultimately to reach a point x = $(x_1, x_2, ..., x_n)$ at which all the partial derivatives of f(x) are equal to zero.

The method described here is known as the **gradient search** procedure. Recall that the gradient of a function f(x) at a point x = x′ is:

$$\nabla f(x') = \left[\frac{\delta f}{\delta x_1}, \frac{\delta f}{\delta x_2}, ..., \frac{\delta f}{\delta x_n} \right] \text{ at } x = x'$$

and the gradient will be used here as an indication of the direction of the fastest rate of increase of the function f(x), viewed from the point x = x'. The gradient method will generate successive points by repeatedly moving in the direction of the gradient at each point.

The next question is *how far to move* in the direction of the gradient. A move from an initial point x^0 all the way to a solution x^* for which $\nabla f(x^*) = 0$ would involve a circuitous route that would require constant re-evaluation of the gradient along the way. Because this would be computationally unreasonable, our method will instead move in a *straight line* in the direction of the gradient, and the distance to the next point will be: *as long as f(x) keeps increasing*. At that new point where f(x) is no longer increasing, the gradient is re-evaluated to determine the next direction to move, a distance for the next move is determined, and the next point is computed. This process repeats until two successive points are essentially the same, or $\nabla f(x)$ is within numerical tolerance of zero at one of the points.

This approach bears a resemblance to the method one might follow when climbing a mountain. At a given point, look around and select the direction of **steepest ascent** in the terrain, and follow that direction until the path is no longer ascending. At this point, look around again and select the direction of steepest ascent, and continue to repeat this process until arriving at a point at which none of the surrounding terrain is ascending. Assuming the mountain is concave, the climber has now reached the peak.

This analogy is only a two variable case in which the two variables represent the horizontal plane and the function value represents the vertical height of the surface of the mountain. Let us now describe this **steepest ascent** process for maximizing an n-variable function.

An initial approximation x^0 is chosen, then successively a point x^{j+1} is found from the current point x^j as follows:

$$x^{j+1} = x^j + d^j \cdot \nabla f(x^j)$$

where d^j specifies the distance to be moved in this iteration.

The value of d^j must be found so as to maximize the function f at the *new* point; therefore, we wish to

$$\text{maximize } f(x^j + d^j \cdot \nabla f(x^j))$$

with respect to d^j. Because all the other variables are now playing the role of constants in this context, we actually are merely faced with the problem of maximizing a function of a single variable. For this, we can take the derivative with respect to d^j, set it equal to zero, and solve for d^j; or use a one dimensional search method such as described in Section 5.2.1. The multivariable steepest ascent algorithm can now be stated succinctly as follows.

5.2.2.1 Multivariable Gradient Search

1. Establish an error tolerance ε. Determine an initial approximation or trial solution $x^0 = (x_1^0, x_2^0, ..., x_n^0)$. Set j = 0.
2. Determine the value of d^j that maximizes

$$f(x^j + d^j \cdot \nabla f(x^j))$$

and use that value of d^j to evaluate x^{j+1}.

3. Compute the next trial solution:

$$x^{j+1} = x^j + d^j \bullet \nabla f(x^j)$$

4. If $|x^{j+1} - x^j| \leq \varepsilon$, then terminate.

$\nabla f(x^{j+1})$ must be very close to zero

5. Set $j = j + 1$ and go to Step 2.

The gradient search always eventually converges to a stationary point as long as $f(x^{j+1}) > f(x^j)$ at every iteration. Note that a line search algorithm finds a *local* optimum. Therefore, it is possible for a naïve line search algorithm to find a solution $f(x^{j+1}) < f(x^j)$, in which case convergence is not guaranteed. Consider the example in Figure 5.8. If the initial distance is long enough, then a search such as a bisection search could easily converge to a worse solution than the initial solution, and the process could conceivably even cycle back to x^j.

It has been observed that the gradient method often *overshoots*. By going as far as possible while $f(x)$ is increasing, excessive zig-zagging toward the optimum typically occurs. Several modifications improve performance (Simmons 1975), but the simplest is to use $0.9d^j$ instead of d^j as the distance. This practice has been observed to double the convergence rate.

It might be pertinent to mention here that not all methods for multi-variable optimization rely on the use of derivatives. There are a number of methods that do not require explicit first derivative information. For example, the gradient vector is composed of n elements each of which measures the slope or the rate of change of the function if we take a small step in each of the coordinate directions. Therefore, one simple method of approximating the gradient at x^j is to perform n additional function evaluations at each

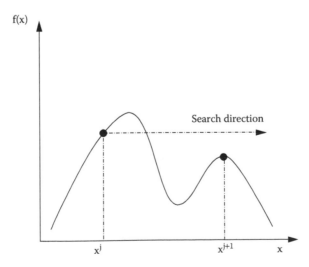

FIGURE 5.8
Potential problem for line searches.

of the points $f(x^j + \delta_i)$ for each i, simply perturbing the i-th component of x^j by some small constant. The terms of the gradient measure the per-unit difference in the function value. This same concept can be extended to approximating the second derivatives of a function. Normally, it is preferable to provide an explicit function for the derivatives. However, if that is not practical, and if function evaluations are not too expensive, the approximation methods may be valuable.

5.2.3 Newton's Method

A criticism that could be made of the gradient search method is that, although the gradient direction is the best direction to move, viewed from the current trial solution, as soon as we begin moving away from the current point, that direction is immediately *not* the best direction any longer. And the farther we move, the worse the chosen direction becomes. The gradient direction is an especially poor choice in the neighborhood of the optimum. Therefore, convergence is not easily predictable.

This behavior is explained by the fact that the gradient search method follows the gradient direction dictated by a linear approximation to f(x) near the current trial solution x^j. Whereas a *straight line* is generally a poor approximation to a nonlinear function, most general nonlinear functions can be reasonably well approximated by a *quadratic function*, in the vicinity of the maximum.

Newton's method is an iterative technique that makes use of this fact by choosing as its next trial solution that point that maximizes the quadratic approximation to f(x). Specifically, given a current trial solution x^j, the next point x^{j+1} is computed as

$$x^{j+1} = x^j + d^j(-H^{-1}(x^j) \bullet \nabla f(x^j))$$

where:
   H(x) is the Hessian matrix of f(x) evaluated at the point x
   $H^{-1}(x)$ is its inverse

The optimizing distance d^j can be chosen just as it was in the gradient search. Convergence occurs when the direction vector becomes close to zero.

Newton's method generally requires fewer iterations for convergence than the gradient search method because it uses a better direction of movement from one point to the next. However, there is little else to recommend this method from a practical standpoint. First, of course, the function f must be twice continuously differentiable, and the Hessian matrix must be nonsingular. The computational effort associated with inverting the Hessian matrix is excessive. (For economy of computation, it is reasonable to use the same inverse for several consecutive iterations. This slows convergence, but simplifies each iteration so much that overall performance is actually improved.)

Even so, the calculations are more extensive than for the gradient search method, and the efficiency diminishes rapidly as the number of variables increases because the matrix H becomes quite large. Moreover, Newton's method may fail to converge in general. The formula for computing a new point x^{j+1} from x^j does not necessarily imply an increase in the function value, for it could be that $f(x^{j+1}) < f(x^j)$. In particular, if the Hessian is positive definite, Newton's method will approximate a quadratic minimum. If it is negative definite, it approximates a quadratic maximum. When the Hessian is indefinite, Newton's method takes us to a saddle point solution of the approximation. Certainly, if f(x) were quadratic and H(x) were negative definite, then the method would converge in one iteration. In general, convergence to a local maximum is guaranteed, and occurs quite rapidly for any smooth, continuous nonlinear function once we get *close enough* to the maximum. However, *close enough* can be a very small region.

5.2.4 Quasi-Newton Methods

The computational demands of repeatedly inverting the $n \times n$ Hessian matrix H motivated the development of a large number of modifications to the original Newton's method. These modifications differ from one another primarily in the way that the second derivatives are approximated from one iteration to the next.

These **Quasi-Newton methods** begin with an arbitrary negative definite approximation to H, or its inverse, and through a succession of improvements, eventually converge to the true matrix H. For example, the methods could begin with $H = -I$, a negative identity matrix at some initial point, x^0. The Newton direction corresponds to a simple gradient direction. We first perform a line search to get a new point, x^1. Then, based on the new point and the function value, we perform a rank 1 update to the matrix H (and H^{-1}) which fits the current points with a quadratic. In so doing, we *correct* the estimate of H in one dimension, but we also maintain a negative definite approximation. This process is repeated using the new estimate of H to perform a line search and get a new maximum at x^2. After n iterations on a negative definite quadratic function, the approximation is exact.

The first such method was introduced by Davidon (1959), and shortly thereafter was improved upon by Fletcher and Powell (1963). The combined technique was known as the **DFP method**. A few years later, minor variations were proposed independently by Broyden (1970), Fletcher (1970), Goldfarb (1969), and Shanno (1970) and these became known collectively as the **BFGS update formula**. This is the method upon which almost all commercial software for nonlinear unconstrained optimization is based. The mathematical foundations and the precise formula typically used for updating the Hessian matrix is given in Beale (1959) and Avriel (1976).

5.3 Constrained Optimization

General nonlinear objective functions with general nonlinear constraints are the subject of this section. The methods to be applied will differ, depending on the nature of the constraints. Equality constraints can be dealt with using the method of **Lagrange multipliers**. Inequality constraints require the more comprehensive Karush–Kuhn–Tucker theory, which is central to the entire subject of mathematical programming. We will conclude with a short discussion of some popularly used techniques.

5.3.1 Lagrange Multipliers (Equality Constraints)

The method of Lagrange multipliers is named after the 18th century French mathematician Joseph-Louis Lagrange, and applies to nonlinear optimization problems with equality constraints, which can be expressed in the form:

$$\text{maximize} \qquad f(x)$$
$$\text{subject to} \qquad g_i(x) = b_i \text{ for } i = 1, \dots, m$$

where $x = (x_1, x_2, \dots, x_n)$.

We wish to find a solution such that each $g_i(x) = b_i$, so we are going to rewrite the original problem as:

$$\text{maximize } F(x, \lambda) = f(x) - \sum_{i=1}^{m} \lambda_i (g_i(x) - b_i)$$

The quantities λ_i are called Lagrange multipliers, and it is clear that if all the equality constraints are met precisely, then $F(x, \lambda) = f(x)$ for any values of $\lambda_1, \lambda_2, \dots, \lambda_m$. We wish to find values of $\lambda_1, \lambda_2, \dots, \lambda_m$ and $x_1, x_2, \dots, x_n$ that maximize $F(x, \lambda)$ and also satisfy $g_i(x) = b_i$ for $i = 1, \dots, m$. Such a solution would solve our original equality constrained problem.

We already know that a necessary condition for an optimum of $F(x, \lambda)$ is that $\delta F/\delta x_j = 0$ for $j = 1, \dots, n$ and $\delta F/\delta \lambda_i = 0$ for $i = 1, \dots, m$. Taking $(m + n)$ partial derivatives of F, with respect to the components of x_j and the λ_i, and setting each equal to zero, we can write the necessary conditions as

$$\frac{\delta f(x)}{\delta x_j} - \sum_{i=1}^{m} \left[\lambda \frac{\lambda_i \delta g_i}{\delta x_j} \right] = 0 \text{ for } j = 1, \dots, n$$

$$g_i(x) - b_i = 0 \text{ for } i = 1, \dots, m$$

We now have a set of $(m + n)$ equations in $(m + n)$ variables, which may be solvable by some iterative technique such as Newton–Raphson. There may be more than one critical point, but if so, the global optimum will be among them.

As a final observation, it is interesting to apply the method of Lagrange multipliers to the standard linear programming problem with constraints expressed as $Ax = b$, and to see that the Lagrange multipliers are precisely equivalent to the dual variables. This is merely a special case of a further generalization which will be examined next.

5.3.2 Karush–Kuhn–Tucker Conditions (Inequality Constraints)

The most general nonlinear programming problem can be defined as

$$\text{maximize} \qquad f(x)$$
$$\text{subject to} \qquad g_i(x) \leq 0 \text{ for } i = 1, \dots, m$$

where $x = (x_1, x_2, \dots, x_n)$. Clearly, *any* mathematical programming problem can be expressed in this form. It is tempting to introduce slack variables and convert all the inequality constraints into equalities, then apply the method of Lagrange multipliers. However, the m extra variables introduce an unwelcome computational expense, and we have more attractive alternatives that we will now consider.

Actually, we do try to extend the idea of Lagrange multipliers by recognizing that if the *unconstrained* optimum of $f(x)$ does not satisfy all the inequality constraints indicated earlier, then when the constraints are imposed, at least one of the constraints will be satisfied as an *equality*. That is, the constrained optimum will occur on a boundary of the feasible region.

This observation suggests an algorithm for solving the problem. We begin by solving the unconstrained problem of maximizing f(x). If this solution satisfies the constraints, stop. Otherwise, we repeatedly impose increasingly larger subsets of constraints (converted to equalities) until either a feasible solution is found via the method of Lagrange multipliers, or until it is determined that no feasible solution exists.

Unfortunately, this method is very computationally demanding (and consequently essentially useless on most problems of practical size), as well as not guaranteeing that a solution found is globally optimal. Still, the Lagrange multiplier idea leads to what are known as the Karush–Kuhn–Tucker conditions that are necessary at a stationary point, corresponding to x and λ, of a maximization problem. The Karush–Kuhn–Tucker conditions can be stated as:

$$\frac{\delta f}{\delta x_j} - \sum_{i=1}^{m} \frac{\lambda_i g_i}{\delta x_j} = 0 \text{ for } j = 1, \ldots, n$$

$$g_i(x) \leq 0 \text{ for } i = 1, \ldots, m$$

$$\lambda_i \geq 0 \text{ for } i = 1, \ldots, m$$

$$\lambda_i g_i(x) = 0 \text{ for } i = 1, \ldots, m$$

The Karush–Kuhn–Tucker conditions correspond to the optimality conditions for linear programming where the λ's represent the dual variables. The gradient of the objective function at the optimal solution, x, can be written as a non-negative linear combination of the gradients (normal vectors) of the active constraints. The second condition states that x must be feasible. The third condition is non-negativity, and the fourth condition corresponds to complementary slackness: λ can be positive only if the corresponding constraint is active ($g_i(x) = 0$). If the i-th constraint is satisfied as a strict inequality, then the i-th resource is not scarce and there is no marginal value associated with having more of that resource. This is indicated by $\lambda_i = 0$.

The Karush–Kuhn–Tucker necessary conditions are also sufficient for a maximization problem if the objective function f(x) is concave and the feasible region is convex. Establishing the convexity and concavity and applying the Karush–Kuhn–Tucker necessary conditions do not yield procedures that are reasonable for direct practical numerical application. However, the Karush–Kuhn–Tucker conditions do form the very foundation of the theory of general mathematical programming, and will be seen again in the next section where—at last—we will see some *efficient* computational methods.

5.3.3 Quadratic Programming

Quadratic programming comprises an area of mathematical programming that is second only to linear programming in its broad applicability within the field of Operations Research. While quadratic objective functions are not as simple to work with as linear objectives, we can see that the gradient of a quadratic function is a *linear* function.

Consequently, the Karush–Kuhn–Tucker conditions for a quadratic programming problem have a simple form that can make solutions to these problems considerably easier to obtain than for general nonlinear programming problems.

The quadratic programming problem can be expressed in the following form:

$$\text{maximize} \qquad z - \sum_{j=1}^{n} c_j x_j + \sum_{j=1}^{n} \sum_{k=1}^{n} d_{jk} x_j x_k$$

$$\text{subject to} \qquad \sum_{j=1}^{n} a_{ij} x_j = b_i \text{ for } i = 1, \ldots, m$$

$$x_j \geq 0 \text{ for } j = 1, \ldots, n$$

The problem can be expressed more succinctly, using matrix notation, as:

$$\text{maximize} \qquad z = c^T x + x^T D x$$
$$\text{subject to} \qquad Ax = b$$
$$x \geq 0$$

where:
  x and c are n-component vectors
  A is an m × n matrix
  b is m × 1
  D is an n × n symmetric matrix.

Several algorithms have been developed to solve certain forms of quadratic functions, and we will describe some of the best known and most widely used ones. Because of the complexity of these procedures, we will give only brief overviews. The curious reader is urged to consult a more advanced reference such as Simmons (1975) or Nash and Sofer (1996) for a deeper appreciation of these methods.

One of the earliest and simplest methods for solving quadratic programs is Wolfe's algorithm (Wolfe 1959), which is still widely used today. In this method, a sequence of feasible points is generated via a modified Simplex pivoting procedure that terminates at a point x^* where the Karush–Kuhn–Tucker conditions are satisfied. Because the Karush–Kuhn–Tucker conditions represent a system of linear equations when the objective function is quadratic, the problem reduces to finding a feasible solution to a system of equations. Wolfe's algorithm uses phase 1 of the Simplex algorithm to find a feasible solution. The complementary slackness conditions are not linear, but the algorithm simply maintains a set of active constraints, and allows only the corresponding λ_i dual variables to be greater than zero. Wolfe's method, like most of the later procedures, moves along an active constraint set.

When D is negative definite, Wolfe's algorithm converges to an optimal solution, or demonstrates infeasibility within a finite number of iterations, assuming that the possibility of infinite cycling due to degeneracy is excluded.

Beale's method (Beale 1959), introduced by E.M.L. Beale as early as 1955, is based on classical calculus rather than on the Karush–Kuhn–Tucker conditions. This method is applicable to any quadratic program of the form described earlier except that Beale does not require D to be negative definite or negative semidefinite (i.e., the objective function need not be concave). Thus, this algorithm will generally yield local optima and the first solution generated will be the global optimum when the objective is concave.

Beale's method partitions matrices and uses partial derivatives to choose pivots until it is no longer possible to improve the objective value by any permitted change in a non-basic variable. Initially, all redundant constraints are eliminated and an initial basic feasible solution is determined via a Phase 1 Simplex process. The matrices are partitioned in such a way that a new system of equations is developed in which the basic variables, along with the associated constraints, are separated from the non-basic variables and their associated constraints. Partial derivatives determine which non-basic variable to increase or decrease.

When an apparent solution is achieved, an examination of the second partial derivative will determine whether the solution is a false optimum or not. If the second partial derivative is positive for some x, then the current solution is a minimum (rather than a maximum). In this case, the objective function can be improved by bringing x into the basis.

A slightly less popular, but more recent and more sophisticated method was originally presented by Lemke (1962). It is applicable to any quadratic problem, but is typically described in terms of solving problems expressed in the form

$$\text{maximize} \qquad z = c^T x - \frac{1}{2} x^T D x$$

$$\text{subject to} \qquad Ax \leq b$$

where D is not only symmetric but also positive definite. This new restriction on D is critical and is used throughout the procedure. Lemke's formulation of the constraints in terms of inequalities rather than equations causes the Karush–Kuhn–Tucker conditions to assume a particularly simple form which is exploited by the algorithm. These constraints also include any non-negativity restrictions.

Lemke's algorithm first formulates the Karush–Kuhn–Tucker conditions for the original problem, then defines a new set of variables, from which a second quadratic program is constructed. This new problem is solved and from its solution is obtained a solution to the original problem. The basic strategy is to generate a sequence of feasible points until a point is reached at which a certain gradient satisfies two specific restrictions. Three situations may arise, each of which is handled differently but results in a matrix being updated via the usual Simplex transformation technique. When the algorithm terminates with an optimal solution to the second quadratic program, the optimal solution to the original quadratic program is constructed based on the definitions of the new set of variables.

Historically, we find that Beale's method is used less extensively in practice than the other two algorithms mentioned here. Computational experiments (Ravindran and Lee 1981) have shown that Lemke's algorithm outperforms Wolfe's and four other lesser-known algorithms. Although Lemke's algorithm can fail to converge, when convergence does occur, it occurs more quickly than in the other methods.

Quadratic programming models are important for a number of reasons. General nonlinear problems with linear constraints are sometimes solved as a sequence of quadratic program approximations. Many nonlinear relations occurring in nature are not quadratic, but can be approximated by quadratic functions and then solved with the methods just described.

However, a wide variety of problems fall naturally into the form of quadratic programs. The kinetic energy of a projectile is a quadratic function of its velocity. The least-squares problem in regression analysis has been modeled as a quadratic program. Certain problems in production planning, econometrics, activation analysis in chemical mixture problems, and in financial portfolio management are often treated as quadratic problems. We will elaborate on this last problem in the following example.

Example 5.3

A classical problem that is often used to illustrate the use of the quadratic programming model is called **portfolio selection**. A portfolio is a collection of assets, all of which have positive present values (called prices) and which also have positive future values that are currently unknown. Analysts often use the term *rate of return on investment* to describe future value as follows:

$$\text{Future value} = \text{Price} \times \left(1 + \text{Rate of return on investment}\right)$$

Future values are positive but certainly may be less than present values (prices).

Rates of return are not known nor are they guaranteed. A very high expected return on an asset is usually accompanied by great variability. The future values can be estimated, but because such estimates are subject to error, there is a risk associated with any portfolio. The risk of a portfolio can be reduced by diversification, the extent of which is determined by the number of assets in the portfolio and the proportion of the total investment that is in each asset. It is generally easier to predict the future value of the portfolio than to predict the future values of the individual assets.

The portfolio manager is responsible for assigning a weight to each asset held in the portfolio. The weight of the i-th asset is the ratio of the dollar amount invested in that asset, divided by the total dollar value of the portfolio. The sum of the weights must be one, and all weights are non-negative. A portfolio p is defined by this set of weights. We will see that these weights determine the portfolio's expected future value as well as the portfolio's risk.

The portfolio manager generally begins his decision making process with

- A fixed amount of money to be invested.
- A list of n assets to invest in.
- The expected return of each asset.
- The variance of each asset return.
- All covariances.

If risk were of no concern, the manager would undoubtedly just invest all the money in the one asset offering the greatest expected return, that is, assigning a weight of 1 to that asset and 0 to all the others, regardless of risk. But risk almost always *is* a consideration, and most investors are risk-averse.

It is desirable to maximize return *and* minimize risk, but in a competitive market, prices fluctuate so that the *safer* investments are more expensive than the *riskier* ones. So, in general, it is not possible to simultaneously achieve both goals of maximizing return and minimizing risk. Instead, we define a class of **efficient portfolios**. A portfolio is said to be efficient if either

- There is no other less risky portfolio having as high a return.

  or

- There is no other more profitable portfolio having as little risk.

Thus, the problem of efficient portfolio selection can be viewed as having primal and dual expressions: to minimize variance subject to a specified expected return, or to maximize expected return subject to a specified variance.

Let n be the number of assets being considered, and let r_i be the expected return on the i-th asset. We will let W denote a column vector of n asset weights, indicating what fraction of the portfolio should be allocated to each asset. We use a variance-covariance matrix V in which diagonal element v_{ii} is the variance of the i-th asset, and the off-diagonal elements $v_{ij} = v_{ji}$ denote the covariance between the i-th and j-th assets. Then, *risk* is defined as the variance σ^2 of the portfolio p as:

$$\text{risk } (p) = \sigma^2(p) = W^T V W$$

The expected return of the portfolio p is given by:

$$E(p) = W^T R$$

where R is a column vector of expected asset returns. So the *portfolio selection problem* can be expressed as

$$\text{minimize} \quad z = W^T V W$$

$$\text{subject to} \quad \sum_{i=1}^{n} w_i r_i \geq P$$

$$\sum_{i=1}^{n} w_i = 1$$

$$w_i \geq 0 \ \text{ for } i = 1, ..., n$$

where P is a desired minimum return on investment.

Equivalently, we could

$$\text{maximize} \quad z = W^T R$$

$$\text{subject to} \quad W^T V W \leq Q$$

$$\sum_{i=1}^{n} w_i = 1$$

$$w_i \geq 0 \ \text{ for } i = 1, ..., n$$

where Q is a desired maximum risk. The first of these formulations is clearly in the form of a quadratic programming problem with linear constraints. The noted economist

Harry Markowitz (1959) is credited with formulating the portfolio selection problem as a quadratic programming model.

5.3.4 More Advanced Methods

Quadratic programming models represent one special case of nonlinear programming problems, but there are many additional methods that have been developed for solving various other special types of nonlinear problems. **Gradient methods** are based on the ideas presented previously in Section 5.2.2, but must include special provisions to restrict the search within the feasible region. One of the best-known of these is the **reduced gradient method** (Wolfe 1967, Lasdon and Warren 1978, Murtagh and Saunders 1978). In particular, the Lasdon and Waren algorithm, GRG2 (Generalized Reduced Gradient) is available in standard spreadsheet packages using the *Solver* tool. It is based on a Simplex scheme, but instead of improving a current solution by a change of *one* non-basic variable at a time, the reduced gradient method simultaneously changes as many non-basic variables as can change and yield an improvement, but at different rates, proportional to their respective partial derivatives.

Unconstrained optimization techniques have also been adapted for **constrained optimization** by the imposition of *penalty functions* or *barrier functions* (Bazaraa et al. 2013).

5.4 Software for Nonlinear Optimization

Linear programming models require only the coefficients for objective and constraint functions, and it is easy to enter this input to software by using well-established input formats that have been in use for many years. By contrast, nonlinear models come in many different forms. In some cases, the model itself may actually be very complicated. In others, the model may be fairly simple, but just does not conform to any particular standard model, and therefore finding and using the *right* software is difficult. This lack of any standard form (resulting from the fact that *nonlinear* programming includes every imaginable form of mathematical programming except linear!) has always made the selection and use of appropriate software cumbersome.

As modeling languages such as AIMMS, AMPL, and GAMS have become more sophisticated, software use has become accessible to a larger community of analysts. Software for nonlinear programming often requires that the derivatives of functions be explicitly entered along with other components of the problem. But even this obstacle has been alleviated by software, which analyzes nonlinear formulae and generates software that can evaluate derivatives (a process much more sophisticated than the symbolic differentiation that is available in some mathematical packages). In light of these advances, why are nonlinear programming problems still considered difficult to solve?

1. In many problems, it is computationally difficult to determine whether or not the objective function is concave (convex) in the feasible region; hence, it is difficult to guarantee convergence to a global optimum.
2. If a method finds a *solution*, it is often difficult to know whether it is local or global.

3. Existence of feasible solutions for a problem having nonlinear constraints is difficult to determine, which means there is no guarantee of finding an initial feasible point (starting solution) even when one exists.

4. Special-purpose software may need to be used in conjunction with more general nonlinear programming optimization packages.

5. Software often employs some variation on Newton or Quasi-Newton methods. This works well if the current point is close to the optimal, but the results are unpredictable when the initial point is far away from the optimal solution.

6. Some algorithms require more global knowledge about their nonlinear functions to give satisfactory performance.

For these reasons, although modeling languages and convenient software interfaces seem to invite a wide audience of users, it must be recognized that for nonlinear programming, friendly software is no substitute for a savvy and mathematically astute analyst. Yet nonlinear programming is a valuable tool for many applications in science, engineering, and finance, and there is a wide selection of powerful and ingenious software available.

Choosing software for nonlinear optimization problems is difficult because no one algorithm is efficient and effective for finding a global optimum for general nonlinear problems. Because no method is invariably superior to others, many software products include a number of methods, with the hope that one of the methods will suffice for a given problem. All methods will typically require repeated computation of the objective function, the gradient vector, and an approximation to the Hessian matrix. For many problems, evaluating the gradient requires more time than evaluating the objective function, and approximating the Hessian matrix can take an enormous amount of processing time as well as memory. It is tempting to seek a technique that does not rely on the Hessian, but such techniques (because they are poorly guided) may require many more iterations and in the end are therefore slower.

Software for nonlinear optimization in general has always been characterized by its variety. Some algorithms seem to perform exceptionally well on problems having constraints that are nearly or mostly linear, while a very different approach may be effective on problems that are highly nonlinear but that have relatively few variables or constraints. And some recent progress has been made by extending interior point methods for linear programming problems to quadratic and even general nonlinear problems.

MINOS is one of several linear and nonlinear optimizers offered within the AIMMS, APMonitor, GAMS, TOMLAB, and AMPL modeling systems and the NEOS Server, but it also can be used as a stand-alone package. The MINOS system is a general-purpose optimizer, designed to find locally optimal solutions involving smooth nonlinear objective and constraint functions. It takes advantage of sparsity in the constraint set, is economical in its use of storage for the reduced Hessian approximation, and is capable of solving large-scale linear and nonlinear programs.

NPSOL is sometimes offered as a companion to MINOS, and the two systems share a number of features such as computer platforms and languages. However, NPSOL is especially designed for dense linear and nonlinear programs, and for small models involving nonlinear constraints or whose functions are highly nonlinear and expensive to evaluate. It does not exploit sparsity (its Hessian is always stored in full form); it requires fewer evaluations of the nonlinear functions than does MINOS; it is more robust than MINOS if constraints are highly nonlinear; and convergence is assured for a large class of problems (particularly some for which MINOS fails to converge).

MATLAB Optimization Toolbox includes a wide variety of methods for linear and nonlinear optimization on various platforms (Beck 2015). The MATLAB language facilitates problem input. Constraints and objective functions must be differentiable. *TOMLAB* is a modeling environment in MATLAB that has a unified input-output format and integrates automatic differentiation. It works with MATLAB solver algorithms as well as other solvers.

SAS Institute, Inc. provides a general nonlinear optimization package that runs on various platforms. SAS offers several techniques including Newton–Raphson, Quasi-Newton, conjugate gradient, Nelder-Mead simplex, hybrid Quasi-Newton, and Gauss–Newton methods, which comprise special routines for quadratic optimization problems. The Quasi-Newton methods use the gradient to update an approximation to the inverse of the Hessian and is applicable where the objective function has continuous first and second derivatives in the feasible region. *SAS OPTMODEL* provides a general nonlinear optimization problem solver. *SAS/OR* handles nonconvex nonlinear optimization problems that may have many locally optimal solutions that are not globally optimal. *SAS/OR* applies multiple global and local search algorithms in parallel to solve difficult optimization problems such as those having discontinuous or non-differentiable functions, to identify global optima.

IMSL libraries comprise an extensive set of subroutines and procedures for a variety of mathematical and statistical purposes that are supported across a wide range of languages as well as hardware and operating system environments including Windows, Linux, and many UNIX platforms. It includes routines to solve nonlinear problems whose size is limited only by the available memory, and is generally successful on problems involving smooth functions.

LINGO modeling language and solver and *LINDO API* combine large-scale linear, nonlinear, and integer optimizers through an interactive modeling environment. The primary underlying technique is a generalized reduced gradient algorithm, and the system incorporates a global solver, multi-start capability, and a quadratic solver.

Gurobi Optimization has a reputation for their robust and high performance software for solving difficult and complex problems. Gurobi products can be embedded in existing development environments or can run in stand-alone mode. They offer advanced implementations of the newest algorithms including parallel algorithms running in innovative shared memory hardware contexts.

IBM CPLEX Optimizer can solve both convex and non-convex quadratic to global optimality. It can find the unique solution to a concave maximization problem and a first-order solution to a non-concave problem. CPLEX has both barrier and simplex algorithms for solving convex quadratic programs and a barrier algorithm for solving non-convex problems. It can also solve problems with convex quadratic constraints.

Frontline Solvers is the developer of MS-Excel Solver that comes with Excel but it is limited in its capability for solving large problems. The company offers a more powerful (premium) solver that works as an add-in to Excel but is capable of solving larger linear and nonlinear problems. The Solver's SDK (software developer kit) can be used with multiple modern programming languages such as C++, Java and Python. Solver uses the GRG nonlinear method for nonlinear optimization.

NEOS Server is a free internet-based service for solving numerical optimization problems including nonlinear problems. It offers several nonlinear constrained programming solvers such as CONOPT, Knitro, MINOS, LOQO among others. These solvers can be accessed by using modeling languages such as AMPL and GAMS.

COIN-OR (COmputational INfrastructure for Operations Research), is an open-source community for the development and deployment of operations research software including nonlinear optimization solvers such as DFO and FilterSD.

Valuable reference material for serious practitioners and analysts can be found in Gill et al. (1981, 1984). These volumes do not necessarily stress the intuitive appeal of the methods discussed, but rather they realistically present the details pertinent to the practical performance of some of the most powerful and advanced methods and implementations. Additional recommended sources of information include Fourer (1998, 2017), Moré and Wright (1993), and Nash (1998).

Finally the Nonlinear Programming Frequently Asked Questions (FAQ) web page offers many resources on nonlinear programming, software and solvers.

5.5 Illustrative Applications

5.5.1 Gasoline Blending Systems (Rigby et al. 1995)

Texaco's most important refinery product is gasoline. Crude oil entering a refinery is distilled and split into various components, which can then be reformed or cracked into lighter compounds that may be of greater commercial value. The resulting stocks (having varied and unanticipated properties) must then be blended to achieve certain quality specifications. The greatest profitability will result if the refinery can maximize its production of higher octane gasoline blends from the available stocks.

Gasoline blend qualities include familiar properties such as octane (measured as research, motor, and road octanes) and lead content, but also other characteristics such as Reid vapor pressure (RVP), sulfur and aromatic contents, and volatilities (the temperatures at which certain percentages of the blend boil away). Other qualities are important because of federal and state agency emission standards. While some properties of gasoline blending can be (and have been for decades) modeled as *linear* optimization problems, it is known that octane, RVP, and volatilities are highly nonlinear functions of volume and weight.

Prior to the late 1970s, gasoline blending was a simple mixture of various stocks, and octane requirements were met by injecting tetraethyl lead into the blend. Blending recipes were based on hand calculations that did not significantly affect the overall economies of the plant. Tetraethyl lead was inexpensive and available in ample supplies, so the octane requirement was not a binding constraint in the model.

However, during the 1970s, governments mandated that lead be phased out of the blending recipe; and by the early 1980s, the federal government also clamped down on volatility specifications for gasoline. These two changes had a drastic effect on the economics of refining, and Texaco responded by developing a nonlinear gasoline blending optimization system. The first version of the system resulted in an estimated annual savings of $30 million, improved quality control, and increased the ability to plan refining operations and market the products, and perform sensitivity analysis on current operations schedules.

The blending model was coded in the GAMS modeling language, and uses MINOS solvers. Subsequent versions of the system allowed additional flexibility in handling blending stocks, increased the number of constraints that could be modeled, and permitted computations to be placed on more sophisticated client server network hardware.

The system is used for both immediate, short-range, and long-range planning. Refinery planners make use of the system to generate the recipe for the next blending operation. For short-term planning purposes, it is important to be able to examine the multi-period model covering the next few days, to ensure that components consumed in today's blends do not render tomorrow's schedule infeasible. And finally, refinery operations planners must anticipate all the activities associated with gearing up plants for gasoline reformulation. Estimates produced by older linear programming planning models must be checked for consistency with larger nonlinear models, and the system allows planners to identify errors and restructure problem formulations where necessary.

5.5.2 Portfolio Construction (Bertsimas et al. 1999)

A large investment firm in Boston manages assets in excess of $26 billion, for clients that include pension funds, foundations, educational endowments, and several leading investment institutions. This firm employed the widely used classical theory of portfolio optimization (Markowitz 1959), in which managers determine the proportion of total assets to invest in each available investment to minimize risk (variability of return) subject to constraints that require the expected total return to meet a certain target. This famous model includes an objective that is a quadratic function of the decision variables and constraints that are linear.

For a variety of reasons, large clients typically subdivide their asset classes and allow each portion to be managed by different analysts who have distinctly unique investment styles. This strategy ensures that the composite return will be a linear combination of the returns resulting from the different investment styles. Because this linear diversification approach is generally accepted by clients, the investment firm applies the technique within its individual funds. Portfolios are partitioned into subportfolios, each characterized by a different investment style. Quadratic optimization can still be used for the multiple subportfolio problem, but the number of decision variables increases dramatically because each subportfolio can conceivably invest in any of the securities available to the composite portfolio. (One of the firm's funds is partitioned into 50 subportfolios.)

A notable advantage of this partitioned portfolio framework is the ability to reduce trading costs by swapping shares internally among subportfolios, thereby often avoiding the costs of trading on the open market. Globally optimizing multiple subportfolios thus makes it possible to sharply increase the turnover within each subportfolio without necessarily increasing turnover for the composite portfolio. This portfolio construction methodology produces funds with good performance, high liquidity, relatively low turnover, use of multiple investment styles, and diversification over time.

The desired diversification that is achieved through multiple subportfolios unfortunately gives rise to certain complications that are not handled within the standard quadratic programming model. With risk management through diversification, the number of different stocks (or other investments) in the portfolio becomes very large, and as the portfolio is rebalanced over time, the number of transactions also grows, resulting in increased custodial fees and transaction costs. These phenomena can sometimes be dealt with by adding a post-processing step to the quadratic optimization phase, simply to prohibit positions and trades smaller than a given threshold. But this firm's strategy specifically included investing in small market capitalization stocks, so merely eliminating small positions would be inconsistent with established investment criteria. Additionally, post-processing that eliminates many small but key positions can interfere with optimization objectives and can violate constraints.

On the basis of these considerations, the investment firm decided to modify its quadratic optimization approach so that it could simultaneously optimize its multiple subportfolios *and* maintain control over the number of positions and transactions in the composite portfolio. These stock position levels and transaction counts are inherently integer-valued quantities, and the quadratic model therefore had to be expanded to include integer components, resulting in a mixed-integer programming model. The solution was implemented using ILOG CPLEX 4.0 as the underlying mixed-integer solver. The mixed integer solution allowed the firm to reduce its average number of different holdings by approximately 50%, and its average number of transactions by about 80%, significantly decreasing its operational costs and trading costs while maintaining essentially the same originally targeted expected returns on investment.

5.5.3 Balancing Rotor Systems (Chen et al. 1991)

Large steam turbine generators, high speed gas turbine engines, and other machinery with flexible rotating shafts must be balanced to reduce vibration. Minimizing vibration is important in extending the life of the machine, improving operating efficiency, and maintaining a safe operating environment. The design and fabrication of rotating machinery has undergone evolutionary changes over time, in particular being influenced by increased energy costs and safety and maintenance concerns. The use of lighter weight materials in rotors and faster rotating speeds necessitates more accurate manufacturing processes, which result in improved balancing characteristics in the rotors.

One of the most popular techniques for flexible rotor balancing treats the rotordynamic system as a linear system in calculating balance correction weights. The primary disadvantage of this approach is that it typically requires a large number of actual trial runs to collect enough data to estimate accurately the required balance corrections. The linear programming approach seems attractive from a computational standpoint, but in many applications, such as for utility companies, the costs of shutdown, installation of trial weights, startup, and data collection are prohibitive.

Using a recently developed nonlinear programming model, it is now possible to determine an optimal system balance without the necessity of trial runs. In place of *actual* trial runs, this new technique requires developing a mathematical model of the system dynamics that can be used to *simulate* rotor response to balance corrections.

The unbalance of a rotor is continuously distributed along the axis of the rotor. However, in the nonlinear model, this continuous distribution is discretized into a finite number of balance planes to which corrections can be applied. Similarly, measurements are taken at only a limited number of points (in some cases, as few as two points is sufficient). The nonlinear optimization process then seeks to find the *unbalance vector* that minimizes a least-squares difference between the adjusted analytical model and the measured experimental model. In the nonlinear solver for this constrained least-squares problem, a search direction is found using a steepest descent method, and a constrained line search is used to determine the step size. Gradients of the objective function, with respect to all the design variables, determine the search direction; and at each gradient evaluation, the rotor model must be solved to obtain the system response. The computations are frequently complicated by ill-conditioned gradients, but normalization procedures are employed effectively against this difficulty. Because the rotor systems being balanced often have multiple possible operating speeds, the optimization objective function includes weights (coefficients) associated with each different operating speed of the rotor, with the largest weights applied to the most critical operating speeds.

In test rigs, significant improvements in vibration levels were observed through the use of this model. And in the numerical computations, convergence to an optimum solution took place in less than a minute of mainframe processing time.

5.6 Summary

Nonlinear optimization models are used for mathematical programming problems in which the objective function and constraints are not necessarily linear. This class of problems is very broad, encompassing a wide variety of applications and approaches to solving the problems. No single algorithm applies equally to all nonlinear problems; instead, special algorithms have been developed that are effective on certain types of problems.

Unconstrained optimization can often be dealt with through the use of calculus to find maximum and minimum points of a function. Constrained optimization typically requires solving systems of equations. As helpful as the mathematical theories are that can be used to describe the characteristics of optimal solutions to nonlinear problems, such insights nevertheless often fail to suggest computationally practical methods for actually finding the desired solutions.

Iterative search techniques are frequently used for nonlinear optimization. A one-dimensional search suffices for finding the optimum value of a function of one variable; at each step, the slope, or derivative, of the function is used to guide and restrict the search. Although such a technique seems much too elementary for a realistic nonlinear optimization problem, single-variable search methods are often incorporated into more sophisticated **multi-variable search** procedures.

For finding the optima of functions of many variables, gradient search methods are guided by the slope of the function with respect to each of the variables. At each step, the method follows the direction indicated by the sharpest improvement from the current point. For this reason, techniques that operate in this way are often referred to as steepest ascent methods. Straight-line searches can be improved upon by using Newton's method, which is based on quadratic approximations to nonlinear functions.

Constrained optimization methods differ depending on the nature of the constraints. The method of Lagrange multipliers is applicable to problems with equality constraints. For problems with inequality constraints, Karush–Kuhn–Tucker theory describes necessary and sufficient conditions for optimality and forms the foundation of general mathematical programming.

Key Terms

BFGS updates
concave function
constrained optimization
convex function
convex region
convex set
DFP method
efficient portfolio
Fibonacci method
global maximum

global minimum
golden section method
gradient search
Hessian matrix
inflection point
Karush–Kuhn–Tucker conditions
Lagrange multipliers
local maximum
local minimum
multivariable search
necessary conditions
Newton's method
one dimensional search
portfolio selection
quadratic programming
Quasi-Newton methods
reduced gradient method
risk
steepest ascent
sufficient conditions
unconstrained optimization

Exercises

5.1   Consider the function $f(x, y) = 3x^2 - 2xy + y^2 + 3e^{-x}$. Is this function convex, concave, or neither? Explain your answer.

5.2   Consider the function $f(x) = x^4 - 8x^3 + 24x^2 - 32x + 16$. Is this function convex, concave, or neither? Explain your answer.

5.3   Consider the following nonlinear problem. Is the feasible region convex?

$$\begin{aligned} \text{minimize} \quad & f(x, y) = x - 2xy + 2y \\ \text{subject to} \quad & x^2 + 3y^2 \le 10 \\ & 3x + 2y \ge 1 \\ & x, y \ge 0 \end{aligned}$$

5.4   Consider the following nonlinear problem. Is the feasible region convex?

$$\begin{aligned} \text{minimize} \quad & f(x, y) = 3x^2 - 2xy + y^2 \\ \text{subject to} \quad & x^3 - 12x - y \ge 0 \\ & x \ge 1 \end{aligned}$$

5.5 Use the one dimensional search described in Section 5.2.1 (also known as a *bisection search*) to find a minimum of the function

$$f(x) = x^4 - 3x^3 + 2x^2 - 2x + 7$$

over the range $1 \leq x \leq 10$. Use $\varepsilon = 0.1$.

5.6 The *golden section* search is similar to the bisection search for one dimensional problems, except that it uses only function values and it does not require calculating derivatives. This is particularly useful when the function does not have first derivatives defined, or when computing the first derivatives is very expensive computationally. Suppose you are given two initial end-points, $a \leq x \leq d$, and the minimum of $f(x)$ is known to lie in this range. Evaluate the function at two points $c = a + 0.618 \cdot [d - a]$, and $b = d - 0.618 \cdot [d - a]$. Note that $a < b < c < d$, but they are not evenly spaced. If $f(b) < f(c)$, then let $[a, c]$ be the new interval, and repeat the calculation. Otherwise, let $[b, d]$ be the new interval. The *magic* aspect of the golden section is that when you have to compute the new interior points between $[a, c]$, you discover that b is precisely 0.618 of the distance between a and c. In other words, you only need to make one additional function evaluation. Similarly, if the new interval is $[b, d]$, then point c is already lined up with one of the new required points.

Use the method of golden section to find the minimum of the function in the previous problem with $\varepsilon = 0.1$.

5.7 Consider the unconstrained problem:

$$\text{minimize } f(x, y) = 3x^2 - 2xy + y^2 + 3e^{-x}$$

Starting from the solution $(x, y) = (0, 0)$, and an initial step length of 2, perform two iterations of the gradient search algorithm to find a minimum. That is, compute the gradient at the point $(0, 0)$, and perform a one dimensional line search to find a minimum along the line. From this new point, perform a second line search.

5.8 Repeat Exercise 5.7, but use Newton's method to find the solution.

5.9 *Rosenbrock's function* is a particularly difficult problem that looks deceptively simple. Consider the unconstrained function:

$$\text{minimize } f(x, y) = 100 (y - x^2)^2 + (1 - x)^2$$

The function has a unique minimum at the point $(1, 1)$ (where $f(1, 1) = 0$). This pathological example has also been called the *banana function*. If you plot the function, it follows a narrow banana-shaped valley from the point $(-1, 1)$ to the minimum $(1, 1)$. Because the valley has quite steep sides, anytime an algorithm tries to follow the downward slope in a straight line, the line almost immediately starts going up, resulting in very short steps. It is very difficult for any algorithm to find the way around the banana.

Try using both a gradient search and Newton's method beginning at the point $(-1, 1)$. Perform several iterations. You will likely observe rather poor progress along a narrow zig-zag path.

5.10 Consider the problem:

$$\begin{array}{ll} \text{minimize} & f(x, y) = x^2 y \\ \text{subject to} & x^2 + y^2 \text{ " } 1 \end{array}$$

Use the method of Lagrange multipliers to express this problem as an unconstrained minimization problem, and solve the problem using both the gradient method and Newton's method.

5.11 Consider the following nonlinear problem with linear constraints:

$$\begin{array}{ll} \text{maximize} & f(x, y) = x^2 y + 2y^2 \\ \text{subject to} & x + 3y \leq 9 \\ & x + 2y \leq 8 \\ & 3x + 2y \leq 18 \\ & 0 \leq x \leq 5 \\ & 0 \leq y \leq 2 \end{array}$$

Solve this problem graphically. Begin at the point $(0, 0)$, and check the gradient. If the Karush–Kuhn–Tucker conditions are not satisfied, you should be able to find an improving direction in the feasible region.

References and Suggested Readings

Avriel, M. 1976. *Nonlinear Programming: Analysis and Methods.* Englewood Cliffs, NJ: Prentice-Hall.

Bazaraa, M. S., H. D. Sherali, and C. Shetty. 2013. *Nonlinear Programming: Theory and Algorithms.* New York: John Wiley & Sons.

Beale, E. M. L. 1959. On quadratic programming. *Naval Research Logistics Quarterly* 6 (3): 227–243.

Beale, E. M. L. 1985. The evolution of mathematical programming systems. *Operational Research Society* 36 (5): 357–366.

Beale, E. M. L. 1988. *Introduction to Optimization.* Chichester, UK: John Wiley & Sons.

Beale, E. M. L., G. C. Beare, and P. Bryan-Tatham. 1974. *The DOAE Reinforcement and Redeployment Study: A Case Study in Mathematical Programming in Mathematical Programming in Theory and Practice.* Amsterdam, the Netherlands: North-Holland

Beck, A. 2015. *Introduction to Nonlinear Optimization: Theory, Algorithms and Applications with MATLAB SIAM.* New York: Society for Industrial & Applied Mathematics.

Beightler, C. S., D. T. Phillips, and D. J. Wilde. 1979. *Foundations of Optimization,* 2nd ed. Englewood Cliffs, NJ: Prentice-Hall.

Beightler, C., and D. T. Phillips. 1976. *Applied Geometric Programming.* New York: Wiley.

Bertsekas, D. P. 2015. *Convex Optimization Algorithms.* Belmont, MA: Athena Scientific.

Bertsekas, D. P. 2016. *Nonlinear Programming,* 3rd ed. Belmont, MA: Athena Scientific.

Bertsimas, D., C. Darnell, and R. Soucy. 1999. Portfolio construction through mixed-integer programming at Grantham. *Interfaces* 29 (1): 49–66.

Broyden, C. G. 1970. The convergence of a class of double-rank minimization algorithms. 1. General considerations (pp. 76–90); 2. The new algorithm (pp. 222–231). *Journal of the Institute of Mathematics and its Applications* 6.

Burley, D. M. 1974. *Studies in Optimization*. New York: John Wiley & Sons.

Chen, W. J., S. D. Rajan, H. D. Nelson, and M. Rajan. 1991. Application of nonlinear programming for balancing rotor systems. *Engineering Optimisation* 17: 79–90.

Cheney, E. W., and D. Kincaid. 1980. *Numerical Mathematics and Computing*. Monterey, CA: Brooks/Cole.

Davidon, W. C. 1959. Variable metric method for minimization. *AEC Research & Development Report ANL-5990 (Rev.)* Lemont, IL: Argonne National Laboratory.

Diwekar, U. 2010. *Introduction to Applied Optimization*. New York: Springer.

Fletcher, R. 1970. A new approach to variable metric algorithms. The *Computer Journal* 13: 317–322.

Fletcher, R. 1987. *Practical Methods of Optimization*, 2nd ed. Chichester, UK: John Wiley & Sons.

Fletcher, R., and M. J. D. Powell. 1963. A rapidly convergent descent method for minimization. *The Computer Journal* 6: 163–168.

Floudas, C. A., and P. M. Pardalos (Eds.). 1992. *Recent Advances in Global Optimization*. Princeton, NJ: Princeton University Press.

Fourer, R. 2017. Software survey: Linear programming. *OR/MS Today* 44 (3): 48–59.

Fourer, R. 1998. Software for optimization: A survey of recent trends in mathematical programming systems. *OR/MS Today* 25 (6): 40–43.

Gay, D. M. 1983. Subroutines for unconstrained minimization. *ACM Transaction on Mathematical Software* 9: 503–524.

Gill, P. E., W. Murray, and M. H. Wright. 1981. *Practical Optimization*. New York: Academic Press.

Gill, P. E., W. Murray, and M. H. Wright. 1984. Trends in nonlinear programming software. *European Journal of Operational Research* 17: 141–149.

Goldfarb, D. 1969. *Sufficient Conditions for the Convergence of a Variable-Metric Algorithm, in Optimization*. London, UK: Academic Press.

Hadley, G. 1974. *Nonlinear and Dynamic Programming*. Reading, MA: Addison-Wesley.

Hooke, R., and T. A. Jeeves. 1961. Direct search solution of numerical and statistical problems. *JACM* 8: 212–229.

Kuhn, H. W., and A. W. Tucker. 1951. Non-linear programming. *Proceedings of the 2nd Berkeley Symposium on Mathematical Statistics and Probability*. Berkeley, CA: University of California Press, Vol. 481–492.

Lasdon, L. S., and A. D. Warren. 1978. Generalized reduced gradient software for linear and non-linear constrained problems. In Greenberg (Ed.), *Design and Implementation for Optimization Software*. Alphen aan den Rijn, the Netherlands: Sijthoff/Noordhoff, pp. 363–397.

Lemke, C. E. 1962. A method of solution for quadratic programs. *Management Science* 8 (4): 442–453.

Liebman, J., L. Lasdon, L. Schrage, and A. Waren. 1986. *Modeling and Optimization with GINO*. Palo Alto, CA: The Scientific Press.

Luenberger, D. G., and Y. Ye. 2008. *Linear and Nonlinear Programming*, 3rd ed., Vol. 116, International Series in Operations Research & Management Science. New York: Springer.

Markowitz, H. M. 1959. *Portfolio Selection, Efficient Diversification of Investments*. New York: John Wiley & Sons.

Markowitz, H. M. 1987. *Mean-Variance Analysis in Portfolio Choice and Capital Markets*. New York: Blackwell.

McCormick, G. P. 1983. *Nonlinear Programming: Theory, Algorithms and Applications*. New York: John Wiley & Sons.

Moré, J. J., and S. J. Wright. 1993. *Optimization Software Guide*. Philadelphia, PA: Society for Industrial and Applied Mathematics.

Murtagh, B. A., and M. A. Saunders. 1978. Large-scale linearly constrained optimization. *Mathematical Programming* 14: 41–72.

Nash, S. G. 1998. Software survey: Nonlinear programming. *OR/MS Today* 25 (3): 36–45.

Nash, S. G., and A. Sofer. 1996. *Linear and Nonlinear Programming*. New York: McGraw-Hill.

Nelder, J. A., and R. Mead. 1965. A simplex method for function minimization. *Computer Journal* 7 (4): 308–313.

Powell, M. J. D. 1964. An efficient method for finding the minimum of a function of several variables without calculating derivatives. *Computer Journal* 7: 155–162.

Ravindran, A., and H. K. Lee. 1981. Computer experiments on quadratic programming algorithms. *European Journal of Operational Research* 8 (2): 166–174.

Rigby, B., L. S. Lasdon, and A. D. Waren. 1995. The evolution of Texaco's blending systems: From OMEGA to StarBlend. *Interfaces* 25 (5): 64–83.

Rosenbrock, H. H. 1960. An automatic method for finding the greatest or least value of a function. *Computer Journal* 3: 175–184.

Schrage, L. 1986. *Linear, Integer and Quadratic Programming with LINDO.* Palo Alto, CA: The Scientific Press.

Shanno, D. F. 1970. Conditioning of Quasi-Newton methods for function minimization. *Mathematics of Computation* 24: 647–656.

Sharpe, W. F. 1963. A simplified model for portfolio analysis. *Management Science* 9: 277–293.

Simmons, D. M. 1975. *Nonlinear Programming for Operations Research.* Englewood Cliffs, NJ: Prentice-Hall.

Wilde, D. J. 1964. *Optimum Seeking Methods.* Englewood Cliffs, NJ: Prentice-Hall.

Wolfe, P. 1959. The simplex method for quadratic programming. *Econometrica* 27 (3): 382–398.

Wolfe, P. 1967. *Methods of Nonlinear Programming in Nonlinear Programming.* Amsterdam, the Netherlands: North-Holland.

6

Markov Processes

Certain complex systems exhibit characteristics that evolve randomly over time. A Markov process is a mathematical model, based on principles developed by the Russian probability theorist A.A. Markov, that allows systems engineers and analysts to describe and predict the behavior of such systems. Probabilities and uncertainties arise in the most diverse applications, and in many cases, Markov analysis provides a framework in which to study the behavior of these systems.

For example, a Markov model was developed for aircraft landing decisions and was used to study data collected from the Pittsburgh Airport. Aircraft arriving at an airport are supplied with information describing the congestion, and based on that information, must decide whether to join the queue of planes waiting to land or to instead fly on to a different airport (Rue and Rosenshine 1985). In a very different context, a Markov decision process framework has been applied in the fishing industry to determine what proportion of a salmon population to catch in a given season, and what proportion to leave and allow to spawn and thus build up the population for the next season (White 1985, 1988).

In the passenger airline industry, decisions must be made continuously by airline bookings managers about how many reservations to accept for a specific flight up until the day of departure. The objective is to maximize passenger revenues while minimizing passenger rejections. A Markov model to assist with this decision process was applied to data from Scandinavian Airlines (Alstrup et al. 1986). And when a fire alarm is received at a fire station, a dispatcher must make decisions about how many fire engines to send out in response to the alarm, to minimize long run average fire losses. A Markov model helps with such decisions, based on the type of alarm and the number of fire engines currently out on calls (Swersey 1982).

Markov analysis has been found to be useful in areas as disparate as population dynamics, inventory management, equipment maintenance and replacement problems, market share analysis, and economic trend analysis. Our study of Markov processes will begin with some preliminary definitions, and we will then investigate the types of analysis that can be performed.

Suppose we let x_t denote some observable system characteristic at time t. The characteristic is seen to change probabilistically as time progresses; therefore, x_t is not known with certainty until time t, and can be thought of as a random variable. The sequence of random variables $x_0, x_1, x_2, ..., x_t, ...$ represents a **stochastic process** in which the value of x_t typically depends on the values of the previous random variables in the sequence. If primarily interested in studying the *changes* in the system, then we may merely index the points in time when significant events occur. (We may even wish to assume that the time between changes is a constant, and label the points in time as 0, 1, 2, ...) Such a process is called a **discrete-time stochastic process**. If, on the other hand, we wish to measure the actual progress of absolute clock time and study the time between transitions, then we have a **continuous-time stochastic process** in which the system is viewed at arbitrary times rather than at discrete instants in time. We will restrict our discussion to discrete-time stochastic processes, and in particular to a special type of process known as a Markov process.

6.1 State Transitions

In the systems we will study, the observed characteristic or condition of the system at any given time is referred to as the **state** of the system. We will assume that there is a finite number of states, numbered 1,..., N, and that at any time the system occupies (or is completely described by) exactly one of these states. When a change occurs in the system, we say that the system makes a **transition.**

A discrete-time stochastic process is called a **Markov process** if a transition from one state to another depends only on the current state of the system and not on previous states that the system may have occupied. More formally, this property can be expressed in terms of conditional probabilities:

$$P(x_{t+1} = s_{t+1} \mid x_t = s_t, x_{t-1} = s_{t-1},..., x_1 = s_1, x_0 = s_0)$$
$$= P(x_{t+1} = s_{t+1} \mid x_t = s_t)$$

The state at time $t + 1$ depends only on the state the system was in at time t and not on the values of any of the random variables $x_{t-1}, ..., x_0$. This is called the **Markov property.** And because each x_t depends only on x_{t-1} and has an effect only on x_{t+1}, the process is sometimes called a **Markov chain.** Since we assume there are finitely many states, the process is called a **finite state Markov chain.**

An additional assumption fundamental to the analysis of Markov processes is that the probability of a transition from any state i to any state j is the same for any time t. That is,

$$P(x_{t+1} = j \mid x_t = i) = p_{ij}$$

is independent of the time index t. The property that a Markov process's transitional behavior does not change over time is called the **stationarity property.**

The probability p_{ij} described earlier is called the **transition probability** of a system changing from state i at some time t to state j at time $t + 1$. Transition probabilities are defined for all states $i, j = 1, 2,..., N$ that the system may occupy, and are usually written as a **transition probability matrix**

$$P = \begin{bmatrix} p_{11} & p_{12} & \cdots & p_{1N} \\ p_{21} & p_{22} & \cdots & p_{2N} \\ \cdot & & \cdot & \\ \cdot & & \cdot & \\ \cdot & & \cdot & \\ p_{N1} & p_{N2} & \cdots & p_{NN} \end{bmatrix}$$

The elements p_{ij} are sometimes called **one-step transition probabilities** because they refer to system changes that can occur directly in one time period. And because the system must be in some state after each transition, each row of probabilities must sum to one; that is,

$$\sum_{j=1}^{N} p_{ij} = 1$$

The values of these transition probabilities define the probability distributions of the Markov chain $\{x_t\}$ and therefore describe the evolutionary behavior of the system.

A Markov process begins at some initial time $t = 0$. If the state x_0 is not known with certainty, then we must specify the probabilities with which the system is initially in each of the N states. We denote this as $P(x_0 = i) = p_i(0)$ for each state i, and we use the vector

$$p(0) = (p_1(0)\ p_2(0)\ \dots\ p_N(0))$$

to describe the **initial probability distribution** for the system.

In summary, a system can be modeled as a Markov process if it has the following four properties:

Property 1: A finite number of states can be used to describe the dynamic behavior of the system.

Property 2: Initial probabilities are specified for the system.

Property 3: Markov property—We assume that a transition to a new state depends only on the current state and not on past conditions.

Property 4: Stationarity property—The probability of a transition between any two states does not vary in time.

It should be noted that the validity of any study using the tools of Markov analysis hinges on the extent to which the Markov and stationarity assumptions are met by the actual system under investigation. We certainly realize that processes involving human choice are often affected, if only subtly, by past experiences and are not based just on a current scenario. Strictly speaking, this violates the Markov property. Furthermore, seasonal variations and political cycles may interfere with the stability or constancy with which probabilistic transitions occur, and therefore the stationarity of the transition probabilities may be questionable. In light of this, we must emphasize the importance of the analyst's understanding of the system being modeled and of the assumptions upon which Markov analysis is predicated. Almost as important as the mathematical model itself is the role that keen judgment plays in applying these procedures and in interpreting the results.

If an analyst determines that a Markov analysis is appropriate for the system being studied, then the techniques for analyzing Markov processes may provide answers to such questions as the following:

- How many transitions (steps) will it likely take for the system to move from some specified state to another specified state?
- What is the probability that it will take some given number of steps to go from one specified state to another?

- In the long run, which state is occupied by the system most frequently?
- Over a long period of time, what fraction of the time does the system occupy each of the possible states?
- Will the system continue indefinitely to move among all N states, or will it eventually settle into a certain few states?

We will look briefly at an example of a very simple Markov chain. Then in the following sections, we will show how questions such as the above can be answered for systems that can be modeled as Markov processes.

Example 6.1

Consider using Markov chains to model changes in weather at a ski resort, and try to use the model to help describe the operation and maintenance of the ski mountain equipment and the ways in which the skiers respond to the weather. Suppose that winter days can be described as either sunny, cloudy, or snowing. We can arbitrarily denote that state 1 corresponds to a sunny day, state 2 corresponds to a cloudy day, and state 3 corresponds to a day with snowfall. Suppose that after studying historical weather patterns in this particular ski location, we believe that if we know the weather condition on any given day, the weather on the next day can be described according to the following transition probabilities:

$$P = \begin{bmatrix} 0.7 & 0.1 & 0.2 \\ 0.2 & 0.7 & 0.1 \\ 0.5 & 0.2 & 0.3 \end{bmatrix}$$

So, for example, the probability that a clear day is followed by a snowy day is $p_{13} = 0.2$, and the probability that if today is cloudy then tomorrow will also be cloudy is $p_{22} = 0.7$. (Note that p_{ii} is the probability of no change from one day to the next.) We realize that the ski season probably lasts only a few months and that these weather patterns certainly do not endure into the summer. Nevertheless, *within* the winter season, these probabilities are stationary.

These one step transition probabilities can be illustrated in a **state transition diagram** in which the nodes of a graph represent system states, and arcs represent possible transitions and are labeled with transition probabilities. Figure 6.1 shows the transition diagram for our example, indicating the one step transitions that can be made. If we are interested in the probability that a certain weather condition will prevail after two days, we can use a two-step **transition tree**, as shown in Figure 6.2.

Suppose that on a given day, the ski area is experiencing sunny weather, and we wish to know the probability that in two days there will again be sunny weather. There are three ways in which a sunny ski resort can, two days later, be sunny again (that is, be in state 1 again): the weather may never change, and this happens with probability (0.7)(0.7) = 0.49; it may change to cloudy then change back to sunny, with probability (0.1)(0.2) = 0.02; or it could snow the next day then return to sunny conditions on the second day with probability (0.2)(0.5) = 0.10. The probability of the second day being sunny is then the sum of these probabilities 0.49 + 0.02 + 0.10 = 0.61. Similarly, the probability of a cloudy day two days after a sunny day is (0.7)(0.1) + (0.1)(0.7) + (0.2)(0.2) = 0.18, and the probability of a snowy day two days after a sunny day is (0.7)(0.2) + (0.1)(0.1) + (0.2)(0.3) = 0.21. Notice that we are assuming that the weather at the ski resort must be in *some*

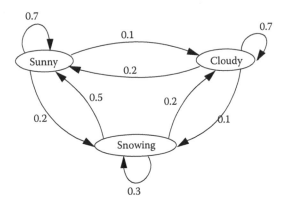

FIGURE 6.1
Transition diagram.

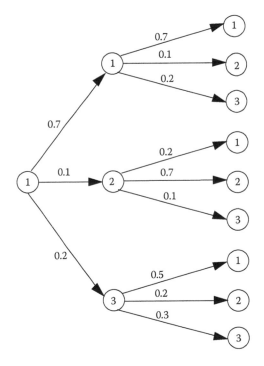

FIGURE 6.2
Transition tree.

state after two days, and indeed the three probabilities sum to one: $0.61 + 0.18 + 0.21 = 1$. Similar trees could be drawn for projecting weather two days beyond a cloudy day or a snowy day.

The transition tree is a handy way of illustrating the pattern of paths through the states, as long as the number of transition periods is small. But this technique becomes quite cumbersome if we want to examine weather behavior over many days. Fortunately, there is a much simpler and more direct way to obtain this information.

Let us denote by $p^{(n)}_{ij}$ the probability of a transition from state i to state j in n steps. Above, we calculated $p^{(2)}_{11} = 0.61$, $p^{(2)}_{12} = 0.18$, and $p^{(2)}_{13} = 0.21$. Our calculation of $p^{(2)}_{11}$ is

$$(0.7)(0.7)+(0.1)(0.2)+(0.2)(0.5)$$
$$=(p_{11})(p_{11})+(p_{12})(p_{21})+(p_{13})(p_{31})$$

which simply accounts for the three possible ways of making the transition from state 1 and back again. This is precisely the inner product of the first row of P with the first column of P; the first row defining probabilities for leaving state 1 together with the first column giving probabilities for re-entering state 1. From this observation, we can generalize that for any i and j, $p^{(2)}_{ij}$ is the inner product of the i-th row of P with the j-th column of P.

$$p^{(2)}_{ij} = \sum_{k=1}^{N} p_{ik}p_{kj}$$

Because this is exactly matrix multiplication, we find that we can compute $P^{(2)} = P^2 = P \cdot P$, and that each element of $P^{(2)}$ is just the two-step transition probability $p^{(2)}_{ij}$. In our example,

$$P^2 = \begin{bmatrix} 0.61 & 0.18 & 0.21 \\ 0.33 & 0.53 & 0.14 \\ 0.54 & 0.25 & 0.21 \end{bmatrix}$$

We can further generalize that n-step transition probabilities can be obtained from

$$P^{(n)} = P^n = P \cdot P^{n-1}$$

and

$$P^{(n)}_{ij} = \sum_{k=1}^{N} P_{ik}P^{(n-1)}_{kj}$$

Then,

$$P^3 = \begin{bmatrix} 0.568 & 0.229 & 0.203 \\ 0.407 & 0.432 & 0.161 \\ 0.533 & 0.271 & 0.196 \end{bmatrix}$$

and from this, for example, we can see that the probability that, if any given day is cloudy, then snowy weather will occur three days later, is 0.161, because $p^{(3)}_{23} = 0.161$.

Table 6.1 shows the matrices P^n for n values of from 1 to 50. We will have occasion to refer again to these computational results when we discuss related topics in Section 6.5.

TABLE 6.1

n-Step Transition Probability Matrices for $1 \leq n \leq 50$

$P^1 =$	0.7000000	0.1000000	0.2000000	$P^{18} =$	0.5135283	0.2972783	0.1891930
	0.2000000	0.7000000	0.1000000		0.5134839	0.2973344	0.1891813
	0.5000000	0.2000000	0.3000000		0.5135191	0.2972900	0.1891906
$P^2 =$	0.6100000	0.1800000	0.2100000	$P^{19} =$	0.5135220	0.2972862	0.1891913
	0.3300000	0.5300000	0.1400000		0.5134963	0.2973187	0.1891846
	0.5400000	0.2500000	0.2100000		0.5135167	0.2972930	0.1891899
$P^3 =$	0.5680000	0.2290000	0.2030000	$P^{20} =$	0.5135184	0.2972908	0.1891904
	0.4070000	0.4320000	0.1610000		0.5135035	0.2973097	0.1891864
	0.5330000	0.2710000	0.1960000		0.5135153	0.2972948	0.1891895
$P^4 =$	0.5449000	0.2577000	0.1974000	$P^{21} =$	0.5135163	0.2972935	0.1891898
	0.4518000	0.3753000	0.1729000		0.5135076	0.2973044	0.1891875
	0.523000	0.2822000	0.1925000		0.5135145	0.2972958	0.1891893
$P^5 =$	0.5316700	0.2743600	0.1939700	$P^{22} =$	0.5135150	0.2972950	0.1891895
	0.4777700	0.3424700	0.1797600		0.5135100	0.2973013	0.1891882
	0.5204000	0.2885700	0.1910300		0.5135143	0.2972960	0.1891892
$P^6 =$	0.5240260	0.2840130	0.1919610	$P^{23} =$	0.5135143	0.2972959	0.1891893
	0.4928130	0.3234580	0.1837290		0.5135114	0.2972996	0.1891885
	0.5175090	0.2922450	0.1902460		0.5135137	0.2972967	0.1891891
$P^7 =$	0.5196013	0.2896039	0.1907948	$P^{24} =$	0.5135139	0.2972964	0.1891892
	0.5015252	0.3124477	0.1860271		0.5135122	0.2972985	0.1891887
	0.5158283	0.2943716	0.1898000		0.5135135	0.2972969	0.1891891
$P^8 =$	0.5170390	0.2928418	0.1901190	$P^{25} =$	0.5135136	0.2972967	0.1891891
	0.5065707	0.3060713	0.1873579		0.5135127	0.2972979	0.1891888
	0.5148541	0.2956029	0.1895428		0.5135135	0.2972970	0.1891890
$P^9 =$	0.5155552	0.2947169	0.1897277	$P^{26} =$	0.5135135	0.2972969	0.1891890
	0.5094927	0.3023785	0.1881286		0.5135130	0.2972976	0.1891889
	0.5142899	0.2963160	0.1893939		0.5135134	0.2972970	0.1891890
$P^{10} =$	0.5146959	0.2958029	0.1895010	$P^{27} =$	0.5135134	0.2972970	0.1891890
	0.5111849	0.3002400	0.1885749		0.5135131	0.2972974	0.1891889
	0.5139631	0.2967290	0.1893077		0.5135133	0.2972970	0.1891890
$P^{11} =$	0.5141982	0.2964318	0.1893697	$P^{28} =$	0.5135134	0.2972970	0.1891890
	0.5121649	0.2990014	0.1888334		0.5135132	0.2972973	0.1891890
	0.5137738	0.2969681	0.1892578		0.5135133	0.2972971	0.1891890
$P^{12} =$	0.5139100	0.2967960	0.1892937	$P^{29} =$	0.5135133	0.2972970	0.1891890
	0.5127325	0.2982842	0.1889831		0.5135132	0.2972972	0.1891890
	0.5136642	0.2971066	0.1892289		0.5135133	0.2972971	0.1891890
$P^{13} =$	0.5137431	0.2970070	0.1892497	$P^{30} =$	0.5135133	0.2972971	0.1891890
	0.5130611	0.2978688	0.1890698		0.5135133	0.2972971	0.1891890
	0.5136008	0.2971868	0.1892121		0.5135133	0.2972971	0.1891890
$P^{14} =$	0.5136464	0.2971291	0.1892242	$P^{31} =$	0.5135133	0.2972971	0.1891890
	0.5132515	0.2976282	0.1891200		0.5135133	0.2972971	0.1891890
	0.5135640	0.2972333	0.1892024		0.5135133	0.2972971	0.1891890
$P^{15} =$	0.5135904	0.2971998	0.1892094	$P^{32} =$	0.5135133	0.2972971	0.1891890
	0.5133617	0.2974889	0.1891491		0.5135133	0.2972971	0.1891890
	0.5135427	0.2972602	0.1891968		0.5135133	0.2972971	0.1891890
$P^{16} =$	0.5135580	0.2972408	0.1892008	$P^{33} =$	0.5135133	0.2972971	0.1891889
	0.5134256	0.2974082	0.1891659		0.5135133	0.2972971	0.1891890
	0.5135304	0.2972757	0.1891935		0.5135133	0.2972971	0.1891889
$P^{17} =$	0.5135392	0.2972645	0.1891959	$P^{34} =$	0.5135133	0.2972970	0.1891889
	0.5134625	0.2973615	0.1891757		0.5135133	0.2972971	0.1891889
	0.5135235	0.2972848	0.1891917		0.5135133	0.2972970	0.1891889

(Continued)

TABLE 6.1 (*Continued*)

n-Step Transition Probability Matrices for $1 \leq n \leq 50$

$P^{35} =$	0.5135133	0.2972970	0.1891889	$P^{43} =$	0.5135132	0.2972970	0.1891889
	0.5135133	0.2972971	0.1891889		0.5135132	0.2972970	0.1891889
	0.5135133	0.2972970	0.1891889		0.5135132	0.2972970	0.1891889
$P^{36} =$	0.5135133	0.2972970	0.1891889	$P^{44} =$	0.5135132	0.2972970	0.1891889
	0.5135133	0.2972970	0.1891889		0.5135132	0.2972970	0.1891889
	0.5135132	0.2972970	0.1891889		0.5135132	0.2972970	0.1891889
$P^{37} =$	0.5135132	0.2972970	0.1891889	$P^{45} =$	0.5135132	0.2972969	0.1891888
	0.5135132	0.2972970	0.1891889		0.5135132	0.2972970	0.1891888
	0.5135132	0.2972970	0.1891889		0.5135132	0.2972970	0.1891888
$P^{38} =$	0.5135132	0.2972970	0.1891889	$P^{46} =$	0.5135132	0.2972969	0.1891888
	0.5135132	0.2972970	0.1891889		0.5135132	0.2972970	0.1891888
	0.5135132	0.2972970	0.1891889		0.5135132	0.2972969	0.1891888
$P^{39} =$	0.5135132	0.2972970	0.1891889	$P^{47} =$	0.5135132	0.2972969	0.1891888
	0.5135132	0.2972970	0.1891889		0.5135132	0.2972969	0.1891888
	0.5135132	0.2972970	0.1891889		0.5135132	0.2972969	0.1891888
$P^{40} =$	0.5135132	0.2972970	0.1891889	$P^{48} =$	0.5135132	0.2972969	0.1891888
	0.5135132	0.2972970	0.1891889		0.5135132	0.2972969	0.1891888
	0.5135132	0.2972970	0.1891889		0.5135132	0.2972969	0.1891888
$P^{41} =$	0.5135132	0.2972970	0.1891889	$P^{49} =$	0.5135132	0.2972969	0.1891888
	0.5135132	0.2972970	0.1891889		0.5135132	0.2972969	0.1891888
	0.5135132	0.2972970	0.1891889		0.5135132	0.2972969	0.1891888
$P^{42} =$	0.5135132	0.2972970	0.1891889	$P^{50} =$	0.5135132	0.2972969	0.1891888
	0.5135132	0.2972970	0.1891889		0.5135132	0.2972969	0.1891888
	0.5135132	0.2972970	0.1891889		0.5135132	0.2972969	0.1891888

6.2 State Probabilities

We have already seen notation to describe the *initial* probability of the system being in each of the possible states. We used the vector

$$p(0) = (p_1(0) \ p_2(0)\ldots p_N(0))$$

where each $p_i(0) = P(x_0 = i)$ is the probability that the system is initially in state i. We can extend this notation and define a **state probability vector**

$$p(t) = (p_1(t) \ p_2(t) \ \ldots \ p_N(t))$$

where $p_i(t)$ is the probability that the system will occupy state i at any time t if the **state probabilities** at time 0 are known.

State probabilities can be defined recursively as follows:

$$p(1) = p(0) \cdot P$$

$$p(2) = p(1) \cdot P = p(0) \cdot P^2$$

$$p(3) = p(2) \cdot P = p(1) \cdot P^2 = p(0) \cdot P^3$$

and, in general,

$$p(n) = p(0) \cdot P^n \text{ for } n = 0, 1, 2, \ldots$$

Returning to our example of weather patterns, suppose that on a certain day at the beginning of a series of weather observations, the weather is sunny. The initial state probability vector is

$$p(0) = (1.0\ 0\ 0)$$

Then, the state probabilities after one day are:

$$p(1) = p(0) \cdot P = (1.0\ 0\ 0) \cdot \begin{bmatrix} 0.7 & 0.1 & 0.2 \\ 0.2 & 0.7 & 0.1 \\ 0.5 & 0.2 & 0.3 \end{bmatrix} = (0.7\ 0.1\ 0.2)$$

After two days, the probabilities are:

$$p(2) = p(1) \cdot P = (0.7\ 0.1\ 0.2) \cdot P$$

but because we have already computed P^2 we can more directly obtain $p(2)$ as:

$$p(2) = p(0) \cdot P^2 = (1.0\ 0\ 0) \cdot \begin{bmatrix} 0.61 & 0.18 & 0.21 \\ 0.33 & 0.53 & 0.14 \\ 0.54 & 0.25 & 0.21 \end{bmatrix} = (0.61\ 0.18\ 0.21)$$

Likewise,

$$p(3) = p(0) \cdot P^3 = (1.0\ 0\ 0) \cdot \begin{bmatrix} 0.568 & 0.229 & 0.203 \\ 0.407 & 0.432 & 0.161 \\ 0.533 & 0.271 & 0.196 \end{bmatrix} = (0.568\ 0.229\ 0.203)$$

$$p(4) = p(0) \cdot P^4 = (1.0\ 0\ 0) \cdot \begin{bmatrix} 0.5449 & 0.2577 & 0.1974 \\ 0.4518 & 0.3753 & 0.1729 \\ 0.5253 & 0.2822 & 0.1925 \end{bmatrix} = (0.5449\ 0.2577\ 0.1974)$$

$$p(5) = p(0) \cdot P^5 = (1.0\ 0\ 0) \cdot \begin{bmatrix} 0.5317 & 0.2744 & 0.1940 \\ 0.4778 & 0.3425 & 0.1798 \\ 0.5204 & 0.2886 & 0.1910 \end{bmatrix} = (0.5317\ 0.2744\ 0.1940)$$

$$p(6) = p(0) \cdot P^6 = (1.0\ 0\ 0) \cdot \begin{bmatrix} 0.5240 & 0.2840 & 0.1920 \\ 0.4929 & 0.3235 & 0.1837 \\ 0.5175 & 0.2922 & 0.1902 \end{bmatrix} = (0.5240\ 0.2840\ 0.1920)$$

If we performed the same calculations under the assumption that on day 1 the weather is cloudy and therefore p(0) = (0 1.0 0), we would find:

$$p(1) = (0\ 1.0\ 0)\ P = (0.2\ 0.7\ 0.1)$$

$$p(2) = (0.33\ 0.53\ 0.14)$$

$$p(3) = (0.407\ 0.432\ 0.161)$$

$$p(4) = (0.4518\ 0.3753\ 0.1729)$$

$$p(5) = (0.4778\ 0.3435\ 0.1798)$$

$$p(6) = (0.4928\ 0.3235\ 0.1837)$$

Now suppose that, instead of actually *observing* the weather on the first day, we assume that on that day it is equally likely to be sunny, cloudy, or snowing; that is, p(0) = (1/3 1/3 1/3). Then,

$$p(1) = (1/3\ 1/3\ 1/3) \cdot P = (0.467\ 0.333\ 0.200)$$
$$p(2) = (1/3\ 1/3\ 1/3) \cdot P^2 = (0.493\ 0.320\ 0.187)$$
$$p(3) = (1/3\ 1/3\ 1/3) \cdot P^3 = (0.503\ 0.310\ 0.187)$$
$$p(4) = (1/3\ 1/3\ 1/3) \cdot P^4 = (0.507\ 0.305\ 0.188)$$
$$p(5) = (1/3\ 1/3\ 1/3) \cdot P^5 = (0.510\ 0.302\ 0.188)$$
$$p(6) = (1/3\ 1/3\ 1/3) \cdot P^6 = (0.511\ 0.300\ 0.189)$$

What we can observe in this particular example is that after several transitions, the probabilities of the system being in given states tend to converge, or become constant, independent of the initial state. (We will see in later sections that not all Markov systems behave in this way.) In our example, however, it appears that after six days, the probability of a sunny day occurring at the ski area is roughly 0.51, the probability of a cloudy day is roughly 0.30, and the probability of a snowy day is roughly 0.19. And these state probabilities hold, regardless of the actual or expected initial weather conditions.

We have no precise way of knowing how long it will take a Markov chain to *stabilize*, as we have seen above; but, if there are not many P entries very near to zero or one, this stabilization will be achieved fairly quickly. For our example, the rows of the matrix P^n become almost indistinguishable from one another for n > 10. (Refer back to Table 6.1.) Thus, after about 10 days, the effects of the initial distribution of weather probabilities will have disappeared. We can assume that since this stabilization appears to occur within 10 days (transition steps), then surely this technique will be of some use in modeling weather patterns during a ski season of, say, 120 days. (Exercise 6.4 provides some further insight into the contrast in rates of convergence.)

When we say that the state probabilities become constant, this does *not* mean that after a long period of time, the system does not change states any longer. Rather, as transitions continue to occur, the system's occupancy of each state is in some sense predictable.

Based on information such as this, we could, in our example, answer such questions as:

- Should we plan more (or less) mid-slope barbecues to entertain skiers on sunny days?
- Do we need better canopies on the chairlifts to shield the skiers from falling snow?
- Should we sell special ski goggles to improve skiers' visibility during cloudy conditions?

In Section 6.5, we discuss just what characteristics a Markov chain must have that will permit us to make this kind of analysis of long-range behavior of the system. We will see that long-term trends can be studied directly without our having to compute state probabilities for huge values of n.

6.3 First Passage Probabilities

In a typical Markov chain, we frequently observe that states are left and re-entered again and again. We have developed a means of computing the probability $p^{(n)}_{ij}$ that a system will leave state i and be in state j after n transitions. But this does not give us any information about whether the system entered state j *at any time before* the n-th transition. Suppose we are specifically interested in the probability that a system leaves state i and enters state j *for the first time* after n steps. This is called the **first passage probability**, and it is clearly related to the n-step probability. However, we must exclude all the ways in which the system may have entered state j *before* the n-th transition. For example, if we know the probability of a cloudy day on the slopes being followed three days later by a sunny day, we realize that this three-step transition can occur in several ways:

cloudy → cloudy → cloudy → sunny
cloudy → cloudy → sunny → sunny
cloudy → sunny → sunny → sunny
cloudy → cloudy → snowy → sunny
cloudy → snowy → snowy → sunny
cloudy → snowy → sunny → sunny
cloudy → sunny → snowy → sunny
cloudy → sunny → cloudy → sunny
cloudy → snowy → cloudy → sunny

The probability $p^{(3)}_{21}$ accounts for *all* of these possible paths. By contrast, the *first passage* probability will account for only those paths in which a sunny day does not occur until the third step. Thus, we want to measure the probability that one of the following paths will occur:

cloudy → cloudy → cloudy → sunny
cloudy → cloudy → snowy → sunny
cloudy → snowy → snowy → sunny
cloudy → snowy → cloudy → sunny

The **first passage probability** $f^{(n)}_{ij}$ is computed as $p^{(n)}_{ij}$ minus the probabilities of the n-step paths in which state j occurs *before* the n-th step. In one transition, $p^{(1)}_{ij} = f^{(1)}_{ij}$. Then for larger n,

$$f^{(2)}_{ij} = p^{(2)}_{ij} - f^{(1)}_{ij}p^{(1)}_{jj}$$

$$f^{(3)}_{ij} = p^{(3)}_{ij} - f^{(1)}_{ij}p^{(2)}_{jj} - f^{(2)}_{ij}p^{(1)}_{jj}$$

.

.

.

$$f^{(n)}_{ij} = p^{(n)}_{ij} - f^{(1)}_{ij}p^{(n-1)}_{jj} - f^{(2)}_{ij}p^{(n-2)}_{jj} - \ldots - f^{(n-1)}_{ij}p^{(1)}_{jj}$$

or, more succinctly,

$$f^{(n)}_{ij} = p^{(n)}_{ij} - \sum_{k=1}^{n-1} f^{(k)}_{ij}p^{(n-k)}_{jj}$$

Returning to our example, we will illustrate $f^{(n)}_{ij}$ for n = 1, 2, and 3.

$$F^{(1)} = P = \begin{bmatrix} 0.7 & 0.1 & 0.2 \\ 0.2 & 0.7 & 0.1 \\ 0.5 & 0.2 & 0.3 \end{bmatrix}$$

$$F^{(2)} = \begin{bmatrix} 0.61-(0.7)(0.7) & 0.18-(0.1)(0.7) & 0.21-(0.2)(0.3) \\ 0.33-(0.2)(0.7) & 0.53-(0.7)(0.7) & 0.14-(0.1)(0.3) \\ 0.54-(0.5)(0.7) & 0.25-(0.2)(0.7) & 0.21-(0.3)(0.3) \end{bmatrix} = \begin{bmatrix} 0.12 & 0.11 & 0.15 \\ 0.19 & 0.04 & 0.11 \\ 0.19 & 0.11 & 0.12 \end{bmatrix}$$

$$F^{(3)} = \begin{bmatrix} 0.568-(0.7)(0.61)-(0.12)(0.7) & 0.229-(0.1)(0.53)-(0.11)(0.7) & 0.203-(0.2)(0.21)-(0.15)(0.3) \\ 0.407-(0.2)(0.61)-(0.19)(0.7) & 0.432-(0.7)(0.53)-(0.04)(0.7) & 0.161-(0.1)(0.21)-(0.11)(0.3) \\ 0.533-(0.5)(0.61)-(0.19)(0.7) & 0.271-(0.2)(0.53)-(0.11)(0.7) & 0.196-(0.3)(0.21)-(0.12)(0.3) \end{bmatrix}$$

$$= \begin{bmatrix} 0.057 & 0.099 & 0.116 \\ 0.152 & 0.033 & 0.107 \\ 0.095 & 0.088 & 0.097 \end{bmatrix}$$

And, just to check our intuition, let us re-examine $f^{(3)}_{21}$. This should be $f^{(3)}_{21} = p^{(3)}_{21}$

- Probability (cloudy → cloudy → sunny → sunny)
- Probability (cloudy → sunny → sunny → sunny)
- Probability (cloudy → snowy → sunny → sunny)
- Probability (cloudy → sunny → snowy → sunny)
- Probability (cloudy → sunny → cloudy → sunny)

$$= 0.407 - 0.098 - 0.098 - 0.035 - 0.020 - 0.004$$

$$= 0.152$$

which is exactly the value we computed earlier as the element $f^{(3)}_{21}$ in the matrix $F^{(3)}$.

6.4 Properties of the States in a Markov Process

Before continuing with our analysis of the long-term behavior of Markov processes, we must define some of the properties of the states that can be occupied by a Markov process. As we will see, the particular patterns with which transitions occur into and out of a state have a great deal to do with the role which that state plays in the eventual behavioral trends of a system.

A state j is **reachable** from state i if there is a sequence of transitions that begins in state i and ends in state j. This is, $p^{(n)}_{ij} > 0$ for some n.

An **irreducible** Markov chain is one in which every state is reachable from every other state. That is, in an irreducible chain, it is not possible for the process to become *trapped* and thereafter to make transitions only within some subset of the states.

A set of states is said to be **closed** if no state outside of the set is reachable from any state inside the set. This means that once the system enters any state in the set, it will never leave the set. In an irreducible chain, all the states constitute a closed set and no subset of the states is closed.

A particularly interesting case arises if a closed set contains only one state. This state i is called an **absorbing state**, and $p_{ii} = 1$. The system never leaves an absorbing state.

A state i is a **transient state** if there is a transition out of state i to some other state j from which state i can never be reached again. Thus, whenever a transient state is left, there is a positive probability it will never be occupied again. And therefore, the long-term probability of a system being in a transient state is essentially zero because eventually, the state will be left and never entered again.

A **recurrent state** is any state that is not transient. In an irreducible finite-state Markov chain, all states are recurrent. A special case of a recurrent state is an absorbing state, from which no other state can be reached.

The various state characteristics just defined can be illustrated by the Markov process whose **one-step transition probability** matrix is given by:

$$P = \begin{bmatrix} 1/3 & 1/3 & 0 & 0 & 0 & 1/3 \\ 0 & 0 & 9/10 & 0 & 0 & 1/10 \\ 0 & 0 & 1/3 & 2/3 & 0 & 0 \\ 0 & 0 & 1/2 & 1/4 & 1/4 & 0 \\ 0 & 0 & 1 & 0 & 0 & 0 \\ 0 & 0 & 0 & 0 & 0 & 1 \end{bmatrix}$$

and which is illustrated by the transition diagram in Figure 6.3. In this example, state 6 is an absorbing (and therefore recurrent) state because there is only one arc out of state 6, and $p_{66} = 1$. States 1 and 2 are transient because after each state is left, it is never re-entered. In the case of state 2, there is no possible way to return. On the other hand, it is possible for state 1 to recur as the system changes from state 1 directly again to state 1; but once a transition is made out of state 1, that state is never entered again. States 3, 4, and 5 are recurrent states.

A state is said to be **periodic** if it is occupied only at times which differ from one another by multiples of some constant greater than 1. In Figure 6.4a, all three states are periodic with period 2; and in Figure 6.4b, all states are periodic with period 4. In general, the period t of a **periodic state** i is the smallest integer such that all transition sequences from state i back to itself take some multiple of t steps, and t > 1.

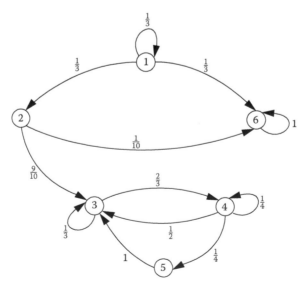

FIGURE 6.3
Transition diagram with transient and recurrent states.

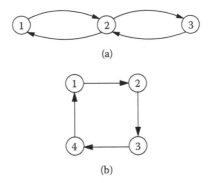

FIGURE 6.4
(a,b) Periodic states.

If a state can be occupied at *any* time, then it is said to have period 1, or to be **aperiodic**. In an irreducible chain, all states are either aperiodic, or are periodic and all have the same period. An irreducible finite-state Markov chain in which all states are aperiodic is called an **ergodic** Markov chain.

In the limit, as time goes to infinity, the occupancy of periodic states never stabilizes because these states are *likely* to be occupied at certain times and yet *cannot* be occupied at other times. Similarly, in the long run, transient states are not of interest because they are eventually left and not re-entered. If we eliminate periodic and transient states from our consideration, and focus on ergodic processes, we find that we can further characterize the limiting behavior of a Markov process. In the following sections, we will describe the calculations for two such behavioral characteristics observed in Markov processes.

6.5 Steady-State Analysis

In Section 6.2, we noticed, while computing successively higher powers of the transition probability matrix, that state probabilities tended to stabilize regardless of the initial state of the system. This behavior is typical of ergodic Markov chains and is found in most practical applications. When we say that state probabilities become *stable*, what we really mean is that as the time parameter n becomes large, $P^{(n+1)}$ is essentially the same as $P^{(n)}$. Furthermore, the rows of $P^{(n)}$ begin to look identical as n grows large, which means that, independent of the initial state of the system, the probabilities of having evolved *into* each state after any large number of steps do not change. Mathematically, if P is the one-step transition probability matrix, then the powers of P approach a matrix $[\mathcal{P}]$ where

$$\mathcal{P} = \lim_{n \to \infty} P^n$$

in which

1. Each row of $\mathcal{P}$ is an identical probability vector called Π, and

$$\Pi_j = \lim_{n \to \infty} p^{(n)}_{ij} \text{ independent of i.}$$

2. All elements of Π are positive.

3. Π has the property that $\Pi\,P = \Pi$; that is,

$$\Pi_j = \sum_{i=1}^{N} \Pi_i p_{ij} \text{ for all } j = 1, 2, \ldots, N$$

4. The elements of Π represent a probability distribution and therefore

$$\sum_{j=1}^{N} \Pi_j = 1$$

The vector Π is called the **steady-state probability vector,** and each Π_j denotes the steady-state probability that the process is in state j after a large number of steps (long after the effects of the initial state have eroded away).

As we cautioned at the end of Section 6.2, when we establish the steady-state (or *long-term* or *equilibrium*) behavior of a system, we are not determining that the dynamic system has finally come to rest in some particular state. On the contrary, transitions continue to occur with exactly the probabilities that governed the transitions early in the chain (this is the stationarity assumption). What the steady state probabilities do tell us can be interpreted in several ways:

- Π_j is the probability that, if we inspect the system at any instant (long after the process has begun), we will find the system in state j.
- Π_j is the percentage of time the system spends in state j in the long run.
- If the Markov chain models the behavior of a large number of entities that all obey the same transition probabilities, then Π_j is the fraction (or proportion) of entities that occupy state j at any given time.

Solving the system of equations $\Pi \cdot P = \Pi$ is much more satisfactory computationally than raising P to higher and higher powers (see Exercise 6.2). However, to obtain the solutions needed, we must note that the rank of the matrix P is $N - 1$, where N is the number of states in the process. (That is, if we add together any $N - 1$ rows, we get the remaining row.) Therefore, we have a system of N dependent equations in N unknowns, and consequently infinitely many solutions. We want to obtain that unique solution for which the unknowns are a probability distribution; therefore, we discard any one of the first N equations in $\Pi \cdot P = \Pi$ and introduce the equation

$$\sum_{j=1}^{N} \Pi_j = 1$$

We now have a system of N independent equations in which we can uniquely solve the N unknowns Π_j for $j = 1, 2, \ldots, N$.

To illustrate this, we can re-examine the daily changes in the weather system at the ski resort. This process is ergodic, so we can apply a steady state analysis:

$$\begin{bmatrix} \Pi_1 & \Pi_2 & \Pi_3 \end{bmatrix} \begin{bmatrix} 0.7 & 0.1 & 0.2 \\ 0.2 & 0.7 & 0.1 \\ 0.5 & 0.2 & 0.3 \end{bmatrix} = \begin{bmatrix} \Pi_1 & \Pi_2 & \Pi_3 \end{bmatrix}$$

The three dependent equations are:

$$\Pi_1 = 0.7\Pi_1 + 0.2\Pi_2 + 0.5\Pi_3$$
$$\Pi_2 = 0.1\Pi_1 + 0.7\Pi_2 + 0.2\Pi_3$$
$$\Pi_3 = 0.2\Pi_1 + 0.1\Pi_2 + 0.3\Pi_3$$

We can arbitrarily choose to discard any one of the three (how about the first one?), and use instead the normalizing equation $\sum \Pi_j = 1$ to obtain the system:

$$0.1\Pi_1 - 0.3\Pi_2 + 0.2\Pi_3 = 0$$
$$0.2\Pi_1 + 0.1\Pi_2 - 0.7\Pi_3 = 0$$
$$\Pi_1 + \Pi_2 + \Pi_3 = 1$$

The solution to this system of simultaneous linear equations is

$$\Pi_1 = 0.5135135$$
$$\Pi_2 = 0.2972973$$
$$\Pi_3 = 0.1891892$$

If we compare these results with the elements in the higher powers of P that we computed in Section 6.2 (as shown in Table 6.1), we find that indeed the value $\Pi_1 = 0.5135135$ appears in all rows of column 1, the value $\Pi_2 = 0.2972973$ appears in all rows of column 2, and the value $\Pi_3 = 0.1891892$ appears in all rows of column 3. However, this pattern does not stabilize until we compute P^n for n values around 26 or 27. (The pattern becomes apparent at about P^{10}, but small changes continue to be evident as we compute successively higher powers, up to about P^{27}, after which no significant changes in P^n occur.)

The computational effort required to raise P to the 27-th power is considerably greater than the effort required to solve the system of three equations. Furthermore, we have no way of knowing in advance just exactly what power of P needs to be computed. Exercise 6.2 will allow you to observe this contrast for yourself. Solving the steady-state equations is clearly the preferred method for determining the steady-state probabilities.

6.6 Expected First Passage Times

We have defined the **first passage time** of changing from state i to state j to be the number of transitions made by a Markov process as it goes from state i to state j for the first time. If $i = j$, then first passage time is the number of steps before the process returns to the same state, and this is called the **first recurrence time**. We will denote the first passage time from state i to state j as T_{ij}.

If the Markov process is certain to go from state i to state j eventually (given that the process ever enters state i), then T_{ij} is a random variable. In Section 6.3, we discussed the first passage probability $f^{(n)}_{ij}$, which is the probability that $T_{ij} = n$.

If a process in state i is not certain to ever reach state j, then

$$\sum_{n=1}^{\infty} f^{(n)}_{ij} < 1$$

Otherwise, the $f^{(n)}_{ij}$ are a probability distribution for the first passage times T_{ij}, and

$$\sum_{n=1}^{\infty} f^{(n)}_{ij} = 1$$

We can then write the **expected first passage times** m_{ij} from state i to state j as

$$E(T_{ij}) = m_{ij} = \sum_{n=1}^{\infty} n \, f^{(n)}_{ij}$$

(If $i = j$, this is called **expected recurrence time**.) Using these results, we could answer such questions as:

- How many days might we expect to wait for snowy weather to become sunny?
- After how many days on the average will snowy weather conditions again be snowy, after possibly changing to sunny or cloudy in the meantime?

From a computational standpoint, obtaining expected first passage times using the previous formula is difficult because we have to compute $f^{(n)}_{ij}$ for all n. However, in the case of expected recurrence time from state i back to itself, we can simply take the reciprocal of the steady-state probability to obtain

$$m_{ii} = \frac{1}{\Pi_i}$$

(For example, if a process is in state i 1/4 of the time during steady-state, then $\Pi_i = 1/4$ and $m_{ii} = 4$. That is, we would expect that an average of four steps are required to return to state i.)

For general i and j, we need a practical way of computing m_{ij}. Suppose a Markov process is currently in state i. Then with probability p_{ij}, the process will make a transition to state j for the first time in *one* step. Otherwise, the first move will be to some state k other than j; and for each $k = 1, \ldots, N$, $k \neq j$, this will happen with probability p_{ik}. In each of these cases, the first passage time will be 1 (the transition from state i to state k) plus the expected first passage time from the state k to state j. Therefore,

$$m_{ij} = (p_{ij})(1) + \sum_{\substack{k=1 \\ k \neq j}}^{N} (p_{ik})(1 + m_{kj})$$

which can be expressed as

$$m_{ij} = \left(p_{ij} + \sum_{\substack{k=1 \\ k \neq j}}^{N} p_{ik} \right) + \sum_{\substack{k=1 \\ k \neq j}}^{N} p_{ik} m_{kj}$$

to obtain

$$m_{ij} = 1 + \sum_{\substack{k=1 \\ k \neq j}}^{N} p_{ik} m_{kj}$$

Thus, m_{ij} is defined in terms of other expected first passage times m_{kj}. Since the p_{ik} are known constants, we simply have a linear equation involving $N - 1$ expected first passage times. However, this formula can be used to obtain an equation describing *each* of the m_{kj} involved in the first equation. The resulting system of $N - 1$ simultaneous linear equations in $N - 1$ unknowns can be solved to obtain unique values for all $N - 1$ expected first passage times into state j.

To answer the question: how many days might we expect to wait before snowy ski conditions change to sunny conditions, we need to compute the expected first passage time

$$m_{31} = 1 + p_{32} m_{21} + p_{33} m_{31}$$

Since m_{31} is defined in terms of m_{21}, we also need the equation:

$$m_{21} = 1 + p_{22} m_{21} + p_{23} m_{31}$$

These two equations can be solved simultaneously to obtain $m_{31} = 2.6316$ and $m_{21} = 4.2105$. Therefore, on the average, it is 2.6316 days before snowy weather first becomes sunny. In the process of finding this result, we also observe that it is an average of 4.2105 days before cloudy conditions become sunny.

If we wish to know the first recurrence time m_{22} to answer the question: after how many days on the average will cloudy weather again become cloudy?, then we solve

$$m_{22} = 1 + p_{21} m_{12} + p_{23} m_{32}$$

To do this, we also need

$$m_{12} = 1 + p_{11} m_{12} + p_{13} m_{32}$$

and

$$m_{32} = 1 + p_{31} m_{12} + p_{33} m_{32}$$

We solve this system to find that $m_{12} = 8.1818$, $m_{32} = 7.2727$, and $m_{22} = 3.3636$. Recall that we can also find m_{22} more quickly as $1/\Pi_2 = 1/.2972973 = 3.3636$, if we have already computed the steady-state probabilities. Thus, cloudy conditions that change to sunny or snowy can be expected to return to cloudy after 3.3636 days.

6.7 Absorbing Chains

The ergodic Markov chains that we have been studying represent processes which continue indefinitely and whose behavior at arbitrary future times is characterized through the steady state analysis presented in Section 6.5. Yet another interesting class of Markov chain

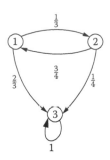

FIGURE 6.5
Absorbing chain.

applications arises when the process arrives, after a finite number of transitions, at some state from which it never leaves. Such a process evolves initially through (or within) a set of transient states (according to one-step transition probabilities), but eventually is certain to leave these states and enter one of the absorbing states. A Markov chain with at least one absorbing state is called an **absorbing chain**. By analyzing the behavior of absorbing chains, we can model processes whose stochastic behavior eventually terminates rather than continuing indefinitely.

Consider the non-ergodic Markov chain depicted in Figure 6.5. In this simple example, it is clear that if the process begins in state 1 or state 2, it may alternate between those two transient states for some time, but eventually a transition will occur—either from state 1 or 2—into state 3 (an absorbing state). Then, since $p_{33} = 1$, this system will never again enter state 1 or state 2. It might be imagined that this Markov model represents the conditions of patients in a hospital ward for trauma victims, in which states 1 and 2 denote critical and serious conditions and state 3 denotes terminal conditions. Whereas critical patients may be upgraded to serious, and serious patients may turn critical, no improvements are made by those classified as terminal.

Steady state conditions for such systems are not determined in the same way as for ergodic chains. If we wish to define steady-state probabilities to describe the situation shown in Figure 6.5, we should recognize that in the long run, the transient states will not be occupied at all, and $\Pi_i = 0$ for all transient states i. In this example, $\Pi_1 = \Pi_2 = 0$. The absorbing state, on the other hand, will always be occupied in the long run, and thus its steady-state probability is 1. In this example, $\Pi_3 = 1$. In a process that has more than one absorbing state (only one of which will ever eventually be occupied), steady-state probabilities do not exist.

There is an interesting distinction between ergodic chains and absorbing chains. While initial conditions do not affect steady-state probabilities in an ergodic chain, the initial state of an absorbing chain has a strong effect on *which* absorbing state is eventually entered. For example, in the transition diagram in Figure 6.6, we can examine the transition probabilities and see that, if the process begins in state 2, it is most likely to be absorbed into state 4; whereas if the process is initially in state 1, then the most likely absorbing state to be entered is state 3. The probability that an absorbing state will be entered is called its **absorption probability**. Absorption probabilities are conditional probabilities, dependent on the initial state of the process. In this section, we will learn how to analyze the ways in which transient states are occupied before an absorbing chain enters an absorbing state, and we will see how to compute the probabilities with which each absorbing state will be entered.

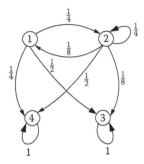

FIGURE 6.6
Absorbing chain with two absorbing states.

We first rearrange the one-step transition probability matrix, if necessary, so that both rows and columns are indexed by transient states first, and then absorbing states. Then, if we have r transient states and N – r absorbing states, the matrix P has the following structure.

$$
P = \begin{array}{c} \text{r rows} \\ \text{N – r rows} \end{array}
\left[\begin{array}{c|c}
\overbrace{Q}^{\text{r columns}} & \overbrace{R}^{\text{N – r columns}} \\ \hline
0 & I
\end{array} \right]
$$

In this form, the submatrix Q contains the one-step transition probabilities from transient states to transient states, and R contains transition probabilities from transient states to absorbing states. The lower-left submatrix of zeros indicates the impossible transitions from absorbing states to transient states, and the identity matrix I indicates the certain transitions from each absorbing state to itself.

The Markov chain in Figure 6.6 has the one-step transition probability matrix

$$
P = \begin{bmatrix}
0 & 1/4 & 1/2 & 1/4 \\
1/8 & 1/4 & 1/8 & 1/2 \\
0 & 0 & 1 & 0 \\
0 & 0 & 0 & 1
\end{bmatrix}
$$

with

$$
Q = \begin{bmatrix} 0 & 1/4 \\ 1/8 & 1/4 \end{bmatrix} \quad \text{and } R = \begin{bmatrix} 1/2 & 1/4 \\ 1/8 & 1/2 \end{bmatrix}
$$

Now, the matrix $(I - Q)$ is always non-singular, so we can obtain its inverse $F = (I - Q)^{-1}$. The matrix F is called the **fundamental matrix** of the Markov chain, and its elements specify the expected number of times the system will be in its various transient states before absorption occurs. More precisely, the element f_{ij} tells us, for a system initially (or currently) in state i, the expected number of times the system will occupy state j before absorption into some state. (The fundamental matrix F does not directly provide any information about *which* absorbing state will eventually be the one that is entered.)

In our example,

$$(I-Q) = \begin{bmatrix} 1 & -1/4 \\ -1/8 & 3/4 \end{bmatrix}$$

and

$$F = (I-Q)^{-1} = \begin{bmatrix} 24/23 & 8/23 \\ 4/23 & 32/23 \end{bmatrix}$$

From this we can determine that if the process begins in state 1, we can expect it to enter state 1 24/23 times and state 2 8/23 times. Therefore, the total expected number of transitions before absorption is $24/23 + 8/23 \approx 1.39$. Similarly from initial state 2, we would expect to occupy state 1 4/23 times and state 2 32/23 times, and to undergo $4/23 + 32/23 \approx 1.565$ transitions before absorption into state 3 or 4. In general, from initial state i, the total number of transitions through transient states before absorption is

$$T_i = \sum_{j=1}^{r} f_{ij}$$

that is, the sum of the elements in the i-th row of the fundamental matrix F. This essentially characterizes the duration of the finite stochastic process.

While the matrix F alone does not indicate which absorbing state will be entered, we can easily obtain **absorption probabilities** by multiplying the fundamental matrix F by the matrix R to obtain the matrix A:

$$A = F_{rxr} \cdot R_{rx(N-r)}$$

The element a_{ij} tells us, for a system initially (or currently) in state i, the probability of the system being absorbed into state j. In our example,

$$A = \begin{bmatrix} 1.043 & 0.3478 \\ 0.1739 & 1.391 \end{bmatrix} \begin{bmatrix} 1/2 & 1/4 \\ 1/8 & 1/2 \end{bmatrix} = \begin{array}{c} \\ \\ \text{transient states} \end{array} \begin{array}{c} \text{absorbing states} \\ \begin{array}{cc} 3 & 4 \end{array} \\ \begin{array}{c} 1 \\ 2 \end{array} \begin{bmatrix} 0.5653 & 0.4347 \\ 0.2608 & 0.7390 \end{bmatrix} \end{array}$$

which tells us that from initial state 1, the absorption probability into state 3 is 0.5653 and into state 4 is 0.4347. And from initial state 2, the absorption probability into state 3 is 0.2608 and into state 4 is 0.7392. (Notice that each row in A sums to 1 because, from any initial state, one of those absorbing states *will* eventually be entered.)

Recall our intuitive observation of Figure 6.6 concerning the effect of initial states on absorption: that from state 2, absorption into state 4 seemed more likely; while from state 1 absorption into state 3 seemed more likely. And indeed, in our calculations earlier, $a_{24} > a_{23}$ while $a_{13} > a_{14}$.

The matrix method for obtaining absorption probabilities can be explained by examining the equations for the individual elements a_{ij}. Given a system in state i, absorption into state j can happen in two ways. There could be a one-step transition into state j that could happen with probability p_{ij}. Otherwise, there could be a one-step transition to some transient state k (where $k = 1, \ldots, r$) for which the probability is p_{ik}, followed by eventual absorption into state j, for which we have the absorption probability a_{kj}. Because one of these will certainly occur, we can compute the absorption probability a_{ij} as

$$a_{ij} = p_{ij} + \sum_{k=1}^{r} p_{ik} a_{kj}$$

This gives us one equation in many unknowns; and if we apply the same formula for all k, we will obtain a uniquely solvable system of equations that will give us all of the absorption probabilities for a system initially in state i.

If we are interested in only one or a few initial states i, solving these systems of equations would probably be simpler computationally than performing the matrix inversion required to obtain the fundamental matrix F and then the matrix multiplication by R. However, if all the absorption probabilities are desired, it is more succinct to note that the formula for *all* the a_{ij} is just the matrix equation

$$A = R + Q \cdot A$$

This can be rewritten as

$$A = (I - Q)^{-1} R = F \cdot R$$

as given earlier.

6.8 Software for Markov Processes

The calculations required for finding steady-state probabilities and expected first passage times are just the standard procedures for solving systems of linear equations. Software for solving linear systems has already been mentioned in previous chapters, and these routines are included in many software products such as the **IMSL** and **SAS** libraries. The **SAS/IML** (Interactive Matrix programming Language) system provides an extensive library of numerical methods for solving linear algebra problems. **LINGO** Integrated Modeling Language and Solvers systems now incorporate matrix functions that support solving linear systems.

Calculating the n-step transition probabilities requires matrix multiplication, which is trivial to implement in any general-purpose programming language, but standard subroutine libraries typically supply this function. There are also several mathematical software packages in which a matrix is the fundamental data object, which may be of particular use in manipulating transition probability matrices.

MATLAB is an integrated software package in which the procedural language is built around the concept of a matrix. The language includes a rich set of matrix functions that allow a user to express algorithms involving matrix operations much more succinctly than would be possible using a general-purpose language. MATLAB versions run on PCs and Macintosh systems, with larger versions for higher performance platforms. Data file formats are compatible across all these platforms and problem sizes are limited only by the amount of memory available on the system. Further details are described in Gilat (2004), and Quarteroni and Saleri (2006).

R is now among the most powerful and commonly used statistical computing, graphics and analytics software. It includes routines for generating Markov chains, computing stationary distributions, calculating empirical transition matrices among others. **R** is a free open source package with a large community of developers and users, and it is compatible with all common operating systems. There are many **R** packages for specific statistical methods and applications, including packages for advanced Markov models. We mention here **DTMCPack** (Nicholson 2013) and the **markovchain** (Spedicato et al. 2017) **R** Packages for basic Markov chains computations.

O-MATRIX is a matrix-based scripting language which originated as an object-oriented analysis and visualization tool for Windows computing environments. Data and procedures are built with a text editor, and computations are expressed via a powerful, but small and easy-to-learn, language in which all operations are performed on matrix objects. This integrated technical computing environment is now aimed at providing high performance capabilities for solving computationally intensive mathematical and engineering problems. O-MATRIX is compatible with a version of MATLAB from MathWorks. See the software's website and user manual for a more detailed description of O-MATRIX.

6.9 Illustrative Applications

6.9.1 Water Reservoir Operations (Wang and Adams 1986)

The operation of a water reservoir is determined by a sequence of decisions concerning the volume of water to be released during a time period. Optimizing the operations involves finding a set of optimal release decisions over successive time periods to maximize the expected total reward associated with long-term operation of the reservoir. This process can be viewed as a Markov system by discretizing reservoir storage volumes into a finite number of states, and treating the release events as transitions among the storage states from one time period to the next.

An optimization method developed for general water reservoir management was applied to the Dan River Issue Reservoir on a tributary of the Yangtze River. A two stage analysis framework involved a real-time model followed by a steady-state model. Each time period was analyzed by using information about the current reservoir storage state and historical information (gathered over a period of 30 years) of inflow into the reservoir. Thus, the Markov process is derived from transition probabilities based on stochastic inflow data, coupled with operational decisions for releasing water from the reservoir.

The objectives in this Yangtze River case were flood control, production of hydroelectric power, and the ability to augment the flow of water during low seasons. Analysts developed operational standards for each activity. The rewards obtained through operation of the reservoir were measured as revenues resulting from electric power production minus penalties to account for failures to meet targeted standards of operation. Average annual rewards computed in this way are the primary performance criterion for this system, but additional indicators of performance include average annual energy production and probabilities of the occurrence of various undesirable scenarios associated with floods and low flow.

The optimal operating strategies derived from the Markov analysis represented significant performance improvements of 14% for average annual reward, and 3.6% for average annual power production, when compared with conventional strategies or the results of deterministic optimization. Steady state optimization yielded increases in both power production and the effectiveness of flood control.

Because of the magnitude of the annual revenues from the operation of such a major reservoir, the modest percentage improvements represent substantial actual increases in economic returns. Furthermore, the analysis undertaken for the Yangtze River project was intended to be used for real-time operational decisions, so it was especially valuable that this optimization method turned out to execute in a reasonable amount of time. This computational efficiency, together with the profitability obtained through the optimization process, indicate that the steady-state Markov analysis is an effective component in this decision system for real-time operation of a multi-purpose, large-scale water reservoir.

6.9.2 Markov Analysis of Dynamic Memory Allocation (Pflug 1984)

Computer operating systems are responsible for the management of computer storage facilities (memories) during execution of programs. Dynamic memory allocation techniques are widely used to promote economical and adaptive use of memory in computing environments where exact storage requirements are not specified in advance. Memory managers respond to requests for memory from executing programs by selecting a free block of memory, allocating it to the requesting program, and eventually taking a released block back and returning it to the pool of available space.

Allocation methods, such as *first-fit, best-fit, worst-fit,* and the *buddy method,* have been used and are well-known to computer scientists; however, little mathematical analysis of the performance of these methods has been done. Instead, simulation results have provided most of the reliable insights into the behavior of systems using these various allocation strategies.

A unique application of a Markov model has been used to describe the essential characteristics of a dynamic memory allocation system, and to provide a means of comparing different allocation methods. The states in this model correspond to storage configurations, and the Markov analysis describes how these configurations vary over time.

More specifically, the state of the process is completely described by two vectors that indicate the ordered sequences of the lengths of allocated and free blocks:

$$x = (x_1, \ldots, x_B) \text{ and } y = (y_1, \ldots, y_B)$$

where:
  x_i is the length of the i-th allocated block
  y_i is the length of the free block following the i-th
  B is the number of allocated blocks

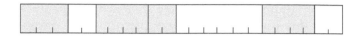

FIGURE 6.7
Memory configuration (B = 4).

Thus, the configuration shown in Figure 6.7 can be described by the vectors

$$x = (3, 4, 2, 4) \text{ and } y = (2, 0, 6, 2)$$

Various assumptions are made in order to permit an analysis using Markov models. In particular, every request for memory is assumed to be followed by a release, so that the number of allocated blocks B remains a constant, and therefore the Markov states are simple to describe.

Furthermore, the probability that a given request is for a block of size i is independent of the current configuration. A one-step transition probability matrix is developed for a *random fit* allocation strategy, which results in a Markov chain that is ergodic. (The set of states for this model is enormous and, for practical reasons, the size of the state space is reduced, at the expense of some loss of *resolution*.)

Steady state probabilities are computed and, for a large memory size, the expected number of free memory locations is close to constant. To study the efficiency and performance of general dynamic memory allocation systems, the methods outlined in this study can be extended to apply to other strategies and to allow for memory requests of arbitrary size.

6.9.3 Markov Models for Manufacturing Production Capability (Foster and Garcia-Diaz 1983)

A Markov analysis has been used to identify the steady-state production capability of manufacturing systems. For specified reliability and maintainability characteristics, the model tracks failures in a multi-unit manufacturing system. Certain assumptions are necessary. First, although the manufacturing system is a continuous-time process, in this case, it is assumed that it can be accurately modeled as a discrete-time Markov chain if the time increments are small enough. Second, the issue of stationarity must be addressed. It is assumed that the probability that a functioning element in the system becomes non-operational in a given interval is independent of the length of time it has been functioning.

In this analysis, the term *failure* refers to any activity, event, or condition causing a decrease in production rate (breakdown, policy change, supply shortage, etc.), while the term *repair* refers to any condition causing an increase in production rate. Each element in the multi-unit system can be classified as either **catastrophic** (one whose failure causes the entire system to shut down immediately), **dependent** (one that cannot be repaired while other elements are in operation), or **independent** (one that can be repaired while other elements are operating).

Three models are developed. For the model consisting of only catastrophic elements, each element has a constant probability of failure and a constant probability of repair. In the model consisting of only dependent elements, all elements have the same probability of failure. The system fails when a specified number of elements have failed, and the probability of system repair is constant over time. Once the system is repaired, all units are functioning again. In the third model, all elements are independent, with the same probability of failure. A specified number of elements can be considered for repair at any given time, and the probability that an element is repaired is constant.

For each of these cases, a transition probability matrix specifies how an initially, fully operational system evolves through stages of element failure, system failure, and repair phases. A steady-state analysis is used to determine such characteristics as the expected steady-state production rate for each specified number of element failures, and the probability of the entire system being down for repairs. As applied to a bottling machine, this analysis led to an optimal repair policy; in particular, it was determined that 2 of the 24 elements should be allowed to fail before the system is shut down for repairs. Under this policy, a system with a peak production capability of 1,200 bottles per minute can achieve an expected production rate of 879 bottles per minute.

6.9.4 Markov Decision Processes in Dairy Farming (Rukav et al. 2014, Suvak et al. 2016)

Dairy and cattle farmers throughout the world face similar problems: namely aiming to produce more and richer milk, handling increases in the price of corn or other feed, varying the diet and conception rate in milk cows, and managing herd growth. Collaborations between researchers in a university mathematics department in Croatia and an enterprising IT solutions provider for agricultural businesses have resulted in a new business environment that can help farmers address the questions earlier. Through the use of Operations Research techniques and analytical software tools, analysts were inspired to apply Markov chains to minimize the expected long term cost of milk production at dairy cow farms.

In this Markov model, each dairy cow is considered to be in a *state* that indicates the quality of her milk (milk fat, lactose, and proteins) and the quantity (with respect to a targeted level considered to be optimal, based on previous studies of lactation). Farmers and veterinarians consider the discrete-time homogeneous Markov chain to be an accurate and appropriate probabilistic model for correctly classifying a dairy cow's transitions from one state to another.

The one step transition probability matrix was constructed by estimating observed and simulated data for particular representative cows over a period of time. The initial distribution puts every cow initially in her *best* state, producing the most favorable quantities and qualities of milk. This and the transition probability matrix describe the transitions of a dairy cow from one lactation state to another over time. However, in the dairy business, the dairy farmer is on site and is therefore in a position to *intervene*, making a decision to take some action when the cow is seen to be in certain states.

By incorporating into the Markov chains the finite set of possible decisions or actions that may be taken in each state, the basic Markov chain is expanded into a broader stochastic model known as a Markov decision process. For example, for a cow in a given state, the farmer may choose to take no action (milk produced is of acceptable quality and quantity), take action to increase lactation (if quantity is too low), take action to improve quality, or replace the dairy cow with a new one who is initialized as being in the *best* state. With any action, there is an associated cost, and our dairy farmer's aim is to minimize cost over the long term.

In the Markov decision process, not every decision is allowable in every state nor with every possible state transition. Because the action decisions are limited, the cow being in only certain states can transition into only particular other states, and the effects of the action decision apply to only particular states. Also the effects of the action decision alter future transitions in only limited ways; steps to improve lactation result in slow results, so that a transition occurs into only a *slightly* better lactation state in a single transition, rather than entering a greatly improved state in only one transition.

With the addition of decision actions that affect transition probabilities, it might seem that the Markov properties have not been preserved. However because the impact of the actions is controlled, analysts are able to build the stochastic model so that each possible decision policy results in a new homogeneous Markov chain which can be analyzed in the conventional way.

6.10 Summary

Markov processes are used to represent systems in which events take place according to specified probabilities. The characteristics of such systems evolve over time, but in many cases, the probabilities themselves can be used to predict some of these system characteristics. Markov analysis provides a framework in which to study the behavior and the emergent properties of these systems.

As events take place in a Markov process, the system makes transitions from one state to another, and these transitions occur according to transition probabilities. If it is known that a system initially is in a given state, then by using these transition probabilities, it is possible to predict the patterns with which the system passes among states and perhaps re-enters states previously occupied by the system.

Some processes involving uncertain or probabilistic transitions exhibit restricted patterns of behavior that tend to dictate the ultimate disposition of the system. However, other systems range more freely among their states indefinitely, and under certain circumstances it is possible to characterize the long-term behavior or status of these systems. Knowing this steady-state behavior of systems is very valuable to analysts in planning or budgeting resources or projecting costs or profits in systems whose events take place in an environment of uncertainty.

Key Terms

absorbing chain
absorbing state
absorption probability
aperiodic state
closed set of states
continuous-time stochastic process
discrete-time stochastic process
ergodic Markov chain
expected first passage time
expected recurrence time
finite state Markov chain
first passage probability
first passage time
first recurrence time
fundamental matrix
initial probability distribution
irreducible Markov chain

Markov chain
Markov process
Markov property
one-step transition probability
periodic state
reachable state
recurrence time
recurrent state
state
state probabilities
state probability vector
state transition diagram
stationarity property
steady state
steady state probability
steady state probability vector
stochastic process
transient state
transition
transition probability
transition probability matrix
transition tree

Exercises

6.1 Suppose

$$P = \begin{bmatrix} 0.080 & 0.184 & 0.368 & 0.368 \\ 0.632 & 0.368 & 0 & 0 \\ 0.264 & 0.368 & 0.368 & 0 \\ 0.080 & 0.184 & 0.368 & 0.368 \end{bmatrix}$$

Obtain the steady state probabilities for the Markov chain whose one-step probabilities are given by P.

6.2 Write a computer program to compute the state probabilities p(50) for a five state system whose initial probabilities are (0.2 0.2 0.2 0.2 0.2) and whose one step transition probabilities are:

$$P = \begin{bmatrix} 0.1 & 0.1 & 0.1 & 0.1 & 0.6 \\ 0.2 & 0.2 & 0.3 & 0.1 & 0.2 \\ 0.2 & 0.2 & 0.2 & 0.2 & 0.2 \\ 0.5 & 0.1 & 0.1 & 0.1 & 0.2 \\ 0.3 & 0.3 & 0.1 & 0.2 & 0.1 \end{bmatrix}$$

by raising P to the 50-th power.

a. Do the elements in your result vector Π sum to 1?

b. After how many steps do the state probabilities cease to change observably?

c. If your state probabilities stabilize, but then exhibit small changes as you continue to compute higher powers of P, how would you explain this?

d. Compare your result with the steady state probabilities you obtain by solving the system $\Pi P = \Pi$ and $\Sigma \Pi = 1$. Do these steady state probabilities sum to 1?

e. Which method of establishing state probabilities takes greater computation time? (Use a timing function on your computer to determine this.)

f. Which method appears to yield more accurate results? (How can you make this determination?)

6.3 Why is the Markov process described by the following transition probability matrix *not* an ergodic process?

$$P = \begin{bmatrix} 0 & 1 & 0 & 0 \\ 0 & 0 & 1 & 0 \\ 0 & 0 & 0 & 1 \\ 1 & 0 & 0 & 0 \end{bmatrix}$$

What do you discover if you try to establish steady state probabilities by solving the steady-state equations for this process?

6.4 Raise the following transition probability matrices to successively higher powers, and note the difference in the number of steps required to reach steady state in each case:

$$P_A = \begin{bmatrix} 0.3 & 0.3 & 0.4 \\ 0.4 & 0.3 & 0.3 \\ 0.5 & 0.2 & 0.3 \end{bmatrix} \qquad P_B = \begin{bmatrix} 0.90 & 0.05 & 0.05 \\ 0.05 & 0.95 & 0 \\ 0 & 0.10 & 0.90 \end{bmatrix}$$

6.5 Try to establish steady state probabilities by solving the steady-state equations corresponding to the Markov system shown in Figure 6.3. Is there any computational difficulty caused by the transient and recurrent states?

6.6 A **doubly stochastic matrix** is one whose row elements sum to 1 *and* whose column elements also sum to 1. Find the steady-state probabilities for the chain whose one-step transition probabilities are given by the doubly stochastic matrix

$$P = \begin{bmatrix} 1/3 & 2/3 & 0 \\ 1/6 & 1/3 & 1/2 \\ 1/2 & 0 & 1/2 \end{bmatrix}$$

In general, for any doubly stochastic matrix, it is true that:

$$\Pi_j = 1/n \text{ for } j = 1, \ldots, n$$

where n is the number of states.

6.7 In a hospital for seriously ill patients, each patient is classified as being either in critical, serious, or stable condition. These classifications are updated each morning as a physician makes rounds and assesses the patients' current conditions. The probabilities with which patients have been observed to move from one classification to another are shown in the following table, where the (i, j)-th entry represents the probability of a transition from condition i to condition j.

	Critical	Serious	Stable
Critical	0.6	0.3	0.1
Serious	0.4	0.4	0.2
Stable	0.1	0.4	0.5

a. What is the probability that a patient who is in critical condition on Tuesday will be in stable condition on the following Friday?

b. How many days on average will pass before a patient in serious condition will be classified as being in stable condition?

c. What is the probability that a patient in stable condition on Monday will experience some sort of reversal and will not become stable again for three days?

d. What proportion of the patient rooms should be designed and equipped for patients in critical condition? In serious condition? In stable condition?

Discuss the validity of the Markov assumption and the stationarity assumption, in the context of this problem.

6.8 Construct a transition probability matrix to model the promotion of high school students through grades 10–12. Ninety-two percent of tenth graders are passed on to the next grade, 4% fail and repeat the tenth grade, and 4% fail and drop out of school. At the end of the eleventh grade, 88% pass to the next grade, 7% fail and repeat, and 5% fail and drop out. Of the twelfth graders, 96% graduate from high school successfully, 3% fail and repeat the twelfth grade, and 1% fail and do not ever complete high school. Students may repeat a grade any number of times, but no student ever returns to a lower grade. Comment on the structure of the transition probability matrix. Of 1,000 students entering the tenth grade, how many are expected to graduate after three years in high school? What other information about the high school students can be obtained from the data given earlier?

6.9 What is the name given to a Markov state that is reachable from the initial state and whose steady-state probability is zero? What is the name given to a Markov state whose steady-state probability is 1?

6.10 What is the interpretation of the element a_{ij} in the matrix A, where A is the product of the matrices F and R? What is the interpretation of the element f_{ij} in the fundamental matrix F of a Markov process?

6.11 Suppose the following one-step transition probability matrix describes a Markov process:

States	1	2	3
1	0.2	0.7	0.1
2	0	0.3	0.7
3	0	0.1	0.9

a. Determine the steady-state probabilities for this process.

b. What is the probability that a first passage from state 2 to state 3 will occur after exactly two transition steps?

c. What is the expected number of transitions that will occur for the system in state 2 to return again to state 2?

6.12 A machine maintenance problem can be modeled as a Markov process. Each day, the machine can be described as either excellent, satisfactory, marginal, or inoperative. For each day the machine is in excellent condition, a net gain of $18,000 can be expected. A machine in satisfactory condition yields an expected $12,000 per day. A marginal machine can be expected to bring a daily net gain of $4,000, and an inoperative machine causes a net loss of $16,000 a day. An excellent machine will be excellent the next day with probability 90%, satisfactory the next day with probability 4%, and marginal with probability 2%, and inoperative with probability 4%. A satisfactory machine will the next day be satisfactory with probability 80%, marginal with probability 12%, and inoperative with probability 8%. A marginal machine will be marginal again the next day with probability 70%, and inoperative with probability 30%. Repairs are made without delay, but only on inoperative machines, and the repairs take exactly one day. (The day long repair process costs $16,000, which accounts for the daily net loss stated earlier.) A machine having undergone repair is 90% likely to be in excellent condition the next day, but in 10% of cases the repairs are ineffective and the inoperative machine will remain out of commission, necessitating a repeat of the repair process on the following day (at an additional cost of $16,000). Find the steady state probabilities for the four states of this machine. Then, assuming that this machine is active (in one of its four states) 365 days per year, find the long-term annual profitability of this machine?

6.13 Given the one step transition probability matrix in the following, compute the expected first passage times from state i to state j, for all i and j.

$$\begin{bmatrix} 0 & 0 & 0.5 & 0.5 \\ 1 & 0 & 0 & 0 \\ 0 & 1 & 0 & 0 \\ 0 & 0.5 & 0.5 & 0 \end{bmatrix}$$

6.14 A computer network is observed hourly to determine whether the network is operational (*up*) or not (*down*). If the network is up, there is a 98% probability that it will be up at the next observation. If it is down, there is a 30% probability that effective repairs will have been completed by the next hourly observation, but a 70% chance that repairs are still in progress and the network is still down the next hour. Analyze the expected first passage times for this computer network. Comment on the performance of the network in general, and in particular interpret and comment on the first passage probabilities. In what type of network environment would the performance of this network be acceptable?

6.15 Customers are often faced with the option of purchasing an extended warranty for a new appliance. Suppose that GalleyKleen dishwashers offer a warranty plan that covers the first three years of ownership of a dishwasher. During the first year of operation, 5% of dishwashers fail. During the second year of operation, 8% fail. And 11% of dishwashers in their third year of service fail. The basic warranty from GalleyKleen covers replacement only when failures occur during the first year. If a failure occurs, a repair during the second year is expected to cost the owner (customer) $150, and during the third year is expected to cost $200. For $80, the customer can purchase an extended warranty that provides free repairs or replacement in case of failures any time within the first three years. Use a Markov model to track the progression of dishwashers through their first three years of service.

   a. Is the extended warranty a good buy for the customer?

   b. By what amount should GalleyKleen increase the sales price of the dishwasher so that the basic (no charge) warranty could be extended to cover three years?

   c. If the basic warranty did cover two years, what is a fair price for the customer to purchase a one-year extension, providing a total of three years of warranty coverage?

6.16 Two companies, one selling Ol' Boy Biscuits and the other selling Yuppy Puppy Pleasers, have cornered the market for dog treats. Each product is packaged to contain a four-week supply of treats, and customers always purchase treats as soon as necessary so as to never run out. For a customer whose last purchase was Ol' Boy, there is a 75% chance of a brand change on the next purchase; and for a customer who most recently bought Yuppy Puppy, there is an 85% chance of a brand change on the next purchase. Ol' Boys are sold at a per unit profit of 60¢, and Yuppy Puppys yield a per unit profit of 70¢.

   a. What proportion of the market is held by each of these two products?

   b. If 30 million customers regularly purchase these products, what are the annual expected profits for each company?

6.17 Consider the assumptions that were made in analyzing the memory allocation processes described in Section 6.9.2. Can you provide arguments that these assumptions are justified in practice?

References and Suggested Readings

Alstrup, J., S. Boas, O. B. G. Madsen, and R. Victor Valqui Vidal. 1986. Booking policy for flights with two types of passengers. *European Journal of Operational Research* 27 (3): 274–288.

Bhat, U. N. 1972. *Elements of Applied Stochastic Processes.* New York: Wiley.

Bhattacharva, R. N., and E. C. Waymire. 1990. *Stochastic Processes with Applications*. New York: Wiley.

Boucherie, R. J., and N. M. van Dijk (Eds.). 2017. *Markov Decision Processes in Practice*. Cham, Switzerland: Springer.

Ching, W.-K., and M. K. Ng. 2006. *Markov Chains: Models, Algorithms, and Applications*. New York: Springer.

Clarke, A. B., and R. Disney. 1985. *Probability and Random Processes for Engineers and Scientists*. New York: Wiley.

Clymer, J. 1988. *System Analysis Using Simulation and Markov Models*. Englewood Cliffs, NJ: Prentice-Hall.

Derman, C. 1970. *Finite State Markovian Decision Processes*. New York: Academic Press.

Feinberg, E. A., and A. Schwartz. 2002. *Handbook of Markov Decision Processes: Methods and Applications*. Boston, MA: Kluwer Academic Publishers.

Flamholtz, E. G., G. T. Geis, and R. J. Perle. 1984. A Markovian model for the valuation of human assets acquired by an organizational purchase. *Interfaces* 14 (6): 11–15.

Foster, J. W., III, and A. Garcia-Diaz. 1983. Markovian models for investigating failure and repair characteristics of production systems. *IIE Transactions* 15 (3): 202–209.

Gilat, A. 2004. *MATLAB: An Introduction with Applications*, 2nd ed. Hoboken, NJ: John Wiley & Sons.

Howard, R. A. 1960. *Dynamic Programming and Markov Processes*. Cambridge, MA: MIT Press.

Isaacson, D., and R. Madsen. 1976. *Markov Chains: Theory and Applications*. New York: Wiley.

Kemeny, J. G., and J. L. Snell. 1976. *Finite Markov Chains*. New York: Springer-Verlag.

Kirkwood, J. R. 2015. *Markov Processes*. Boca Raton, FL: CRC Press.

Leung, C. H. C. 1983. Analysis of disc fragmentation using Markov chains. *The Computer Journal* 26 (2): 113–116.

Nicholson, W. 2013. DTMC: Suite of functions related to discrete-time discrete-state Markov Chains., *R-Package manual*. Available at: ttps://CRAN.R-project.org/package=DTMCPack. (Accessed on May 19, 2018).

Norri, J. R. 1998. *Markov Chains*. Cambridge, UK: Cambridge University Press.

Pflug, G. Ch. 1984. Dynamic memory allocation—A Markovian analysis. *The Computer Journal* 27 (4): 328–333.

Quarteroni, A., F. Saleri, and P. Gervasio. 2014. *Scientific Computing with MATLAB and Octave*, 4th ed. Heidelberg, Germany: Springer.

Ross, S. 1993. *Introduction to Probability Models*, 5th ed. New York: Academic Press.

Ross, S. 1995. *Stochastic Processes*, 2nd ed. New York: Wiley.

Rue, R. C., and M. Rosenshine. 1985. The application of semi-Markov decision processes to queuing of aircraft for landing at an airport. *Transportation Science* 19 (2): 154–172.

Rukav, M., K. Strazanac, N. Suvak, and Z. Tomljanovic. 2014. Markov decision processes in minimization of expected costs. *Croatian Operational Research Review* 5 (2): 247–257.

Sericola, B. 2013. *Markov Chains: Theory and Applications*. New York: Wiley.

Sheskin, T. J. 2010. *Markov Chains and Decision Processes for Engineers and Managers*. Boca Raton, FL: CRC Press.

Spedicato, G. A., T. S. Kang, S. B. Yalamanchi, M. Thoralf, D. Yadav, N. C. Castillo, and V. Jain. 2017. The markovchain Package: A package for easily handling discrete Markov chains in R. Available at: https://cran.r-project.org/web/packages/markovchain/vignettes/an_introduction_to_markovchain_package.pdf. (Accessed on May 19, 2018).

Stewart, W. J. 1995. *Introduction to the Numerical Solution of Markov Chains*. Princeton, NJ: Princeton University Press.

Stewart, W. J. (Ed.). 1991. *Numerical Solution of Markov Chains*. New York: Marcel Dekker.

Suvak, N., Z. Tomljanovic, K. Strazanac, and M. Zekic-Susac. 2016. Markov chains and dairy farming in Croatia. *OR/MS Today* 43 (2): 30–32.

Swersey, A. J. 1982. A Markovian decision model for deciding how many fire companies to dispatch. *Management Science* 28 (4): 352–365.

Taylor, H., and S. Karlin. 1984. *An Introduction to Stochastic Modeling*. New York: Academic Press.

Walters, C. J. 1975. Optimal harvest strategies for salmon in relation to environment variability and production parameters. *Journal of the Fisheries Research Board of Canada* 32 (10): 1777–1784.

Wang, D., and B. J. Adams. 1986. Optimization of real time reservoir operations with Markov decision processes. *Water Resources Research* 22 (3): 345–352.

White, D. J. 1985. Real applications of Markov decision processes. *Interfaces* 15 (6): 73–83.

White, D. J. 1988. Further real applications of Markov decision processes. *Interfaces* 18 (5): 55–61.

Winston, W. L. 2004a. *Introduction to Probability Models: Operations Research Volume II*. Belmont, CA: Brooks/Cole-Thomson Learning.

Winston, W. L. 2004b. *Operations Research: Applications and Algorithms,* 4th ed. Boston, MA: Brooks/Cole.

7

Queueing Models

We have already been introduced to several modeling tools, such as linear programming, network models, and integer programming techniques, which allow us to **optimize** systems. Other techniques, such as Markov analysis, allow us to **observe** and **analyze** the probable behavior of systems over time; and the information gained from these observations may be used indirectly to modify or improve system performance. In this chapter, we will study further mechanisms by which we may observe and characterize the performance of systems. In particular, we will concentrate on the wide variety of systems whose elements include **waiting lines (queues)**, and we will study how such waiting lines interact with other activities or entities in the system toward achieving certain goals for system throughput.

The study of systems involving waiting lines traces its origin to work done many decades ago by A.K. Erlang. Working for the Danish telephone company, this mathematician developed techniques to analyze the waiting times of callers in automatic telephone exchanges. In such systems, waiting is caused by a lack of resources (not enough servers), and system designers must develop ways to balance the value of customer convenience against the cost of providing servers.

Waiting lines inherently create inconvenience, inefficiency, delay, or other problems. Waiting lines represent people waiting for service, machines waiting for a repairman, parts waiting to be assembled, and so on; and these situations cost time and money. Of course, waiting lines can be virtually eliminated by simply adding lots of servers, repairmen, and assembly stations, but this can be very expensive. To make intelligent decisions about how many servers to hire, or how many workstations to build, we must first understand the relationship between the number of servers and the amount of time spent in the queue, so that we can evaluate the trade-off between the various costs of servers and queues.

In this chapter, we will study systems that are simple enough to be modeled analytically and precisely using **queueing models**. These methods have been found to be surprisingly successful in estimating the performance of many kinds of systems. Unfortunately, despite the popularity of analytical techniques, they may be too cumbersome to use (or technically inapplicable) for modeling some very complex systems. These more difficult systems can often be analyzed using **simulation**; therefore, in Chapter 8 we will study the techniques of using computers to simulate the operation of complex systems.

7.1 Basic Elements of Queueing Systems

A queueing system consists of a flow of **customers** into and through a system, who are to be processed or dealt with by one or more **servers**. If there are fewer customers than servers, customers are handled immediately and some servers may be idle. If there is an excess of customers, then they must wait in a line or **queue** until a server becomes available. After being served, the customer leaves the system.

Waiting lines and systems involving waiting lines are so pervasive in real life that it is not at all surprising that the analysis of the operation of such systems forms a major subfield of Operations Research. We expect to see waiting lines at, for example, the bookstore, grocery store, restaurant, bank, gas station, and hospital emergency room. You may even have to *queue up* to visit your professor or to pick up a parcel from the postal service. Queues also form for telephone calls, which must wait for an available circuit in order to complete a connection. Messages or data packets may have to be queued at a processing node of a computer network before they can be forwarded on toward their destinations. Airplanes must wait to take off or land on a particular runway. Manufactured items on an assembly line may have to wait to be worked on by a line worker, a programmed robot, or other machine. And computer programs often wait in a queue to be executed in a large central computing facility. All of these represent systems whose inefficient or improper operation could cause inconvenience, economic loss, or even safety risks. Therefore, engineers and decision analysts are keenly interested in understanding and improving the operation of queueing systems.

The principal elements in a queueing system are the customer and the server. Queues arise only as a result of the servers' inability to keep pace with the needs of the customers. From the customers' point of view, there should be as many servers as there are customers at any given time, but this of course is not economically feasible. It would not make sense to hire enough bank tellers or build enough drive-through teller stations to handle the peak load of customers because, obviously, most of those tellers would be idle during most of the business day. Customers therefore expect to wait some reasonable amount of time for service. The meaning of a *reasonable wait* varies with the context.

If a customer arrives and sees that all the queues look very long, the customer may decide not to wait at all (known as *balking*), and the system loses a customer. A customer in one queue may perceive that a different queue is moving more quickly, so he may abandon his position in the first queue and join the apparently more advantageous one (known as *jockeying*). Or a customer may wait in a line, become discouraged at the slow progress, and leave the queue and the system (known as *reneging*).

This type of behavior certainly complicates the analytical study of queueing systems. Furthermore, customers differ in their perception of queueing patterns. What seems to be a hopelessly long wait to one customer may not seem so to another. A family with several small children in tow might find a 20-minute wait for a seat in their favorite restaurant intolerable, whereas a group of adults might be willing to enjoy conversation during a lengthy wait. An airplane with a nearly empty fuel tank may gladly endure a short wait for a local runway rather than fly to an alternate airport some distance away, whereas an anxious bank customer may change lines several times in the possibly vain hope of receiving more prompt service. Circumstances and personalities strongly influence systems involving human customers.

For our purposes, customers are characterized primarily by the time intervals that separate successive arrivals. (Arrival rates will be discussed in the next sections.) In more complex systems, customers may arrive in groups, such as a group of people wishing to be served together in a restaurant, or a busload of tourists arriving at a museum. Often, the group is treated as a single customer, and these are called *bulk queues*.

Another key characteristic of a queueing system is the type and length of service required by each customer. We will confine our studies to cases in which each customer requires the same type of service, but the server may take a different amount of time for each customer. Human behavior again becomes a factor here. If the server is a machine, it may take exactly the same amount of time for each customer service. More generally, however, the time required for service may be random and will vary from one customer to the next. Moreover, it is easy to envision a human server (a bank teller or an air traffic controller, for example) who sees the queue becoming long, becomes nervous, and makes mistakes, causing the service to take longer. On the other hand, a more adroit server may observe the crowded condition and work more quickly and efficiently. Or an immigration officer at a border crossing may ask fewer questions when the lines get long. Our discussion of service rates, in the next section, will attempt to account for these various considerations.

The following characteristics summarize the main elements of queueing systems.

1. The pattern of **customer arrivals** is typically described by a statistical distribution involving uncertainty.

2. The lengths of **service times** (and therefore the departure times for each customer) likewise are described by a statistical distribution.

3. The number of identical servers (sometimes called **channels** because in some of the earliest systems studied, the servers were information paths) operating in parallel is an important characteristic. If there is more than one server, each may have its own queue or all servers may select customers from a single queue. In more general systems, such as assembly lines, the *customer* (the item being manufactured) may pass through a series of queues and service facilities. The most general systems include both series and parallel queues and are termed **network queues**.

4. The method by which the next customer is selected from the queue to receive service is called **queue discipline**. The most common queue discipline is *first-in, first-out* (FIFO), in which the customer selected for service is the one that has been in the queue for the longest time. Customers could also be selected at random, or according to certain priority schemes such as highest-paying customer, the most urgent customer, or the customer requiring the shortest service time.

5. In some systems, there is a maximum number of customers allowed in the queue at one time; this is called the **system capacity**. If a system is at capacity, new arrivals are not permitted to join the system. This could occur in a drive-in bank where the queue of cars is not allowed to extend into the street, or in a computer network where the buffer space can contain only a certain number of queued data packets. In a bottling plant, there is a certain amount of space between the filling station and the packing lines. When the space fills up, the filling station must be shut down.

6. A final factor in characterizing a queueing system is the population or source from which potential customers are generated. This **calling source** may be finite or infinite in size. In a bank, for example, the calling source would be assumed infinite for all practical purposes because it is unlikely that all possible customers would ever be in the bank and that no others could conceivably arrive. On the other hand, in a computer system with a relatively few number of authorized users, it is certainly possible that all users might be logged on at some time and that there could be no new arrivals. A finite calling source thus can have an effect on the rate of new arrivals.

Once a queueing system has been modeled by specifying all of these characteristics, it may be possible for an analyst to learn a great deal about the behavior of the system by answering questions such as the following.

- How much of the time are the servers idle? Servers usually cost money, and under-utilized servers might need to be removed.

- How much time does a customer expect to spend waiting in line? And is this a reasonable amount of time, considering the context? Is it likely that customers are being lost due to long queues?

- What is the average number of customers in the queue? Should servers be added in order to try to reduce the average queue length?

- What is the probability that the queue is longer than some given length at any point in time?

These are questions facing system designers who must try to get optimal utilization from their service facilities while providing an acceptable level of convenience or safety for customers. Keep in mind that queue analysis normally occurs *before* a system is built. A primary purpose of queueing theory may be to determine how many service facilities (such as operating rooms or checkout counters) to build before it is too late or too costly for modifications to be undertaken. The remainder of this chapter presents some tools for answering just such questions as these.

7.2 Arrival and Service Patterns

7.2.1 The Exponential Distribution

In queueing systems, we generally assume that customers arrive in the system at random and that service times likewise vary randomly. Our intuitive notion of *random* is closely associated with the **exponential distribution** of lengths of time intervals between events; that is, intervals between arrivals or durations of services. Suppose we generate a random number n of arrival times over some fixed time period T by selecting n numbers from a *uniform* distribution from 0 to T. This process coincides with our intuitive idea of independent random events. It can be shown that the distances between these points are exponentially distributed.

The assumption underlying the exponential distribution is that the probability of an arrival occurring in any small interval of time depends only on the length of the interval and not on the starting point (time of day, week, etc.) of the interval or on the history of arrivals prior to that starting point. The probability of an arrival in a given time interval is unaffected by arrivals that have or have not occurred in any of the preceding intervals.

Restating these properties of the exponential distribution in terms of service times: the duration of a given service does not depend on the time of day (e.g., it does not depend on how long the service facility has been in operation), nor on the duration of preceding services, nor on the queue length or any other external factors.

Note that the *stationary* and *memoryless* properties that we observe here are precisely the assumptions that we made in order to model processes using Markov analysis, as discussed in the preceding chapter. In fact, we will return to exactly these same ideas in developing our analytical queueing models in the next section.

The **exponential density function** (sometimes called the negative exponential density function) is of the form

$$f(t) = \lambda e^{-\lambda t}$$

where $1/\lambda$ is the mean length of intervals between events. Therefore, λ is the rate at which events occur (the expected number of occurrences per unit time).

That is, $f(t)$ represents the probability of an event occurring within the next t time units. The curves shown in Figure 7.1 illustrate the shape of the exponential distribution for different values of the parameter λ, shown as λ_1, λ_2, and λ_3 on the vertical axis. Because the area under each of these curves must be one (as for any probability density function), a larger value of λ implies a more rapid decrease and asymptotic convergence to zero. As indicated by the exponential distribution curves in the figure, the most likely times are the small values close to zero, and longer times are increasingly unlikely. The exponential distribution times are more likely to be *small* than above the mean. However, there will occasionally be very large times.

We should mention here that there are clearly some cases that are *not* represented by the exponential distribution function. A machine (or even a human) service facility that performs each service with the identical constant service time yields a service distribution in which all service times are essentially equal to the mean service time, which is inconsistent with the exponential distribution. The exponential distribution also precludes cases where a customer arrives but does not join the queue because he sees another customer arrive just ahead of him, or when the server tries to speed up as the queue length increases. It is easy to imagine other scenarios that cannot be correctly or realistically described by the exponential distribution; however, experience in system modeling has shown that many systems exhibit a typical pattern of random, independent arrivals of customers, most of whom can be served in a short length of time while relatively few require longer service. For example, most bank customers arrive to conduct simple transactions that can be handled quickly, while the few irregular cases require more time. In a hospital emergency room, a large number of the arriving cases require relatively simple first-aid, while serious trauma cases requiring longer attention are less frequent. Thus, the assumption of exponentially distributed interarrival times and service times has been found in many practical situations to be a reasonable one.

An exponential interarrival distribution implies that the arrival process is Poisson distributed. If the interarrival times are exponential, then the number of arrivals per unit time is a Poisson process. A **Poisson distribution** describes the probability of having precisely n arrivals in the next t time units as:

$$\text{Probability }\{X(t) = n\} = \frac{(\lambda t)^n e^{-\lambda t}}{n!}$$

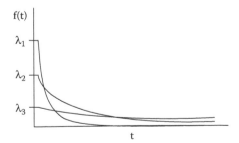

FIGURE 7.1
Exponential distribution.

Notice that when n = 0, the probability

$$\text{Probability } \{X(t) = 0\} = e^{-\lambda t}$$

is precisely the exponential distribution that the *next* arrival will *not* occur until after t time units. However, in describing queueing systems, people often refer to **Poisson arrivals** and **exponential service times** because it seems more natural to describe customer arrivals by stating how many *arrivals per unit time*, whereas service times are more conveniently described by stating the *duration of the service*.

7.2.2 Birth-and-Death Processes

Because of our assumption that interarrival and service times are exponentially distributed, this class of queueing models can be viewed as special cases of continuous time Markov processes. When such a system initially begins operation, performance measures are strongly affected by the system's initial conditions (its initial state) and by how long the system has been in operation. Eventually, however, the state of the system becomes independent of initial conditions, and we say the system has reached steady-state. Our queueing models will deal primarily with a steady-state analysis of the queueing system.

To facilitate our development of formulae for performance analysis of queueing systems in steady-state, we will illustrate the system by using a particular type of transition diagram known as a **birth-and-death process** model. The states in this system are characterized by the number of customers in the system, and thus correspond to the set of non-negative integers. This number includes the number in queue plus the number in service. The term *birth* refers to a customer arrival, and the term *death* refers to a departure.

Only one birth or death may occur at a time; therefore, transitions always occur to the *next higher* or *next lower* state. The rates at which births and deaths occur are prescribed precisely by the parameters of the exponential distributions that describe the arrival and service patterns. In queueing theory, the mean customer arrival rate is almost universally denoted by λ and the mean service rate (departure rate) is denoted by μ, where λ and μ are the exponential distribution parameters.

We can illustrate all the possible transitions using the rate diagram shown in Figure 7.2. An arrival causes a transition from a state i into state i + 1, and the completion of a service changes the system's state from i to i − 1, for a given i. No other transitions are considered possible. Using this diagram, we can now begin to derive the formulae that describe the performance of simple queueing systems.

FIGURE 7.2
Rate diagram for a birth-and-death process.

7.3 Analysis of Simple Queueing Systems

7.3.1 Notation and Definitions

Although we will concentrate on only the simplest of queueing models, we will make use of a notational scheme that was developed by Kendall and Lee and that is commonly used to describe a variety of types of queueing systems (Lee 1966). The system characteristics are specified by the symbols

$$A / B / C / D / E / F$$

where:
A and B are letters that denote the interarrival time distribution and the service time distribution, respectively.
C is a number that denotes the number of parallel servers or channels.
D and E denote the system capacity and size of the calling source, respectively.
F is an abbreviation identifying the queue discipline.

The codes used to denote arrival and service patterns are as follows:

M for exponential (**M**arkovian) interarrivals and service times

D for constant (**d**eterministic) times

E_k for **E**rlang distributions with parameter k

GI for **g**eneral **i**ndependent distribution of interarrival times

G for **g**eneral service times

The code for queue discipline may be FCFS (first-come, first-served), SIRO (service in random order), or any other designated priority scheme.

So, for example, a queueing system described as

$$M / M / 1 / \infty / \infty / FCFS$$

is a single server system with exponential arrivals and departure patterns, infinite queue capacity and calling source, and a first-come, first-served queue discipline. This is the type of system we will study most thoroughly in this chapter. Of course, a variety of combinations of characteristics can be defined, but only a relatively small number of systems have been solved analytically.

Beyond this seemingly cryptic notation describing the essential characteristics of a queueing system, we need some additional definitions and notation to describe various performance measures. Determining these performance measures is, after all, our primary reason for creating analytical models of queueing systems. The following abbreviations are used almost universally and can be found in any textbook on queueing analysis.

λ = Arrival rate (expected number of arrivals per unit time).

μ = Departure rate for customers of each server in the system (expected number of customers completing service and departing the system per unit time).

s = Number of parallel servers.

$\rho = \lambda/s\mu$ = Utilization factor of the service facility (the expected fraction of time the servers are busy); sometimes called *traffic intensity*. Note that $\rho < 1$ in order for the system to reach steady-state; otherwise, the customer load on the system grows increasingly large without bound.

p_n = Steady state probability of the system being in state n, that is, of there being exactly n customers in the system. (Recall from our study of steady state Markov processes that this may be interpreted as the fraction of the time the system has exactly n customers.)

L = Expected number of customers in system.

L_q = Expected number of customers in queue; mean length of queue.

W = Expected waiting time for each customer in the system (includes time spent in queue as well as service time).

W_q = Expected waiting time spent in queue.

Certain relationships have been established between L, L_q, W, and W_q. **Little's formula** (Little 1961) states that $L = \lambda W$ and also that $L_q = \lambda W_q$. Also, because expected waiting time in the system equals expected time in queue plus expected service time, we have the formula $W = W_q + 1/\mu$. Therefore, if we can compute any one of these four performance measures, we can use these relationships to compute the other three. But in order to do this, we need a way to compute the probabilities p_n. For this, we return to our birth-and-death process diagrams.

7.3.2 Steady State Performance Measures

If we consider any state in the rate diagram shown in Figure 7.2, and assume the system to have reached steady state, then it must be true that the mean rate at which transitions are made *into* the state must equal the mean rate at which transitions are made *out of* the state. In fact, we can write an equation for each state that expresses this fact and that accounts for all possible ways in which transitions into and out of the state can occur. The set of equations that results from doing this are called **flow balancing equations**, and we can write them in the following way.

First consider state 0, the simplest case, because there is only one path leading in and one path leading out. The mean rate out of state 0 is the probability of being in state 0 times the rate of **departures** from state 0, $p_0\lambda$. The mean rate in is the probability of being in state 1 times the rate of transitions from state 1 to state 0, $p_1\mu$. Therefore, the equation for state 0 is

$$p_0\lambda = p_1\mu$$

For all other states, there are two arcs leading in and two arcs leading out. Still, the *rate-in = rate-out* principle holds and we can write for state 1:

$$\text{Rate in:} \quad p_0\lambda + p_2\mu$$

$$\text{Rate out:} \quad p_1\lambda + p_1\mu$$

$$\text{therefore} \quad p_0\lambda + p_2\mu = p_1\lambda + p_1\mu$$

And similarly for state 2,

$$p_1\lambda + p_3\mu = p_2\lambda + p_2\mu$$

and for state n

$$p_{n-1}\lambda + p_{n+1}\mu = p_n\lambda + p_n\mu$$

By first solving the state 0 equation for p_1 in terms of p_0, we can proceed to the state 1 equation and solve for p_2 in terms of p_0, and successively solve for all p_n as follows:

$$p_1 = \left(\frac{\lambda}{\mu}\right)p_0$$

$$p_2 = \left(\frac{\lambda}{\mu}\right)p_1 = \left(\frac{\lambda}{\mu}\right)^2 p_0$$

$$p_3 = \left(\frac{\lambda}{\mu}\right)p_2 = \left(\frac{\lambda}{\mu}\right)^2 p_1 = \left(\frac{\lambda}{\mu}\right)^3 p_0$$

.

.

.

$$p_n = \left(\frac{\lambda}{\mu}\right)p_{n-1} = \left(\frac{\lambda}{\mu}\right)^n p_0$$

So, for the birth-and-death process model in which all arrivals are characterized by the parameter λ and all departures by the parameter μ, any of the p_i can be computed in terms of the parameters λ and μ and the probability p_0. To obtain the value of p_0, we just observe that the sum of all the p_i must equal to one:

$$p_0 + p_1 + p_2 + \ldots + p_n + \ldots = 1$$

Then,

$$1 = p_0 + \left(\frac{\lambda}{\mu}\right)p_0 + \left(\frac{\lambda}{\mu}\right)^2 p_0 + \ldots + \left(\frac{\lambda}{\mu}\right)^n p_0 + \ldots$$

$$= p_0\left[1 + \left(\frac{\lambda}{\mu}\right) + \left(\frac{\lambda}{\mu}\right)^2 + \ldots + \left(\frac{\lambda}{\mu}\right)^n + \ldots\right]$$

The series in square brackets converges, if $(\lambda/\mu) < 1$, to the quantity

$$\frac{1}{1-(\lambda/\mu)}$$

Therefore,

$$1 = p_0\frac{1}{1-(\lambda/\mu)}$$

and

$$p_0 = 1 - \left(\frac{\lambda}{\mu}\right)$$

More intuitively, you may also recall that $\rho = \lambda/s\mu$ is the probability of the service facility being in use at any given time. For a single service facility in which the same parameter μ characterizes service times (as in Figure 7.2), regardless of the number of customers requiring service, we let s = 1. Therefore, $\rho = \lambda/\mu$ is the probability of a busy service facility, and thus $1 - (\lambda/\mu) = 1 - \rho$ is the probability of an idle service facility. This is exactly what is meant by the probability of there being zero customers in the system, so it is reasonable that $p_0 = 1 - (\lambda/\mu)$.

We can now express all of the system state probabilities in terms of λ and μ as follows:

$$p_n = \left(\frac{\lambda}{\mu}\right)^n p_0$$

$$p_n = \left(\frac{\lambda}{\mu}\right)^n \left[1 - \left(\frac{\lambda}{\mu}\right)\right] \quad \text{or}$$

$$p_n = \rho^n(1-\rho)$$

Our original purpose in developing these formulae was so that we could compute system performance measures such as the expected number of customers in the system L, and the expected amount of time each customer spends in the system. By defining the expected number of customers, we know

$$L = \sum_{n=0}^{\infty} (n \cdot p_n)$$

$$= \sum_{n=0}^{\infty} [n \times \rho^n (1-\rho)]$$

$$= (1-\rho) [1\rho^1 + 2\rho^2 + \ldots + n\rho^n + \ldots]$$

$$= (1-\rho) [1 + \rho + \rho^2 + \rho^3 + \rho^4 + \ldots][\rho + \rho^2 + \rho^3 + \rho^4 + \ldots]$$

$$= (1-\rho) \left(\sum_{n=0}^{\infty} \rho^n\right) \left(\sum_{n=0}^{\infty} \rho^n - 1\right)$$

$$= (1-\rho) \left(\frac{1}{(1-\rho)}\right) \left(\frac{1}{(1-\rho)} - 1\right)$$

since

$$\sum_{n=0}^{\infty} x^n = \frac{1}{(1-x)} \quad \text{for } |x| < 1$$

Therefore,

$$L = \frac{\rho}{1-\rho}$$

From this we can use Little's formula $L = \lambda W$ to obtain the expected time in the system W. Because $W = W_q + 1/\mu$, we can then compute the expected time a customer spends in queue, W_q. And from this we can obtain the expected queue length L_q using $L_q = \lambda W_q$.

Example 7.3

Suppose computer programs are submitted for execution on a university's central computing facility, and that these programs arrive at a rate of 10 per minute. Assume average run-time for a program is five seconds, and that both interarrival times and run-times are exponentially distributed. During what fraction of the time is the CPU idle? What is the expected turnaround time of a job in this system? What is the average number of jobs in the job queue?

This system is assumed to be an M/M/1 queueing system with $\lambda = 10$ jobs per minute and $\mu = 12$ jobs per minute. We will also assume that job queues may become arbitrarily long and that there is an infinitely large user population. Since $\rho = 10/12 < 1$, the system will reach steady-state and we can use the formulae developed earlier to answer these questions. Since the utilization factor $\rho = 5/6$, the CPU will be idle 1/6 of the time, or for 10 seconds out of every minute. (Since 10 jobs each take an average of five seconds, the CPU is busy for 50 seconds each minute.)

Turnaround time is defined to be waiting time plus execution time, which we call W. We know $L = \rho/(1 - \rho) = (5/6)/(1/6) = 5$, and from this we can calculate, using Little's formula, $W = L/\lambda = 5/10 = 1/2$ minute. The average queue length is $L_q = \lambda W_q$. Since we have just computed W to be $1/2$ minute, we can use the formula $W_q = W - 1/\mu = 1/2 - 1/12 = 5/12$ minutes (or 25 seconds) for expected waiting time spent in the queue. Then the average number of jobs in the queue is $L_q = 10 \cdot 5/12 = 4^1/_6$ jobs. (Because the job queue itself occupies some computer memory, this tells how much space is typically devoted to this system function, and it also indicates how many jobs are experiencing delay.)

Now suppose we want to know the probability that the number of jobs in the system becomes 4 or more. This can be calculated as 1 minus the probability that there are fewer than 4 (i.e., 0, 1, 2, or 3) customers in the system:

$$\text{Probability } [\geq 4 \text{ jobs}] = 1 - \left[p_0 + p_1 + p_2 + p_3 \right].$$

We know

$$p_0 = 1 - \frac{\lambda}{\mu} = 1 - \frac{5}{6} = 0.1667$$

$$p_1 = \rho p_0 = \left(\frac{5}{6}\right)\left(\frac{1}{6}\right) = \frac{5}{36} = 0.1389$$

$$p_2 = \rho^2 p_0 = \left(\frac{5}{6}\right)^2\left(\frac{1}{6}\right) = 0.1158$$

$$p_3 = \rho^3 p_0 = \left(\frac{5}{6}\right)^3\left(\frac{1}{6}\right) = 0.0965$$

Therefore,

$$\text{Probability } [\geq 4 \text{ jobs}] = 1 - [0.1667 + 0.1389 + 0.1158 + 0.0965] = 0.48225$$

In general, the probability that there are at least k jobs in the system is given by:

$$1 - \sum_{n=0}^{k-1} p_k = 1 - \sum_{n=0}^{k-1} \rho^n (1-\rho)$$

$$= 1 - \frac{(1-\rho)(1-p^k)}{(1-\rho)}$$

$$= \rho^k$$

In our example, $\rho = 5/6$, and $k = 4$, so the probability of at least four jobs in the system is $(5/6)^4 = 0.48225$.

The formulas developed earlier are valid only in queueing systems that eventually reach steady-state. Our underlying assumption that the arrival rate λ be less than the service rate μ is sufficient to guarantee that the system will stabilize. Notice that as ρ approaches 1, both W and W_q become large. Clearly, for $\rho > 1$, arrivals are occurring faster than a constantly busy service facility can keep up with the demand. When $\rho = 1$, the sequence is undefined. However, if we look back at the original state equations, we discover that $(\lambda/\mu) = 1$ implies that $p_0 = p_1 = p_2 = p_3 = \ldots$ There are infinitely many states, all equally likely, which means that the actual probability of being in any given state is zero in the limit.

We have also assumed that the system has an infinite capacity. If this were not the case, then arriving customers would occasionally encounter a full system, and although they would *arrive at* the system according to the arrival parameter λ, they would not be permitted to *join* the system at rate λ. Thus, the effective arrival rate would not be constant and would vary in time, according to whether the system is at capacity. For this case, the p_n formulae remain valid as before. However, if we let N denote the system capacity, the steady-state equation for state N is simply

$$p_{N-1}\lambda = p_N\mu$$

and

$$p_N = \left(\frac{\lambda}{\mu}\right) p_{N-1} = \left(\frac{\lambda}{\mu}\right)^N p_0$$

just as before. However, there is no state $N + 1$. We now have a finite set of states whose probabilities must sum to 1:

$$p_0 + p_1 + p_2 + \ldots + p_N = 1$$

Then,

$$p_0 \left[1 + \left(\frac{\lambda}{\mu}\right) + \left(\frac{\lambda}{\mu}\right)^2 + \left(\frac{\lambda}{\mu}\right)^3 + \ldots + \left(\frac{\lambda}{\mu}\right)^N \right] = 1$$

The series in the brackets is a geometric progression that sums to

$$\frac{1 - (\lambda/\mu)^{N+1}}{1 - (\lambda/\mu)}$$

when $\lambda \neq \mu$; therefore,

$$\frac{p_0\left(1-(\lambda/\mu)^{N+1}\right)}{\left(1-(\lambda/\mu)\right)} = 1$$

or

$$p_0 = \frac{1-\rho}{1-\rho^{N+1}}$$

When $\lambda = \mu$, we get $\lambda/\mu = 1$ and $p_0 = 1/(1+N)$.

Other system measures can be computed as before. It can be shown that, when $\rho \neq 1$:

$$L = \frac{\rho\left[1-(N+1)\rho^N + N\rho^{N+1}\right]}{(1-\rho)\left(1-\rho^{N+1}\right)}$$

When $\rho = 1$, $L = N/2$. Moreover, if a customer arrives when the system is full (with probability p_N), the customer will not enter the system. Therefore, the *effective* arrival rate λ_e is

$$\lambda_e = \lambda\left(1 - p_N\right)$$

$$W = \frac{L}{\lambda_e}$$

$$W_q = \frac{L_q}{\lambda_e}$$

$$W = W_q + \frac{1}{\mu}$$

These formulas are still valid when $\lambda > \mu$. As the arrival rate increases relative to the service rate, the system just loses more customers. It is interesting to note that, even in a saturated system in which the arrival rate is greater than the service rate, there is always still some probability p_0 that the system will be empty and the server will experience some idle time.

We have derived performance measures for single-server (M/M/1) systems. For multiple server systems (where $s > 1$), the actual service rate depends on the number of customers in the system. Obviously, if there is only one customer present, then service is being rendered at rate μ. But if there are two customers present, and $s \geq 2$, then the system service rate is 2μ. Likewise, if $s = 3$, then the system service rate is 3μ. However, if there are s service facilities, the maximum system service rate is $s\mu$, even if there are more than s customers. This is illustrated by the rate diagram in Figure 7.3.

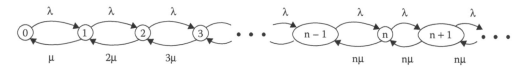

FIGURE 7.3
Rate diagram for multiple servers (s = n).

The results obtained from a birth-and-death process model for an M/M/s system, for s > 1, differ from our previous results because service rates are sensitive to the current customer load. Under the assumption that $\rho < 1$, that is $\lambda < s\mu$, it can be shown that:

$$p_0 = \frac{1}{\sum_{n=0}^{s-1} \frac{(\lambda/\mu)^n}{n!} + \frac{(\lambda/\mu)^s}{s!} \frac{1}{1-(\lambda/s\mu)}}$$

and

$$p_n = \begin{cases} \dfrac{(\lambda/\mu)^n p_0}{n!} & \text{for } 0 \le n \le s \\[3ex] \dfrac{(\lambda/\mu)^n p_0}{s!s^{n-s}} & \text{for } n > s \end{cases}$$

The expected number in queue can be shown to be:

$$L_q = \frac{p_0 (\lambda/\mu)^s \rho}{s!(1-\rho)^2}$$

where $\rho = \lambda/s\mu$ and, as before,

$$L_q = \lambda W_q$$

$$W = W_q + \frac{1}{\mu}$$

$$L = \lambda W = L_q + \frac{\lambda}{\mu}$$

7.3.3 Practical Limits of Queueing Models

In the previous sections, we attempted to give a very brief introduction to the simplest Queueing models and to the mathematical foundations underlying these models. When the systems being studied are more complex, we find that the analytical approach to modeling their behavior grows dramatically in complexity. And for some systems, no analytical models have been developed that are applicable at all.

In selected applications, decomposing or partitioning the system may be a reasonable approach. For example, in a very complex multiple server system, we might simplify the analysis by just considering the servers that create the bottlenecks in the system, and in that way decompose the system into more manageable parts.

The problem of scheduling operating rooms in a hospital provides another good example. We may be able to learn much about the system by treating the operating rooms as a set of parallel identical servers. However, in reality the operating rooms are *not* identical; each one could be characterized by the unique equipment installed in the room. Therefore, it might be useful to partition the patients into several parallel streams of customers, each requiring a different group of identical operating theaters.

There are many examples of systems that involve certain characteristics of queueing models, but whose components do not fit into the standard roles of *customers* and *servers* as we have defined them, or whose customers and servers do not behave according to the simple paradigms to which the equations of analytical queueing models are applicable. In such systems, although we cannot directly and easily *compute* system characteristics such as average queue length, average time spent in a queue, server utilization, and so on, we might instead be able to *write a computer program* to play out and mimic the behavior of the entities in the system. By enacting this behavior under the control of a computer program, we can then also *observe* the performance of the *simulated* system, and, within the computer program, accumulate enough information about the system to then quantify the observed system characteristics. When this is done properly and skillfully, the results may be just as valuable and reliable as the analytical results that can be directly derived for simpler systems.

Although queueing analysis and simulation are often used to accomplish similar goals, the two approaches are quite different. Chapter 8 describes how to develop a simulation of a system, how to cause *events* to happen in the simulated system, how to make observations and gather data about the effects of these events in the simulated system, and thereby infer how the real system being modeled would behave and what some of its essential characteristics are.

7.4 Software for Queueing Models

Application tools for queueing and simulation studies are abundant. While there are a number of software products specifically designed for analytical modeling of queueing-systems, many queueing systems are studied through simulation, and in that case analysts can choose from a variety of software packages and languages for general simulation, as will be described in Chapter 8.

Matlab from **MathWorks** includes queueing blocks such as queues, servers, and switches as part of its discrete-event simulation engine **SimEvents**.

GNU Octave, a Scientific Programming Language that is similar to, and largely compatible with, Matlab also has a queueing package that provides functions for queueing networks and Markov chains analysis. It can be used to compute steady state performance measures for open, closed and mixed networks with single or multiple job classes. Furthermore, several transient and steady state performance measures for Markov chains can be computed, such as state occupancy probabilities, and mean time to absorption. Both discrete and continuous time Markov chains are supported.

R has a freely available queueing package called **Queuecomputer** that implements computationally efficient methods for simulating queues with arbitrary arrival and service times (Ebert 2017).

We have noticed in recent years that there are a number of web based systems that accept input from users to certain queueing models via a web browser, solve the problem and display the solution back on the browser without the need for the user to install any software. A quick web search can easily find systems such as **Solutions to Queueing Systems**, and **Queueing Theory Calculator**. Hlynka (2017) maintains a comprehensive list of queueing theory software.

7.5 Illustrative Applications

7.5.1 Cost Efficiency and Service Quality in Hospitals (Green 2002)

An important factor in evaluating cost efficiency in hospitals is the average percentage of beds that are occupied in a hospital, known as the *average bed occupancy level*. This measure has been for many years the criterion that most often determines bed capacity decisions in U.S. hospitals. The original aim of occupancy level targets has been to manage the supply of hospital beds, limiting the number of unoccupied beds and thereby controlling costs. A widely adopted occupancy target is 85%. Even with such high occupancy levels, health policy planners, government officials, and hospital administrative decision makers have reacted to this figure with the perception that there were too many hospital beds (an average of say 15% unoccupied beds), a situation which they interpreted as a costly and therefore unpopular over-supply of medical resources.

The issue of facility utilization is a complex one. Major changes in health care networks, insurance plans, shorter hospital stays, more outpatient procedures, and fewer inpatient admissions all require a careful reconsideration of efficiency in hospital resource utilization. But based on the measure of average bed occupancy levels, decision makers continue to infer that there is an excess of hospital beds which contributes to high and ever-increasing health care costs. Decisions to reduce the total number of beds in a hospital, or in a specialized unit within a hospital, were almost always based on the long-established measure of average occupancy level targets, which seemed simple to understand and easy to compute, but which inevitably influenced human nature and led to reductions in bed count. Pressure to be more cost-efficient led some managers and administrators to set occupancy targets higher than 90%.

However, a closer look into the broader decision process reveals that the well-meaning but narrow focus on cost efficiency often detracts from much-needed attention to patient service performance quality. Analysts began to look more seriously at the delays experienced when patients cannot be provided with an appropriate bed as needed, either upon initial arrival at a hospital unit or for example when a patient is transitioning from a surgery unit into a recovery unit. Queueing analysts have studied bed unavailability in particular at intensive care units (ICUs) at hospitals in New York state (Green 2002). These hospital units serve the most critically ill patients, and the cost per day is several times as much as for regular inpatient units because of the technology and highly skilled staff required for treatment and monitoring patients. The analysts began with the assumption that standards and practices should be in place that provide the ability to place patients in beds appropriate to their needs without unsafe or unreasonable delays; that is, the purpose of a hospital or unit is to provide appropriate and timely care to patients. Little data had been accumulated to support this study, and furthermore, it was quickly recognized that there were several ill-defined and poorly understood issues and concepts raised by the analysts that directly affected the quality of patient care, and that deserved some preliminary discussion before decision factors could be analyzed.

First of all, the *beds* themselves are characterized as being of a special type officially approved for a specific need, such as for inpatient use, outpatient use, fitted with telemetry or other technological capabilities. Certain specialized beds may be certified but not necessarily staffed with appropriately trained personnel; thus, availability of such a bed depends on the specific needs of an arriving patient.

Next is the question of what is meant by the term *occupied* bed. Hospital census is typically taken at midnight for billing purposes. But a census at that hour of the day usually measures the lowest occupancy of the day. Patients may arrive or depart for day treatment procedures. This *day use* may not be shown as contributing to the average bed occupancy level, even though each patient did in fact occupy a bed for the duration of the stay. In addition, few procedures are scheduled for weekends and holidays, which may have the effect of shifting the patient load to other days. Thus, daily average occupancy level is not a simple and perfect indicator of bed utilization.

For their study of cost efficiency and service quality, analysts proceeded to construct an M/M/s queueing model to describe systems having a single queue, patients arriving according to a Poisson distribution, exponential service times (bed occupancy times), unlimited queue length, and s identical servers (beds appropriate to the hospital or unit). This model was chosen because of its simplicity and tractability, as system performance problems need to be solved quickly; and this model required only data that was already publicly available.

Measures of performance efficiency and effectiveness in such systems include the probability that an arrival has any wait; that is, the probability of delay. These systems have the property that greater occupancy levels cause longer delays for service; and also relatively small increases in occupancy level can cause very large increases in delays. In general, larger service systems can operate at higher utilization levels than smaller ones in achieving the same level of delays, and indeed it was noted that smaller hospitals or units such as might be located in rural areas may need to operate using lower occupancy levels in order to provide good service. When system utilization is high, queueing delays are highly sensitive to even temporary increases in arrival rates. Rural units are often small units, so are particularly stressed by surges in the arrivals.

When the ICUs in the state of New York were analyzed, the average occupancy levels were 75% in the units studied. This could be seen as indicating under-utilization of the beds in these units, but the results of applying the M/M/s queueing study shed some light on the delays associated with this average occupancy level. The probability of delay depends on size s (number of beds) and server utilization. The queueing study revealed that over 90% of the ICUs currently had too few beds and were unable to achieve a proposed goal of a 1% probability of delay. And at half of the ICUs being studied, beds were over-utilized and even the less ambitious goal of a 10% probability of delay could not be reached. So, even at the relatively low average occupancy level, these facilities were far from being able to achieve low probability of delays for their patients.

We have seen that patient service quality should include avoiding delays in providing the patient with a bed. To emphasize the importance of the delay probabilities, let's consider the consequences of there not being a bed available. It has been observed that there are increasing numbers of hospitals turning away arriving patients in ambulances, diverting them to other hospitals, due to lack of available beds. There are increases in the time patients spend in emergency rooms and hallways while waiting for a bed. Bed unavailability is a reality, and long delays and over-crowding are reported routinely. Patients experiencing delays may become agitated or uncomfortable, or they may be in urgent need of critical treatment. Therefore, not surprisingly, there is no standard definition of an *acceptable* delay.

In the study of the New York ICUs, average length of stay was nearly 18 days, so if a delay is encountered, it is likely to be a long delay (awaiting the departure of an existing patient). This could lead to high incidence of ambulance diversions or even to adopting the practice of bumping a current occupant to a different unit (likely one with fewer critical care resources, less telemetry, or lower level of staffing). Clearly, delays can result in poorer service and care of patients. But adding more beds to enlarge the facility is costly, and the acquisition of *excess* beds is often criticized even though patient service may be improved.

The quantitative results of an M/M/s queueing study provided health experts and managers valuable data to support decisions about increasing or decreasing the size of the facility. But in addition, fruitful discussions produced several common sense ideas that also proved effective and relatively inexpensive. Whereas certain ICUs may need more beds in order to deliver good service performance, there may be other units in the same hospital that have too many beds. It may be possible to reallocate beds among units as needed. Another practical solution may be to keep *overflow* beds on hand and hope that they can be staffed when the need materializes. As a further example, the Province of Ontario in Canada has a province-wide system known as CritiCall that keeps track of ICU bed utilization in 150 hospitals. When ambulances are looking for available beds, they call CritiCall and take their patients to the recommended location that has available beds. This improves bed capacity by increasing the relative size of the individual hospitals and spreading the risk of a shortage of beds.

Cross-training nurses or re-purposing selected telemetry equipment could also be possible, especially in large hospitals; done considerately, temporarily increasing the staff might increase patient satisfaction and peace of mind. Management should creatively investigate various options for improving operational efficiency. Goals should respect service performance standards and clinical care quality rather than cost alone.

Queueing analysis has been shown to be one very important aspect of evaluating and improving the efficiency and effectiveness of health care systems. Analysis of delays and the needs for equipment and staff can also be applied in many other areas such as telecommunications, airlines, and agencies providing police, fire, ambulance and other emergency services. In each case, analytical results of queueing studies can help to identify problems and establish a balance in cost of service vs. customer delays and consequences.

7.5.2 Queueing Models in Manufacturing (Suri et al. 1995)

The application of queueing models to the analysis of manufacturing processes began as early as the 1950s, but the extent of use of analytical queueing tools was at times limited, and it varied with trends in the manufacturing industry itself. Queueing models are used to study discrete manufacturing systems, in which products *flow* through a sequence of machines or workstations, where they are worked on either individually or in batches of individual pieces (as contrasted with a continuous flow process such as oil refining). Products must wait to be worked on at each station, so the entire process can be viewed as a **network of queues**. In the 1960s, it was shown that, under appropriate assumptions, the performance of general networks of queues can be predicted just by using the simple formulae that apply to individual queues.

Automated manufacturing systems were traditionally designed and tooled to produce a single product, an approach appropriate only for high-volume production. During the 1970s, the advent of programmable machines made it possible to tailor any machine to make any product within its range of capabilities, instead of being physically limited to a single product. These so-called **flexible manufacturing systems** allowed profitability with lower volume productions. Flexible manufacturing systems usually consisted of numerically controlled automated machines that operated with the support of a materials handling system to move the products between machines.

Interactions among the entities in complex flexible manufacturing systems made it very difficult to predict the performance of these systems. One approach was the very expensive and time consuming process of building a simulation model to predict performance, identify bottlenecks, and analyze the complicated dynamics of such systems. A much more efficient approach was the use of queueing models for analyzing flexible manufacturing systems, and it was during this time that the first queueing software packages were developed for manufacturing systems. These packages were comparatively easy to learn and to use, and the models could be developed and analyzed in a fraction of the time that would have been needed to do a simulation-based study.

As manufacturing strategies matured during the era of flexible manufacturing systems, researchers discovered new and more effective ways to use queueing analysis to predict, manage, and improve performance, giving attention to such issues as resource utilization, queueing bottlenecks, lead times, and productivity. Numerous software packages were available to support such analyses. Companies, such as Alcoa, IBM, Pratt and Whitney, DEC, and Siemens, all reported using queueing packages to achieve improvements in their manufacturing processes. Pratt and Whitney was prepared to spend up to six weeks developing a simulation model to study a preliminary design of a proposed new manufacturing division, but instead turned to the use of queueing software to get the answers it needed much more quickly. Similarly, IBM initiated a progressive project for manufacturing printed circuit boards. An experienced analyst realized that a simulation model of the 200 machines and 50 products would be prohibitively time consuming. A convenient and sufficiently powerful queueing package provided the means of developing the factory model easily within the time available for the analysis.

Throughout the 1980s, manufacturing companies used performance measures based on efficiency of equipment utilization and on cost reduction. Factories had always been sensitive to set-up costs, and attempted to manage such costs by running large lot sizes. However, queueing studies revealed that large lot sizes contribute to long lead-times. By the mid-1980s, there was a shift away from traditional cost reduction and quality improvement objectives toward a strategy of lead-time reduction. Simple techniques collectively known as *just-in-time* scheduling strategies became the vogue, and offered vastly improved productivity over more complex automated systems.

The 1990s saw a new emphasis on speed and bringing new products into a time-competitive market (which coincidentally also contributed to increased quality and improvements in costs), but this new emphasis presented new challenges to managers and analysts. Queueing models turned out to be just the right analytical tool: as manufacturers worked on ways to reduce lead-time, they discovered how much time their products spent waiting in queues. In some cases, it was not unusual for parts to spend up to 99% of their time, not *being* processed, but rather *waiting* to be processed. Thus, reducing the time spent waiting in queues was the most effective approach to managing lead-times. Once again, queueing theory provided the ideal analytical tools.

More recently, Ingersoll Cutting Tool Company began to analyze its manufacturing systems with the goal of reducing set-ups, and quickly discovered that their complex manufacturing processes offered thousands of opportunities for such reductions. Unable to study each possibility, management decided to try to identify just those critical set-ups that could contribute to reducing lead-times by 50% or more. For this analysis, they selected a software package to develop a manufacturing model based on queueing network theory. In just two months, they created a model of a large factory process and were able to make specific recommendations not only to reduce specific set-ups but also to manufacture smaller lot sizes and thereby reduce lead-times. This experience demonstrates the applicability and effectiveness of queueing-based decision software in manufacturing.

7.5.3 Nurse Staffing Based on Queueing Models (De Véricourt and Jennings 2011)

The analytical tools of Operations Research have long been used to study and improve health care delivery systems, principally through better management of available resources such as facilities, staff, and supplies in order to provide healthcare services to patients. A great deal of study has addressed specifically how best to manage and utilize the critical skills of nurses. One very important aspect of managing nursing care focuses on establishing appropriate **nurse staffing levels**; that is, determining just how many nurses should be on staff at any given time in medical units such as hospitals and clinics. A commonly used guideline for nurse staffing is to use the ratio of nurses to patients, which is to set a minimum number of nurses that should be on staff in a hospital that currently has a given number of patients. Advocates of the ratio policy concept seem to believe that good patient outcomes can be achieved simply by increasing nursing ratios, although this position deserves further formal investigation.

California, for example, made a move toward enacting laws to enforce nurse-to-patient ratios in hospitals. But in practice, state policies for nurse staffing levels were influenced by the state health department, hospital administrators, and representatives of nurses unions, all of whose conflicting perspectives on the issue required negotiated compromises not consistent with any documented unbiased analysis. Managing patients' diverse and often unpredictable needs and controlling delays in nurses' response to patients is actually quite a complex problem, and simple ratio guidelines are now known to be inadequate in practice.

Recent research has involved the development of a queueing model to help establish policies for nurse staffing that are efficient and that also meet performance expectations for medical units. In this approach a medical unit is modeled as a closed M/M/s//n/FCFS queueing system in which s nurses serve a fixed-size population of n patients. The exponential arrival and service distributions are commonly used in hospital capacity planning and policy making, and these are considered to be valid assumptions for purposes of this particular study. Arrivals into the system represent patients who will be either in a stable state or in a needy state. A stable patient becomes needy after an exponentially distributed time interval with mean $1/\lambda$. A needy patient is served in first-come first-served order by a nurse who attends to the patient for a service duration that is exponentially distributed with mean $1/\mu$, whereupon the patient reverts to a stable condition.

The purpose for developing this model is to determine how many nurses should be present in the medical unit at any time. In this context, the performance of the medical unit is defined in terms of the **probability of excessive delay**, that is, the probability that the delay between a patient becoming needy and the arrival of a nurse to administer care exceeds a given time threshold T. (It is generally agreed that excessive delays are related to the possibility of adverse events.)

The M/M/s//n/FCFS queueing model permits a sophisticated calculation of the likelihood that a needy patient waits for a time longer than T before being attended to by a nurse. The mathematics underlying this result represents an advanced and elegant extension of the single server probabilities of waiting, which were presented in Section 7.3 of this book, and the calculations are based just on the values of s, n, λ, and μ. Nurse staffing, in this context, consists of finding the minimum staffing level that guarantees a certain bound on the probability of excessive delays.

The queueing model and related analyses in this research are aimed at providing safe service for patients. It is recognized that in any given medical unit, it would be desirable to introduce various additional considerations. For example, nurses may be qualified and certified to offer different levels of care, nurses may have different types and amounts of experience, nurse's service to a needy patient may have to be interrupted in favor of a more urgent need and replaced by a different nurse who completes the service, and so on. All of these complications contribute to the difficulty of analyzing nurse staffing policies, and certainly illustrate the need for guidelines that improve broadly upon the simple ratio rules that have been used in many nurse staffing applications.

The robust queueing system derived in this research provides a framework within which the previous variations can be considered and in which some of the underlying statistical assumptions concerning patient transitions and service times can be relaxed. Experience with this and related queueing models will inevitably raise new issues in healthcare coordination that will require healthcare system decision makers to address new questions, such as:

- How does a given time threshold value T ultimately influence the actual quality of care offered in specific types of medical units? And at what cost?

- What kind of response time constitutes an acceptable level of safety for various types of patients?

- What are appropriate scheduling policies for assigning individual nurses to specific shifts and duties, given a particular level of nurse staffing in a medical unit?

Queueing analysis provides valuable analytical tools that can be used to design effective and efficient healthcare facilities and services. However, more comprehensive studies to assess performance characteristics in actual or proposed health care systems often make use of Simulations, a topic that will be introduced and discussed in Chapter 8 of this book.

7.6 Summary

Queueing models provide a set of tools by which we can analyze the behavior of systems involving waiting lines, or queues. Queueing systems are characterized by the distribution of customers entering the system and the distribution of times required to service the customers.

In the simplest models, these arrival and service patterns are most often assumed to be Poisson arrivals and exponential service times. By viewing queueing systems as Markov birth-and-death processes, and solving flow balancing equations that describe the flow of customers into and out of the system, it is then straightforward to measure the performance characteristics of the system at steady state. These performance criteria include the expected amount of time the customer must wait to be served, the average number of customers waiting in a queue, and the proportion of time that the service facility is being utilized.

For more complicated queueing systems involving different statistical distributions of arrivals and departures, or complex interactions among multiple queues, or multiple servers, the applicability of analytical queueing models may be limited. In such cases, analysts often find that simulation is a more practical approach to studying system behavior.

Key Terms

arrivals
birth-and-death process
calling source
channels
customers
departures
exponential distribution
flow-balancing equations
network queues
Poisson distribution
queue
queueing model
servers
service times
simulation
system capacity

Exercises

7.1   Cars arrive at a toll gate on a highway according to a Poisson distribution with a mean rate of 90 miles per hour. The times for passing through the gate are exponentially distributed with mean 38 seconds, and drivers complain of the long waiting time. Transportation authorities are willing to decrease the passing time through the gate to 30 seconds by introducing new automatic devices, but this can be justified only if under the old system the average number of waiting cars exceeds five. In addition, the percentage of gate's idle time under the new system should not exceed 10%. Can the new device be justified?

7.2 A computer center has one multi-user computer. The number of users in the center at any time is ten. For each user, the time for writing and entering a program is exponential with mean rate 0.5 per hour. Once a program is entered, it is sent directly to the ready queue for execution. The execution time per program is exponential with mean rate of six per hour. Assuming the mainframe computer is operational on a full-time basis, and neglecting the possibility of down-time, find

a. The probability that a program is not executed immediately upon arrival in the ready queue

b. Average time until a submitted program completes execution

c. Average number of programs in the ready queue

7.3 The mean time between failures of a computer disk drive is 3,000 hours, and failures are exponentially distributed. Repair times for the disk drive are exponentially distributed with mean 5.5 hours, and a technician is paid $15.50 per hour. Assuming that a computing lab attempts to keep all drives operational and in service constantly, how much money is spent on wages for technicians in one year?

7.4 Printer jobs are created in a computing system according to a Poisson distribution with mean 40 jobs per hour. Average print times are 65 seconds. Users complain of long delays in receiving their printouts, but the computing lab director will be willing to purchase a faster printer (twice as fast as the present one) only if it can be demonstrated that the current average queue length is four (or more) jobs, and only if the new printer would be idle for at most 20% of the time. Will the lab director be able to justify the acquisition of the new printer?

7.5 Computer programs are submitted for execution according to a Poisson distribution with a mean arrival rate of 90 miles per hour. Execution times are exponentially distributed, with jobs requiring an average of 38 seconds. Users complain of long waiting times. Management is considering the purchase of a faster CPU that would decrease the average execution time to 30 seconds per job. This expense can be justified only if, under the current system, the average number of jobs waiting exceeds five. Also, if a new CPU is to be purchased, its percentage of idle time should not exceed 30%. Can the new CPU be justified? Explain all considerations fully. Make the necessary calculations, and then make an appropriate recommendation to management.

7.6 Customers arrive at a one-window drive-in bank according to a Poisson distribution with mean 10 per hour. Service time per customer is exponential with mean five minutes. The space in front of the window, including that for the car in service, can accommodate a maximum of three cars. Other cars can wait outside this space.

a. What is the probability that an arriving customer can drive directly to the space in front of the window?

b. What is the probability that an arriving customer will have to wait outside the designated waiting space?

c. How long is an arriving customer expected to wait before starting service?

d. How many spaces should be provided in front of the window so that at least 20% of arriving customers can go directly to the area in front of the window?

7.7 Suppose two (independent) queueing systems have arrivals that are Poisson distributed with $\lambda = 100$, but one system has an exponential service rate with $\mu = 120$

while the other system has $\mu = 130$. By what percentage amount does the average waiting time in the first system exceed that in the second system?

7.8   Jobs are to be performed by a machine that is taken out of service for routine maintenance for 30 minutes each evening. Normal job arrivals, averaging one per hour, are unaffected by this lapse in the service facility. What is the probability that no jobs will arrive during the maintenance interval?

Suppose the average service time is 45 minutes. How long do you think the system will take to recover from this interruption and return to a steady-state? Will it recover before the next evening? Does the recovery take a substantial part of the 24-hour day, so that the system essentially never really operates in a steady-state mode?

7.9   Fleet vehicles arrive at a refueling station according to a Poisson process at 20-minute intervals. Average refueling time per vehicle is 15 minutes. If the refueling station is occupied and there are two additional vehicles waiting, the arriving vehicle leaves and does not enter the queue at this facility. What percentage of arriving vehicles do enter this facility? What is the probability that an arriving vehicle finds exactly one vehicle being refueled and none waiting in the queue?

7.10 Customers arrive according to a Poisson distribution with mean six per hour to consult with a guru who maintains a facility that operates around the clock and never closes. The guru normally dispenses wisdom at a rate that serves ten customers per hour.

   a. What is the expected number of customers in the queue?

   b. If there are three chairs, what is the probability that arriving customers must stand and wait?

   c. What is the probability that the guru will actually spend more than ten minutes with a customer?

   d. An idle guru naps. How long in a typical day does this guru nap?

       Infrequently, but at unpredictable times, the guru himself takes off and climbs a nearby mountain to recharge his own mental and spiritual resources. The excursion always takes exactly five hours.

   e. How many chairs should be placed in the waiting room to accommodate the crowd that accumulates during such an excursion?

   f. Customers seeking wisdom from a guru do not want their waiting time to be wasted time, so they always want to bring an appropriate amount of reading material, in case of a wait. What is the normally anticipated amount of waiting time?

7.11 A bank, open for six hours a day, five days a week, gives away a free toaster to any customer who has to wait more than ten minutes before being served by one of four tellers. Customer arrivals are characterized by a Poisson distribution with mean 40 per hour; service times are exponential with mean four minutes. How many toasters does the bank expect to have to give away in one year of 52 weeks?

7.12 Select a system in your university, business, or community that involves queues of some sort, and develop a queueing model that describes the system. Identify the customers and servers. Observe the system and collect data to describe the arrival and service patterns. Apply the appropriate queueing formulae presented in this chapter to quantify the performance characteristics of this system. Are your computed results consistent with your observations?

References and Suggested Readings

Allen, A. O. 1980. Queueing models of computer systems. *IEEE Computer* 13 (4): 13–24.

Bolling, W. B. 1972. Queueing model of a hospital emergency room. *Industrial Engineering* 4: 26–31.

Bose, S. 2001. *An Introduction to Queueing Systems*. Boston, MA: Kluwer Academic Publishers.

Boucherie, N., van Dijk, and N. M. van Dijk. 2010. *Queueing Networks: An Analytical Handbook*. New York: Springer.

Bunday, B. D. 1986. *Basic Queueing Theory*. Baltimore, MD: Edward Arnold.

Chaudhry, M. L., A. D. Banik, A. Pacheco, and S. Ghosh. 2016. A simple analysis of system characteristics in the batch service queue infinite-buffer and Markovian service process. *RAIRO-Operations Research* 50 (3): 519–551.

Ebert, A. 2017. Computationally efficient queue simulation R package. *User Manual*. Available at: https://cran.r-project.org/package=queuecomputer. (Accessed on May 19, 2018)

Gautam, N. 2012. *Analysis of Queues: Methods and Applications*. Boca Raton, FL: CRC Press.

Gilliam, R. 1979. An application of queuing theory to airport passenger security screening. *Interfaces* 9: 117–123.

Green, L. V. 2002. How many hospital beds? *Inquiry* 39 (4): 400–412.

Gross, D., and C. M. Harris. 1998. *Fundamentals of Queueing Theory,* 3rd ed. New York: John Wiley & Sons.

Hassin, R. and M. Haviv. 2003. *To Queue or Not to Queue: Equilibrium Behavior in Queueing Systems.* New York: Springer.

Haviv, M. 2013. *Queues: A Course in Queueing Theory.* New York: Springer.

Hlynka, M. 2017. A comprehensive list of queueing theory software maintained by Myron Hlynka, Professor of the University of Windsor, Last Modified November 2017. https://web2.uwindsor.ca/math/hlynka/qsoft.html.

Jain, R. 1991. *The Art of Computer System Performance Analysis*. New York: Wiley.

Kleinrock, L. 1975. *Queueing Systems*. Vol. I: Theory. New York: John Wiley & Sons.

Kleinrock, L. 1976. *Queueing Systems*. Vol. II: Computer Applications. New York: Wiley Interscience.

Knuth, D. E. 1981. *The Art of Computer Programming,* 2nd ed. Vol. 2. *Seminumerical Algorithms*. Reading, MA: Addison-Wesley.

Kobayashi, H. 1978. *Modeling and Analysis: An Introduction to System Performance Evaluation Methodology*. Reading, MA: Addison-Wesley.

Lee, A. M. 1966. *Applied Queueing Theory*. Toronto, ON: Macmillan.

Little, J. D. C. 1961. A proof for the queuing formula: L= λ W. *Operations Research* 9 (3): 383–387.

Medhi, J. 1991. *Stochastic Models in Queueing Theory*. San Diego, CA: Academic Press.

Pawlikowski, K. 1990. Steady-state simulation of queueing processes: A survey of problems and solutions. *ACM Computing Surveys* 22 (2): 123–170.

Prabhu, N. U. 1997. *Foundations of Queueing Theory*. New York: Springer.

Ravindran, A., D. T. Phillips, and J. J. Solberg. 1987. *Operations Research: Principles and Practice,* 2nd ed. New York: John Wiley & Sons.

Reiser, M. 1976. Interactive modeling of computer systems. *IBM Systems Journal* 15 (4): 309–327.

Saaty, T. L. 1983. *Elements of Queueing Theory: With Applications*. New York: Dover Publications.

Suri, R., G. W. W. Diehl, S. de Treville, and M. J. Tomsicek. 1995. From CAN-Q to MPX: Evolution of queuing software for manufacturing. *Interfaces* 25 (5): 128–150.

Tanner, M. 1995. *Practical Queueing Analysis*. New York: McGraw-Hill Companies.

Tijms, H. C. 2003. *A First Course in Stochastic Models*. New York: John Wiley & Sons.

Véricourt, F. D., and O. B. Jennings. 2011. Nurse staffing in medical units: A queueing perspective. *Operations Research* 59 (6): 320–1331.

Walrand, J. 1988. *An Introduction to Queueing Networks*. Englewood Cliffs, NJ: Prentice-Hall.

8

Simulation

Simulation is the process of studying the behavior of an existing or proposed system by observing the behavior of a model representing the system. Simulation is the imitation of a real system or process operating over a period of time. By simulating a system, we may be able to make observations of the performance of an existing system, hypothesize modifications to an existing system, or even determine the operating characteristics of a nonexistent system. Through simulation, it is possible to experiment with the operation of a system in ways that would be too costly or dangerous or otherwise infeasible to perform on the actual system itself. This chapter introduces simulation models and describes how they can be used in analyzing and predicting the performance of systems under varying circumstances.

8.1 Simulation: Purposes and Applications

Simulation has traditionally been viewed as a method to be employed when all other analytical approaches fail. Computer simulations have been used profitably for several decades now, and simulation seems to have outlived its early reputation as a *method of last resort*. Some systems are simple enough to be represented by mathematical models and *solved* with well defined mathematical techniques such as the calculus, analytical formulas, or mathematical programming methods. The simple queueing systems discussed in Chapter 7 fall into this category. Analytical methods are clearly the most straightforward way to deal with such problems. However, many systems are so complex that mathematical methods are inadequate to model the intricate (and possibly stochastic) interaction among system elements. In these cases, simulation techniques may provide a framework for observing, predicting, modifying, and even optimizing a system.

The use of a computer makes simulation techniques feasible. Information obtained through observing system behavior via simulation can suggest ways to modify a system. And while simulation models remain very costly and time consuming to develop and to run on a computer, these drawbacks have been mitigated significantly in recent times by faster computers and special purpose simulation languages and software products. Indeed, simulation packages have become so widely available and easy to use, and simulation itself has such an intuitive appeal and seems so simple to understand, that a word of caution is in order.

Simulation languages and packages are as easy to misuse as to use correctly. Computer outputs produced by simulation packages can be very impressive. Particularly when other analytical approaches to a problem have been unsatisfactory, it is tempting to embrace whatever *output* is obtained through a sophisticated simulation process. Nevertheless, there is a great deal to be gained through successful simulation. Proper use of simulation methodology requires good judgment and insight and a clear understanding of the limitations of the simulation model in use, so that valid conclusions can be drawn by the analyst. This chapter presents some guidelines that should be helpful in developing the ability to understand and build simulation models. The advantages that may be derived from the use of simulation include:

1. Through simulation it is possible to experiment with new designs, policies, and processes in industrial, economic, military, and biological settings, to name a few. In the controlled environment of a simulation, observations can be made and preparations can be made to deal appropriately with the outcomes predicted in the experiment.

2. Simulation permits the analyst to *compress* or *expand* time. For example, collisions in a particle accelerator may occur too rapidly for instruments to record, while erosion in a riverbed may take place too slowly to permit any effective intervention in the process. By simulating such processes, a time control mechanism can be used to slow down or speed up events and place them on a time scale that is useful to human analysts.

3. While a simulation may be expensive to develop, the model can be applied repeatedly for various kinds of experimentation.

4. Simulation can be used to analyze a proposed system or experiment on a real system without *disturbing* the actual system. Experimentation on real systems, particularly systems involving human subjects, often causes the behavior of the system to be modified in response to the experimentation. Thus, the system being observed is then not the original system under investigation; that is, we are measuring the wrong system.

5. It is often less costly to obtain data from a simulation than from a real system.

6. Simulations can be used to verify or illustrate analytical solutions to a problem.

7. Simulation models do not necessarily require the simplifying assumptions that may be required to make analytical models tractable. Consequently, a simulation model may well be the most realistic model possible.

Application areas that have been studied successfully using simulation models are numerous and varied. Problems that are appropriate for simulation studies include:

- *Activities of large production, inventory, warehousing, and distribution centers*: To determine the flow of manufactured goods
- *Operations at a large airport*: To examine the effects of changes in policies, procedures, or facilities on maintenance schedules, hangar utilization, or even runway throughput
- *Automobile traffic patterns*: To determine how to build an interchange or how to sequence traffic lights at an existing intersection
- *Computer interconnection networks*: To determine the optimum capacity of data links under time varying data traffic conditions
- *Meteorological studies*: To determine future weather patterns

The process of building a simulation of a system is not entirely unlike the process of creating other types of models that have been discussed in this book. The **problem**

formulation phase of a simulation study involves defining a set of objectives and designing the overall layout of the project. **Building a model** of the actual system being studied involves abstracting the essential features of the system and making basic assumptions in order to obtain first a simple model, then enriching the model with enough detail to obtain a satisfactory approximation of the real system. Albert Einstein's advice that things should be made *as simple as possible, but not simpler* might be augmented by the complementary advice that a model need be only complex enough to support the objectives of the simulation study. Real objects and systems have a variety of attributes (physical, technical, economic, biological, social, etc.). In the process of modeling, it is not necessary to identify *all* system attributes, but rather to select just those that efficiently and specifically contribute to the objectives of the model and serve the needs of the modeler or analyst. (For example, if we were studying the structural properties of certain materials to be used in an aircraft, we would include such attributes as tensile strength and weight. And although we might also know the cost or reflectivity of the materials, these latter attributes do not contribute directly to the structural model at hand.) If unnecessary detail and realism are incorporated into the model, the model becomes expensive and unwieldy (although perhaps correct) and the advantages of simulation may be lost. Various types of **simulation models** are discussed in Section 8.2.

The analyst must then **collect data** that can be used to describe the environment in which a system operates. These data may describe observable production rates, aircraft landings, automobile traffic patterns, computer usage, or air flow patterns, and may be used later in experimentation. Extensive statistical analysis may be required in order to determine the distribution that describes the input data and whether the data are homogeneous over a period of time.

Coding the simulation often involves developing a program through the use of simulation languages or packages, as described in Section 8.4.

Verification of the simulation is done to ensure that the program behaves as expected and that it is consistent with the model that has been developed.

Validation tests whether the model that has been successfully developed is in fact a sufficiently accurate representation of the real system. This can be done by comparing simulation results with historical data taken from the real system, or by using the simulation to make predictions that can be compared to future behavior of the real system.

Experimental design is closely related to the original objectives of the study and is based on the nature of the available data. Once the nature and extent of the experimentation is fully defined, the **production** phase begins. Simulation *runs* are made, and system analysis is performed. In some cases, an **optimization** algorithm is coupled with the simulation model to find the optimal values for certain variables in the simulation that would produce optimal values for certain performance measures. For example, finding the optimal resource levels that would maximize throughput subject to some constraints such as allocated budget. This is known as **simulation–optimization**. Upon completion of these phases, final reports are made of observations and recommendations can be formulated.

Although we will not fully discuss all of these phases, we will look more carefully now at some specific techniques for creating **discrete simulation** models. We will also discuss the design of simulation experiments, the use of the results, and some of the software systems and languages that are commonly used as tools in developing simulations.

8.2 Discrete Simulation Models

A computer simulation carries out *actions* within a computer program that represent activities in some real system being modeled. The purpose of the simulation is to make observations and collect statistics to better understand the activity in the simulated system and possibly to make recommendations for its improvement.

Simulations can be categorized as either **discrete** or **continuous**. This distinction refers to the variables that describe the *state* of the system. In particular, the variable that describes the passage of time can be viewed as changing continuously or only at discrete points in time. In models of physical or chemical processes, for example, we might be interested in monitoring continuous changes in temperature or pressure over time, and in that case a continuous simulation model would be appropriate. These models generally consist of sets of differential equations; the rate of change in each variable is dependent on the current values of several other variables. Examples include process control systems, the flight of an airplane, or a spacecraft in orbit continuously balancing the forces of gravity, velocity, and booster rockets.

On the other hand, in queueing systems, events such as customer arrivals and service completions occur at distinct points in time, and a discrete event simulation model should be chosen. Continuous simulation will be mentioned again in Section 8.4, and the topic is thoroughly discussed in many of the references cited at the end of this chapter, particularly Roberts et al. (1994). We will concentrate on discrete simulation models throughout the remainder of this chapter.

8.2.1 Event-Driven Models

A simulation model consists of various components and entities. The dynamic objects in the system are called **entities**. Other main components include processes, resources and queues. In a customer queueing system, for example, the entities may be the customers. Each entity possesses characteristics called **attributes**. The attributes of a customer include the customer's arrival time and the type of service required by the customer. The servers would be characterized by the type of service they render, the rate at which they work, and the amount of time during which they are busy. Queue attributes would include the queue length and the type of service for which the queue is designated. Some attributes such as type of service required or work rate are set at the beginning of the simulation, while other attributes are assigned and updated as the simulation proceeds.

The **system state** is defined by the set of entities and attributes, and the state of the system typically changes as time progresses. Processes that affect the system state are called **activities**. An activity in a queueing system may be a customer waiting in line, or a server serving a customer.

Any activity in a simulation will eventually culminate in an **event**, and it is the occurrence of an event that actually triggers a change in the system state in a discrete simulation model. For this reason, certain discrete simulation models are referred to as **event-driven models**. Although other views such as *process oriented* simulation and *object oriented* simulation are found in some of the languages that will be described in Section 8.4, the *event-driven* view is probably the most widely used discrete simulation approach.

To track the passage of time in a simulation model, a simulation **clock** variable is initially set to zero and is then increased to reflect the advance of simulated time. The increment may be fixed or variable. One such time advance mechanism calls for repeatedly increasing the clock by a fixed unit of time, and at each increment, checking the system to determine whether any event has occurred since the last increment. The disadvantage of this mechanism is the difficulty in selecting an appropriate interval for the clock increment. If the interval is too small, a great deal of uninteresting and inefficient computation occurs as the clock is advanced repeatedly and no events have taken place. If the interval is too large, several events may have occurred during the interval and the precise ordering of events within the interval is not registered, since all these events are assumed to have taken place at the end of the interval. In this way, key information may be lost. Because systems are not necessarily uniformly *eventful* throughout the duration of the simulation (i.e., there will be busy times and quiet times), it is virtually impossible to choose the *correct* or *best* interval for incrementing the simulation clock throughout the entire simulation.

An alternative, and more popular, time advance mechanism is to allow the simulation clock to be advanced only when an event actually occurs. The bookkeeping required to maintain a list of events that will be occurring, and *when* they will occur, is straightforward. The mechanism checks the list to determine the *next* event, and advances the clock to the time of that event. The event is then registered in the simulation. This variable increment mechanism is efficient and easy to implement.

An effective way to learn just exactly what a computer simulation does is to work through a simulation manually. In the following example, we will perform an event driven simulation of a queueing system.

Example 8.2.1

The system we will simulate is one in which the customers are computer programs that are entered into a system to be executed by a single central processing unit (CPU), which is the service facility. As a computer program enters the system, it is either acted upon immediately by the CPU or, if the CPU is busy, the program is placed in a *job queue* or *ready queue* maintained in FIFO (first in first out) order by the computer's operating system.

The service facility (the CPU in this case) is always either busy or idle. Once in the system, the customer (computer program in this case) is either in a queue or is being served. The queue is characterized by the number of customers it contains. The status of the server, the customers, and the queue collectively comprise the *state* of the queueing system, and the state changes only in the event of an arrival or departure of a customer. The input data for this simulation example are given in Table 8.1.

The first program arrives for execution at time 0. This event starts the simulation clock at 0. The second program arrives four time units later. The third customer arrives one time unit later at clock time 5, and so forth. Program execution times are two, three, five, and so on, time units. A quick glance at the arrival and service times shows that in some cases a program is executed completely before its successor arrives, leaving the CPU temporarily idle, whereas at other times a program arrives while its predecessors are still in execution, and this program will wait in a queue.

Table 8.2 shows the clock times at which each program enters the system, begins execution, and departs from the system upon completion of execution. Notice that the CPU is idle for two time units between Programs 1 and 2, for three time units between Programs 5 and 6, and for five time units between Programs 6 and 7. Program 9 arrives

TABLE 8.1

Arrival and Service Times

Customer Number	Arrival Time	Length of Service
1	0	2
2	4	3
3	5	5
4	9	1
5	10	2
6	18	2
7	25	3
8	26	4
9	32	5
10	33	1

TABLE 8.2

Simulation Event Clock Times

Customer Number	Arrival Time	Time Execution Begins	Time Execution Completes
1	0	0	2
2	4	4	7
3	5	7	12
4	9	12	13
5	10	13	15
6	18	18	20
7	25	25	28
8	26	28	32
9	32	32	37
10	33	37	38

just exactly as its predecessor is completing, so there is no CPU idle time nor does Program 9 have to join the queue and wait. Programs 3, 4, 5, 8, and 10 must wait in the queue before beginning execution. Table 8.3 shows the chronological sequence of events in this simulation.

The primary aim of a simulation is to make observations and gather statistics. In this particular example, we will be interested in determining the average time programs spend in the system (*job turnaround time*), the average time spent waiting, the average number of programs in the queue, and the amount or percent of time the CPU is idle. We return to this example in Section 8.3.1 to illustrate making these observations.

Before continuing, however, we should note that the single server queueing system we have just observed fails in several respects to match the M/M/1 model developed in Section 7.3. First of all, arrivals and service times were given deterministically in table form rather than being drawn from the more typical Poisson and exponential distributions. Second, the system was tracked through only ten customers and over a period of only 38 time units (probably a short time relative to the life of the system). Thus, because of the deterministic customer and service behavior and the short duration of the simulation, it would be unjustifiable to claim that these results are in any way *typical* of the

TABLE 8.3

Chronological Sequence of Events

Clock Time	Customer Number	Events
0	1	Arrival and begin service
2	1	Departure
4	2	Arrival and begin service
5	3	Arrival and wait
7	2	Departure
	3	Begin service
9	4	Arrival and wait
10	5	Arrival and wait
12	3	Departure
	4	Begin service
13	4	Departure
	5	Begin service
15	5	Departure
18	6	Arrival and begin service
20	6	Departure
25	7	Arrival and begin service
26	8	Arrival and wait
28	7	Departure
	8	Begin service
32	8	Departure
	9	Arrival and begin service
33	10	Arrival and wait
37	9	Departure
	10	Begin service
38	10	Departure

normal operation of the system. The most common way to overcome these deficiencies is to generate numerical values representing a large number of customers with random arrival patterns and service times. We require that these random values be representative of events and activities that occur in the real system. One mechanism for doing this is described in the following.

8.2.2 Generating Random Events

In a discrete event simulation, once an event of any type has been simulated, the most important piece of information we need to know, in order to advance the simulation is: how long until the next event? Once a customer has arrived, we need to know when the next arrival will occur so that we can *schedule* that event within the simulation. Similarly, upon completion of a service or upon arrival of a customer to an idle server, we need to know the length of time this next service will take so that we can *schedule* this customer's departure from the system.

If we are assuming that interarrival times and service times come from some particular probability distributions, then we must have a mechanism within the simulation program to generate the lengths of these intervals of time and therefore to generate the *next events* in the simulated system. The general procedure will be first to generate a random number from the uniform distribution, to apply a mathematical transformation to the uniform deviate to obtain a random number from the desired distribution, and then to use this random number in the simulation (perhaps as the interval of time until the next event).

A great deal of effort has been put into the study and development of computer programs to generate *random* numbers. Truly random numbers are typically obtained from some physical process, but sequences of numbers generated in this way are unfortunately not reproducible. **Pseudorandom numbers** are numbers that satisfy certain statistical tests for *randomness* but are generated by a systematic algorithmic procedure that can be repeated if desired. The purpose of generating pseudorandom numbers is to simulate sampling from a continuous uniform distribution over the interval [0,1].

The most frequently implemented algorithms belong to the class of *congruential generator methods*. These generators are fully described in books by Knuth (1981), Graybeal and Pooch (1980), Banks et al. (1984), Marsaglia (2003) and most introductory texts on simulation; and they are almost always available in any computer installation through simple subroutine calls. Because of the easy accessibility of these pseudorandom number generators, it is doubtful that a simulation analyst would need to develop software from scratch for this purpose. Yet, from a practical standpoint, analysts are encouraged to heed the following *warning*. Because almost every computer system offers at least one means of generating uniform random variates, most computer users employ these capabilities with faith, assume their correctness, and feel happy with the results. Nevertheless, blatantly bad random number generators are prevalent and may fail some of the standard theoretical or empirical statistical tests for randomness, or may generate strings of numbers exhibiting detectable regular patterns (Marsaglia 1985, Park and Miller 1988, Ripley 1988, L'Ecuyer 1990).

Although many simulation models appear to work well despite these defects in the stream of random numbers, there have been simulation studies that yield totally misleading results because they are more sensitive to the quality of the generators. And although such failures are rare, they can be disastrous; therefore, researchers are still actively investigating better ways to generate random numbers.

In any case, it is quite unlikely that a simulation analyst would need to develop his own software for this purpose. Instead we will discuss how to *use* a uniform deviate from the interval [0,1] to produce a random number from an exponential distribution, thus simulating a sampling from an exponential distribution.

A commonly used method for doing this, called the **inverse transform method**, can be applied whenever the inverse of the cumulative distribution function of the desired distribution can be computed analytically.

Recall that the probability density function for the exponential distribution is given by:

$$f(x) = \begin{cases} \lambda e^{-\lambda x} & \text{for } x \geq 0 \\ 0 & \text{for } x < 0 \end{cases}$$

The corresponding cumulative distribution function is given by:

$$F(x) = \int_{-\infty}^{x} f(t)dt = \begin{cases} 1 - e^{-\lambda x} & \text{for } x \geq 0 \\ 0 & \text{for } x < 0 \end{cases}$$

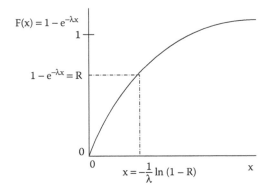

FIGURE 8.1
Inverse transform method.

Figure 8.1 illustrates that the range of F(x) is the interval (0,1), and suggests that uniform random numbers from (0,1) can be transformed into exponentially distributed numbers as follows. Let R denote the uniform random number from (0,1), and set F(x) = R. Then, x = F⁻¹(R) and x can be solved in terms of R by evaluating:

$$F(x) = 1 - e^{-\lambda x} = R$$

$$e^{-\lambda x} = 1 - R$$

$$-\lambda x = \ln (1 - R)$$

$$x = -\frac{1}{\lambda} \ln (1 - R)$$

This formula is called a random variate generator for the exponential distribution. It is often simplified by replacing (1 − R) by R, since both R and (1 − R) are uniformly distributed on (0,1), to obtain the generating formula

$$x = -\frac{1}{\lambda} \ln R$$

Therefore, whenever a simulation program requires a sample from an exponential distribution, R is obtained from a standard pseudorandom number generator, and x is computed by this formula and used in the simulation.

The inverse transform method is not necessarily the most efficient method, but it is straightforward and can be used to generate deviates from a variety of statistical distributions other than the exponential. Unfortunately, for most distributions (including the normal), the cumulative probability function does not have a closed form inverse. In particular, the distribution may be derived from an empirical study of the actual system. In practice, the distributions may not fit any of the theoretical functions. For example, consider a server who can perform several different types of service depending on the customer's need (doctor, bank teller, mechanic). Each *type* of service has a non-zero minimum required time plus a random variable time. However, when all types of service are aggregated, the resulting distribution is likely to be multi-modal, and very non-standard.

In these situations, it is common to approximate the cumulative distribution by a piecewise linear function, and then to apply the inverse transform method using linear

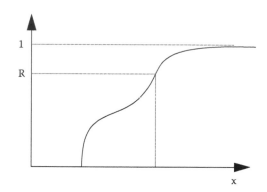

FIGURE 8.2
Example of non-standard cumulative service time distribution.

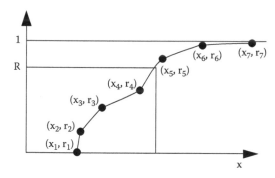

FIGURE 8.3
Piecewise linear approximation of non-standard cumulative service time distribution.

interpolation on each segment. Consider the example in Figure 8.2. We can construct a piecewise linear approximation for this cumulative distribution, as shown in Figure 8.3.

Internally, this distribution can be stored in table form, as in Table 8.4. Then, when we want to generate a random variate from this distribution, we select a uniform random number r, then search for the entry r_i in the table such that $r_i \leq r < r_{i+1}$. The corresponding service time is obtained using standard linear interpolation:

TABLE 8.4

Piecewise Linear Approximation of Cumulative Service Time Distribution

Inflection Point	x Service Time	F(x) Cumulative Probability
1	x_1	$r_1 = 0$
2	x_2	r_2
3	x_3	r_3
4	x_4	r_4
5	x_5	r_5
6	x_6	r_6
7	x_7	$r_7 = 1.0$

$$x = (r - r_i)/slope_i + x_i$$

where

$$slope_i = \frac{(r_{i+1} - r_i)}{(x_{i+1} - x_i)}$$

Clearly, by using a large number of piecewise linear segments, it is possible to achieve any desired level of accuracy. However, there is a trade-off between accuracy and the time required to search repeatedly through large tables.

For further and more general discussions of this method, see the books by Ravindran et al. (1987), Schmeiser (1980), and Law (2007). These references also contain descriptions of other methods, such as the rejection method, the composition method, a derivation technique, and approximation techniques.

8.3 Observations of Simulations

Now that we have discussed some of the techniques for generating the events that push a simulated system through time, let's consider what observations can be made during the simulation that would help to characterize or understand the system being studied.

8.3.1 Gathering Statistics

Because we are currently concerned primarily with the simulation of queueing systems, it is reasonable that the information we would like to obtain from the simulation is just exactly the same type of information that we would calculate with analytical queueing formulae, if we could (i.e., if we had a queueing system in steady-state with known distributions describing arrival and departure patterns). In particular, we might like to determine the average time a customer spends in the system and waiting, the average number of customers in the queue, and the utilization factor of the service facility.

We can return to Example 8.2.1 and show how such information can be gathered. It is important to realize, however, that as we determine these measures of system behavior, we are doing so only for the specific system with the particular arrivals and departures given in the table, and only for the particular time interval covered by these events. No generalization can be drawn about *typical* behavior of the system over the long term. (If it is desirable to make inferences about the steady state characteristics of a simulated system, then a number of issues need to be considered. We will return to this subject after we work through our Example.)

8.3.1.1 Average Time in System

For every customer i, compute

$$T_i = \text{Time spent in the system}$$

$$= \text{Time of service completion} - \text{Time of arrival}$$

Then accumulate the sum of these T_i and divide by the number of customers N:

$$\text{Average time in system} = \frac{\left(\sum_{i=1}^{N} T_i\right)}{N}$$

In the simulation, initialize the sum to zero; then at every service completion event, compute the T_i for this customer and add it to the sum. At the end of the simulation, perform the division. In the Example 8.2.1, we can obtain the T_i from Table 8.2 and compute the sum

$$2 + 3 + 7 + 4 + 5 + 2 + 3 + 6 + 5 + 5 = 42$$

Then the average time in the system for these ten programs is $42/10 = 4.2$ time units.

8.3.1.2 Average Waiting Time

For every customer i in the system, compute

$$W_i = \text{Waiting time} = \text{Time service begins} - \text{Arrival time}$$

$$= \text{Time in system} - \text{Service time}$$

Then accumulate the sum of these W_i and divide by the number of customers N:

$$\text{Average waiting time} = \frac{\left(\sum_{i=1}^{N} W_i\right)}{N}$$

In the simulation, initialize the sum to zero; then at every event corresponding to service beginning (or a departure event), compute the W_i for this customer and add it to the sum. At the end of the simulation, perform the division. In our example, from Table 8.2 again, we obtain the waiting time sum:

$$0 + 0 + 2 + 3 + 3 + 0 + 0 + 2 + 0 + 4 = 14$$

Then the average waiting time for these ten programs is $14/10 = 1.4$ time units.

8.3.1.3 Average Number in Queue

If we let L_i denote the number in the queue during the i-th time interval, then over U time units,

$$\text{Average queue length} = \frac{\left(\sum_{i=1}^{U} L_i\right)}{U}$$

Rather than making an observation of L_i at every *time unit*, it is more practical to observe the queue length at every *event*, and to multiply that queue length by the number of time units that have elapsed since the most recent event that affected the queue length.

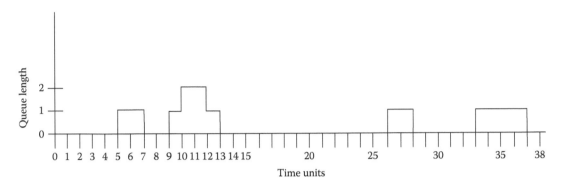

FIGURE 8.4
Queue length.

This bookkeeping requires that we maintain a time duration only for every queue length that occurs during the simulation. In our example, Figure 8.4 charts the queue length during each of the 38 intervals of time. Note that the queue length in this case is always either 0, 1, or 2:

$$\text{Queue length} = 0 \text{ for 26 time units}$$

$$= 1 \text{ for 10 time units}$$

$$= 2 \text{ for 2 times units}$$

Then, over $U = 38$ time units,

$$\text{Average queue length} = \frac{[0 \cdot 26 + 1 \cdot 10 + 2 \cdot 2]}{38}$$

$$= \frac{14}{38}$$

$$= 0.368$$

8.3.1.4 Server Utilization

Upon every event, determine the service facility status (busy or idle) and record it. Then,

$$\text{Server utilization factor} = \frac{\text{Number time units busy}}{\text{Total number time units}}$$

As illustrated in Figure 8.5, our CPU is busy executing programs during 28 time intervals and is idle during 10 time intervals. Therefore, the

$$\text{Server utilization factor} = \frac{28}{38} = 0.74$$

and the

$$\text{Percentage idle time} = \frac{10}{38} = 0.26$$

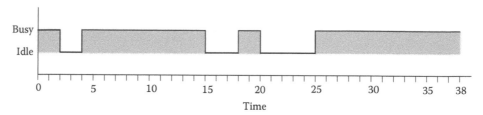

FIGURE 8.5
CPU (server) utilization.

Note that the same information on server utilization can be obtained from Figure 8.5 by computing the shaded area under this step function.

Observations such as these allow us to make judgments concerning, for example, the advisability of acquiring an additional CPU to reduce waiting time. In this example, with an average waiting time of 1.4 time units (a fairly small fraction of average execution time), a queue that is empty more often than not, and a CPU that is idle 26% of the time, it seems unlikely that an additional CPU would be warranted in a general purpose computing environment.

8.3.2 Design of Simulation Experiments

In designing a system simulation in which events are to be generated randomly (rather than introduced into the system deterministically), several questions arise:

- How to start the simulation?
- What to measure?
- What data to gather?
- How long to run the simulation?
- How to recognize whether the system has reached equilibrium?
- How many simulation runs to make?
- What recommendations to make concerning modification of the system being simulated?

We do not necessarily intend to offer answers to these questions, but rather merely to raise the issues that must be considered by the analyst or system designer.

Once a simulation program is developed, and is ready to run, the initial system status must be determined. It may be reasonable to initialize a system as having idle servers and no customers, and let customers begin to arrive randomly. Eventually, there will emerge a certain pattern of queue utilization and service utilization, but when exactly does the *real* pattern emerge? For example, when a bank opens its doors at 9 a.m., it makes sense to assume that the system is empty when it starts. However, if we are simulating a hospital, we probably should run the simulation in *start-up* mode for several days before enough patients have been accumulated so that the apparent simulated demand on the system becomes realistic.

When is it then appropriate to begin observing the system and **collecting data** about queue length, waiting time, and server utilization? It is not valid to start collecting statistics until the system has reached its *steady state*, but this point is difficult to identify precisely. How long should the simulation run after initialization? It would be useful to somehow acquire information in advance (perhaps from previous similar simulations) that describes the system after the initial irregular system behavior patterns have disappeared. It is, however, often difficult to know this in advance.

The questions of what to measure and what data to gather depend of course on the original purpose of performing the simulation. In simulating waiting line systems, there are several obvious performance criteria of common interest. In simulating more general systems, a great deal of data is potentially available; yet gathering all this data is costly and may complicate the simulation program. The efficiency of a simulation may depend on the clear-headed analyst's decision to measure only the behaviors that are relevant to the study.

Perhaps the most important, and most expensive, question is how long to run a simulation, and how to know when additional computation is not going to yield additional information. Recall from our study of Markov and queueing systems that not all systems ever reach a steady state. Some display periodic behavior or other unstable patterns. It may be difficult to know in advance whether the system being simulated is guaranteed to reach a steady state. If it does eventually stabilize, we know that the length of time it takes for this to occur depends on the initial conditions. The only way to make the decision of how long to run a simulation is to gather data, accumulate performance measures (such as average queue length), and compare these measures with those measured earlier in the simulation. When they cease to change significantly, it might be reasonable to surmise that a steady state has been reached. (It could of course be a temporary phenomenon. How can we know for sure?)

Once a simulation program is developed and all the design parameters for a single run have been established, the next question is how many runs to make. Presumably there will be some statistical variations in the system performance measures obtained during each run. So how many samples do we need in order to be confident that we have captured the reality of the system being modeled? Do we make a fixed number of runs? Or enough runs that the variance in outcomes is acceptably low? And if we make numerous runs, should they all cover the identical span of simulated time, or should the simulated time intervals be varied or shifted? Should the various runs involve different system parameters? (For example, we might wish to compare the performance of a computer system with one CPU with that of a system having two CPUs, each with 60% of the speed of the single CPU.) To make such comparisons, it is likely that a battery of experiments would have to be performed for each case. Based on such observations, recommendations could be made for alternative systems having different strengths, advantages, or costs.

8.4 Software for Simulation

Simulation studies can be facilitated by a wide variety of software packages and languages. Specialized computer programming languages have been introduced over the past several decades to assist simulation analysts in the development and use of simulations of real systems. Simulation models have also been implemented in various traditional general purpose programming languages such as Fortran, C, C++, Java, and Python, among others. Simulations developed directly in general purpose, high level languages often execute more efficiently than those implemented in specialized simulation languages. However, most analysts find that it is much more efficient and beneficial to use a special simulation language rather than to try to develop a simulation program from scratch. A competent analyst may lack extensive programming skill, and may prefer to concentrate instead on the system being modeled, using the most convenient tools possible.

Various criteria will determine an analyst's choice of a simulation language or software package. A first consideration is likely to be the analyst's own *programming capabilities* and whether a given language is easy to learn and complements the analyst's skills and experience. A non-programmer may choose a language that is easier to learn, has greater built-in support, and provides less flexibility, whereas a highly skilled programmer may be more adept at learning a language that gives him more *power* and *flexibility*; and this additional control permits the ability to model unusual systems in specialized applications. The nature of the system being modeled can also influence the choice of software; some systems allow the user to add customized subroutines to model non-standard types of activities.

Most simulation software products provide *automatic* mechanisms for collecting statistics, generating reports, and even debugging the simulation. Additional considerations may include the standardization or portability of the language among machines and other software environments. In this section we examine some of the features of several widely used simulation languages and software packages. Further information may be found in published surveys of simulation software (Swain 2017), which review and chart the capabilities of many software packages for simulation.

One of the earliest languages for simulating the dynamic performance of systems is **GASP** (**G**eneral **A**ctivity **S**imulation **P**rogram), a collection of Fortran subroutines developed by Pritsker (1974). GASP subroutines support the development of event-driven and continuous simulations and require subroutines for system initialization, time advance mechanisms, scheduling future events, random variate generators, routines to collect statistics, and report generators. Most of these are supplied as a part of the GASP package; however, the programmer must create a main (driver) routine and fill in the details of initialization and event management.

SIMSCRIPT (**SIM**ple-**SCRIPT**) was developed at the RAND Corporation during the 1960s and was originally an easy to use, Fortran based, system for discrete event simulation and modeling. Over time, the system underwent numerous revisions, evolving into a high level language available for most platforms, and capable of supporting event driven and process oriented simulations, with extensions for continuous simulations and can run on different platforms (Russell 1993, Rice et al. 2005).

GPSS (**G**eneral **P**urpose **S**imulation **S**ystem) was originally developed at IBM around 1960, and because of its early origin and ease of learning, was among the most widely used simulation languages, especially in the 1970s and 1980s (Schriber 1974, Gordon 1978, Solomon 1983). GPSS was then succeeded by **GPSS/H** with additional features and is typically applied to general queueing analysis, manufacturing, and scheduling and can be used under Windows platforms. The most modern version of GPSS is **aGPSS** which comes with a graphical user interface and can run on both Windows and Macintosh (See Schriber [1993] and Hendriksen [1993]).

ARENA is currently among the most widely used software packages for discrete-event simulation. It started as a command language SIMAN and then SIMAN became the engine around which ARENA was built as its Graphical User Interface which allows the user to drag and drop simulation objects and libraries. It includes input analysis and output analysis tools as well as simulation optimization add-in (OptQuest).

It is a general modeling language that is mostly used for process modeling and it works under Windows only at this point.

AnyLogic is a modeling environment that is written in Java and has become very popular in the recent years especially as it allows building not only discrete-event models but also agent based models and system dynamics models. Hybrid models of these three paradigms can also be developed using this object oriented modeling environment which runs on Windows, Mac and Linux.

FlexSim is another object oriented process modeling environment with an outstanding 3D animation capability. It is written in C++ and allows users to augment their models with C++ functional code if necessary. It includes ExpertFit for statistical distribution fitting and has analysis tools.

SAS is a widely used software system that provides support for large models and complex simulation experiments. The graphical user interface offers convenient tools for creating, executing and analyzing simulations. It integrates with **SAS Simulation Studio** for source data and for presenting results of simulation studies. It runs on Windows, Mac and Linux workstations.

Many simulation tools and development environments have emerged over the past 50 years and we recognize that we cannot explore all of them in this book. Additional examples of such packages include AutoMod, ExtendSim, ProModel, Simio, SimProcess, Simula8, and Witness. The reader is referred to the regularly published surveys of simulation software (Swain 1997, 1999, 2001, 2003, 2005, 2007, 2009, 2011, 2013, 2015, 2017), which review and chart the capabilities of many software packages for modeling and simulation. Many simulation languages and software packages had emerged and vanished over the years as more efficient, convenient, and powerful packages were developed. The reader can compare the surveys published by Swain over the years to see which ones have been removed from the list, and which are still on.

Furthermore, a list of open source and commercial simulation packages can be found on the Internet by running a quick web search. For example, a comprehensive list of simulation software is available on Wikipedia at https://en.wikipedia.org/wiki/List_of_discrete_event_simulation_software.

In recent years, cloud computing has started to change the long standing tradition of having software installed locally on computers. Instead, software is installed in the *cloud* on servers hosted by computing farms, and users access the software application via web browser most of the time. It seems that this phenomenon is on its way to becoming the standard for most software systems, including simulation software. An example of this trend is **ClouDES**, a web based, cloud deployed, discrete event simulation platform developed at the Virginia Modeling, Analysis and Simulation Center (VMASC) at Old Dominion University. This cloud based system can be accessed via a web browser without the need to install simulation software on the client side. Instead, users can develop models and execute them through a web browser.

Because simulation studies have traditionally had the reputation of consuming a great deal of computer time for execution of large simulations, substantial amounts of research have been devoted to the development of technologies to use parallel computers to increase the execution speed of simulation programs. This effort has been accelerated and made easier by advances in computer technology where more processing power can now be packed in smaller computers, making it possible to run multiple parallel processors on the same machine. Despite the significant progress made in computer hardware and software, developing simulation models is still a time consuming process for developers and analysts starting with the conceptual development, to implementation and finally validation of simulations.

8.5 Illustrative Applications

8.5.1 Finnish Air Force Fleet Maintenance (Mattila et al. 2008)

The aircraft in Finland's Air Force fleet typically require several hours of maintenance for every hour of flight activity. Depending on the type of aircraft, maintenance involves a variety of policies and procedures, task times, workforce personnel, and equipment, parts and materials handling. Maintenance system performance has an effect on aircraft availability, which is defined to be the number of aircraft that can be used in flight missions. Therefore, an understanding of all aspects of aircraft maintenance decision making is essential to measuring and maintaining the operational capability of the fleet. Analysts determined that discrete event simulation was the most appropriate analytical tool for studying the fleet maintenance system and its effect on aircraft availability.

An adequate model of a maintenance system for flight operations must address the types and number of aircraft, planned and unplanned maintenance, air bases and repair and maintenance sites, levels of maintenance staff, and scenarios for normal flight mission assignments as well as missions involving conflict and hence increased exposure to damage. The fleet in this case consisted of F-18 Hornet fighters, Hawk Mk 51 jet trainers, and certain other aircraft used in transportation, surveillance, and flight training. Various Air Force operational units have facilities for basic inspections, routine maintenance, or specialized shops for more complex tasks or repairs.

The goal of aircraft maintenance during normal peacetime is to preserve the long term operational readiness of the fleet. Enough aircraft must be available for routine training and reconnaissance missions. Everyday maintenance varies greatly based on the type of aircraft, but consists generally of preflight inspections, inspections following completed missions, periodic scheduled maintenance based on accumulated hours in flight, and component replacements or repairs.

Maintenance during conflict conditions must respond to incidences of specific battle damage in which repairs often require unique skills, materials, or replacement parts not needed under normal conditions. The goal of this type of maintenance is restoring failed or damaged aircraft and returning them to mission-capable status as quickly as possible. In order to achieve this goal, it may be judicious to reduce or suspend normal periodic maintenance in favor of keeping aircraft available for high intensity operations. Conflict conditions may include relocation or decentralization of maintenance facilities, and increased durations of various maintenance or repair tasks.

The simulation model for the Finnish Air Force fleet was designed to be capable of representing all the possible events related to different types of aircraft, and including the expected durations and frequencies of each type of maintenance, and materials or part requirements. The task requirements for time, personnel, and materials are estimated or represented by probability distributions. Maintenance operations are simulated under both normal and conflict scenarios. The time advance mechanism in the simulation must account for time spent by an aircraft awaiting maintenance or in transit to an appropriate maintenance site; time waiting for materials, spare parts, or tools, and time waiting for available maintenance mechanics or other crew.

In this simulation, the Finnish Air Force provided much of the actual flight operation and maintenance data and statistical parameters needed for the study. In cases where data were not directly available, analysts interviewed Air Force expert personnel who cooperated ably by offering estimates, insights, and suggestions that contributed significantly to an accurate and successful simulation project. The simulation was developed with ARENA software based on the SIMAN language. The implementation of the model was validated by running it using actual input data and parameters, and observing that the simulation yielded results that were consistent with outcomes that had been observed in actual operational performance in the past. Following validation, numerous simulation experiments allowed analysts to vary the inputs and parameters, and predict system performance under conditions not yet experienced but that may face the Air Force in future operations.

In this simulation project, analysts created a tool that helped study how maintenance resources, policies and operating conditions influenced aircraft availability. This tool facilitated forecasting aircraft availability, analyzing and planning maintenance resource requirements, and studying the feasibility of making modifications to periodic maintenance programs and changes to other operational parameters. For example, planners and representatives from various levels within the Air Force wanted to maximize the operational capability of the fleet under conflict conditions. To this end, the simulation provided information on the expected number of aircraft available and the maximum number of flight missions that can be conducted during conflict conditions. In particular, simulation results revealed that a maintenance system sized for normal operating conditions is likely to encounter difficulties in conducting the maintenance needed during conflict, even if battle damage is small. Simulation studies can guide the Air Force in planning how maintenance resource needs can be met in conflict scenarios, including possibilities of suspending certain normal maintenance activities temporarily.

In addition to meeting the goals discussed earlier, unanticipated benefits accrued from the simulation project. Creating the maintenance model required extensive cooperation and discussions involving the simulation researchers, flight experts, mission specialists, and maintenance professionals. This interaction contributed significantly to a more thorough understanding of the entire fleet command, and it opened up potential for improved communication among personnel throughout the Air Force. The simulation model has been shared with other units in the Finnish Air Force for related maintenance studies. And the simulation also was found to be useful for training purposes, in which graduates of the Finnish Air Force Academy learned to use the simulation and applied it to their own projects in various other areas of study.

8.5.2 Simulation of a Semiconductor Manufacturing Line (Miller 1990)

Turnaround time is often defined to be the elapsed time from start to completion of a manufacturing process. Turnaround time may be more important in semiconductor fabrication than in any other industry because the longer a device is in the fabrication process, the greater the opportunity for contamination. And even in strict clean room environments, particulate contamination onto wafer surfaces over time has a negative effect on product yields.

Variation in the time between steps in the fabrication process is also a source of lower product yields, because certain sequential processes performed minutes apart produce very different results from the same processes performed hours apart, just simply because the physical properties of the materials change over time.

Slow turnaround also means delays in recognizing problems on the assembly line because the functional characteristics of the manufactured devices cannot be tested until the fabrication of the circuits is complete. The correctness of large numbers of items in progress therefore may be unknown, pending completion of initial manufacturing lots.

Just as important as turnaround time is **throughput**, defined to be the number of manufactured items completed per unit time. Semiconductor manufacturing facilities cost hundreds of millions of dollars to build, equip, and operate, and it is essential to obtain maximum utilization of these resources to attain a competitive cost per wafer.

Assembly line loading, the amount of work in progress, affects both turnaround time and throughput. Standard throughput analysis techniques suggest that heavy line loading (to maximize throughput) ensures that the expensive tools and other manufacturing resources never starve for work. On the other hand, queueing theory analysis demonstrates that turnaround time is minimized by having minimal line loading, as this will eliminate the time spent in queues waiting for manufacturing resources.

These conflicting indications make it difficult to determine the most advantageous level of work in progress. Wafer fabrication involves hundreds of different tools and the manufacturing process associated with each tool depends on many variables. Because of the complexity of the semiconductor manufacturing process, one of IBM's facilities found that analytical methods of analysis were inadequate. Analysts there turned instead to the development and use of a simulation model to analyze their stochastic, discrete event system.

Wafer products manufactured in this assembly line required more than 300 processing steps on 100 different tools. The average turnaround time in the original system was not adequate to support the requirements of new product development. It was therefore desirable to cut this time in half, but using only the fabricator's existing tools, human resources, and control capabilities. Thus, the only allowable modifications were to center around assembly line scheduling policies to achieve the desired turnaround time and throughput.

Early in the study, it was discovered that critical data about the system were either not available or outdated. This then necessitated a systematic analysis and review of current processes and tools, flow times, equipment capabilities, and reliabilities, that resulted in an extensive database which would prove to be of immense value both during and after the simulation study. (The importance of having accurate and up to date information about any system being studied cannot be over-emphasized.)

The simulation model had to accurately represent such key characteristics as process flows, tool capabilities and options, tool failures, rework levels, process yields, operators, priority rules, lot sizes, and storage areas for work in progress. An initial attempt to use a generic, pre-developed simulation package proved unsatisfactory in representing all of these details, and did not allow customized logic needed for this study. The requirements of this project were met when the analysts chose the Systems Modeling Corporation's SIMAN simulation language which is currently known as ARENA (refer to Pegden et al. [1991] for a readable introduction to SIMAN and Kelton et al. [2014] for ARENA).

In this study, the experimental frame defined key parameters describing processes, resources (tools), routings and layouts, scheduling policies, and stochastic events. Most of the information required for the experimental frame structures was obtained from the database developed for this simulation. Not only are input parameters specified in the experimental frame, but also output statistics such as queue time, queue length, tool and operator utilization, throughput, and yield. Depending on the process being described and the desired output, experimental frames in this simulation study contained from several hundred to tens of thousands of entries.

The model frame contained all the control logic necessary to describe the manufacturing process, including the movement of wafers through the hundreds of operations and their associated tools, as well as subsidiary activities such as the transporting of wafer lots between operations and the storage and queueing of lots waiting for resources. The model was run on both personal computers and mainframes, but extremely large experimental frames were not well-supported on PCs due to memory limitations.

Simulation experiments were performed to analyze line-scheduling policies, line-loading levels, and lot priorities. The most significant finding was that a 30% reduction in line loading (from current levels) would produce a 17% reduction in turnaround time, with no deterioration to line-throughput performance. This improvement was achievable with no additional tooling, staffing or change in product mix—a surprise to many analysts who did not believe that line scheduling policies alone could lead to major performance enhancements without additional investments in resources.

Further scrutiny of simulation results revealed a number of other (minor) inefficiencies such as bottleneck points and lot-sizing levels, which could be remedied to obtain certain secondary improvements to the system.

Almost all of the recommendations made by the analysts on the basis of the simulation results were implemented, and over a six month period, line turnaround times improved 25%, while throughput rates increased slightly and the number of operators assigned to the line decreased. The study also fostered several advantageous side effects, including improved manufacturing process descriptions, better information for planning, and more thorough measurements and reporting capabilities, as well as identifying improvements that could be made in the future in case it became desirable to acquire additional resources or make further line-scheduling policy changes.

This successful simulation project provided insights into general semiconductor manufacturing performance in addition to the specific information about the actual semiconductor line modeled. It serves as an illustration of the ability of simulation techniques to profitably analyze complex real-world applications.

8.5.3 Simulation of Eurotunnel Terminals (Salt 1991)

In December 1990, Britain and France were linked by a tunnel that was built by a consortium of companies working cooperatively to construct this underground/undersea link. Eurotunnel is the company responsible for operating the tunnel.

Two separate tunnels actually carry two distinct types of rail traffic. High speed passenger service provides connections between London, Paris, and other major European cities. Shuttle trains carry cars and other vehicles whose drivers and passengers accompany their vehicles between Folkestone in the United Kingdom and Coquelles (near Calais) in France. These vehicles pass through immigration, customs, and security checks upon entering a terminal, and drive away immediately upon arrival at their destination.

To optimize procedures at the terminals, it was first necessary to fully understand the pattern of day to day activities in each terminal. It was decided that a simulation model of a terminal would provide the most valuable basis for studying how a terminal handles the predicted demand. This study began with an interesting process of selecting the appropriate simulation tools. The final product was to be placed directly in the hands of management, and needed to be developed quickly and within existing guidelines and standards.

Several languages were considered on the basis of their various strengths. SIMULA was favored because of its object oriented approach, but the SIMGRAPHICS package in

SIMSCRIPT II.5 was attractive because of the graphics presentation capabilities that would appeal to the managers who would ultimately be using the system.

The winning contender was MODSIM II, an object oriented language that also fully supports process based simulations. The analysts noted that MODSIM II supports multiple active methods and multiple inheritance, both of these being popular language features among proponents of object oriented programming. The language was easy to modularize, and also had a completely integrated graphics package. In short, MODSIM II was deemed to offer a practical combination of object oriented power and a good user interface.

A simulation of the Folkestone terminal was developed to model the flow of vehicle traffic through queues and service facilities to pay tolls, pass British and French customs and immigration, undergo security checks, and eventually to be placed on a shuttle train. The goal of this phase of simulation was to establish expected queue lengths and throughput times, and estimate the adequacy of overflow parking lots and waiting areas. Vehicles are classified as *tourist vehicles* or *heavy goods vehicles* and these two categories are tracked through the system via separate service facilities.

Vehicles are the **objects** that are acted upon by various **methods** for paying tolls and passing through checkpoints. Some methods deliver constant time service, while others (such as security) have service times modeled with exponential distributions (because most security checks are brief, but a few are much more extensive and require a longer time). Each service facility has the capability to reject a vehicle, so that the vehicle is removed from the system and not passed on to the next service facility. The simulation provides information on average queue lengths and average waiting times for vehicles.

Animated output and presentation graphics were used successfully in giving comprehensible output to managers, but were also helpful during the program debugging stages. The original simulation was developed on a DEC station, but networked so that managers can easily access the simulation from their own desktops with output delivered to their local printers.

8.5.4 Simulation for NASA's Space Launch Vehicles Operations (Kaylani et al. 2008)

For over three decades, NASA's Space Shuttle had been the only Reusable Launch Vehicle (RLV) used to deliver cargo to space. Almost a decade prior to the end of the Space Shuttle program in 2011, NASA started evaluating options and approaches for replacement programs that were more effective in terms of cost, reliability, safety and availability. It was well understood that it was necessary to study and compare future competing designs consistently to improve upon the Space Shuttle's cost, performance and turnaround time before pursuing the large undertaking of a new RLV. Previous estimates of the Shuttle's operational performance proved overly optimistic, when NASA predicted originally 50–100 flights per year at $6 million per flight. These estimates were off by an order of magnitude for the flight rate and by two orders of magnitude for cost (the Shuttle flew five to ten times a year at a cost of about $600 million per flight). One of the problems of most estimates was that they tended to assume best-case scenarios and failed to take into account factors that can cause operations to take longer, flights to be delayed and costs to increase.

As simulation emerged as a viable tool to model complex systems, many industries, including NASA, started using it to make more accurate predictions. Discrete Event

Simulation (DES) has been widely used for studying processes and has been frequently used in many NASA studies, including those for the Space Shuttle (Mollaghasemi et al. 2000, Cates et al. 2001, 2002). In order to compare RLV design alternatives fairly and consistently, NASA funded the development of the Generic Simulation Environment for Modeling Future Launch Operations (GEM-FLO) to predict processing turnaround times and other effectiveness criteria and to support making key business and program decisions. The primary motivation behind the development of GEM-FLO is to reduce the time and effort required to study the different system designs using simulation.

The underlying simulation model was developed using ARENA discrete event modeling software and was generically designed to be easily configured for the specific characteristics of each proposed RLV and the underlying processes needed for their operations. It accepts design characteristics and operational inputs from the user, and uses them to configure a simulation model that properly reflects the ground processing flow and requirements of that RLV. For example, every RLV is expected to start with mission planning and go through ground processing, vehicle integration, launch, mission execution, and landing. Each vehicle is expected to have multiple Flight Hardware Elements (FHEs), such as orbiters, boosters and fuel tanks. For a certain vehicle design, the number of FHEs, the necessary processing facilities and flow are entered by the user via a graphical interface and the simulation model is configured accordingly. The elements are then expected to merge into an integrated vehicle at an integration facility according to a specific flow and requirements before it moves to the next stage. Process information for all stages that a certain vehicle must go through is defined by a user who is expected to be involved in the vehicle design but not necessarily a simulation expert.

There is a trade-off between how generic and how detailed a model can be; the more detailed the requirements are, the less generic the model will be. In this application of DES, however, RLVs have common core processes that do not deviate drastically from each other, and a generic model can account for variant designs. For example, in case of the Space Shuttle, the solid rocket boosters, which are one of the FHEs, fall into the ocean after they burn out and then they go through a retrieval process. On the other hand, if a new RLV concept uses boosters that fly back on their own as a hypothetical example, we can still consider that there is a retrieval process but it uses different times and resources (instead of falling into the ocean and taking certain amount of time for divers to retrieve them, they land on a runway and take a different amount of time and resources for example).

When the simulation model is executed, it provides a number of performance measures including operations turnaround time, expected flight rate, and resource utilizations, thus enabling users to fairly assess multiple future vehicle designs using the same generic tool. Of course there is a limit to how refined the granularity of a generic model can be; if a model must be very detailed, then it might be more effective to develop separate models for each RLV instead of one generic model for all of them.

Since simulation validation of future systems is in general challenging due to nonexistence of historical data, the output produced by GEM-FLO from the ARENA software was validated using the Space Shuttle historical data. GEM-FLO was used by several NASA programs including the Next Generation Launch Technology (NGLT) Program, the Orbital Space Plane (OSP) Program, and the Crew Exploration Vehicle (CEV) Program.

8.6 Summary

Simulation techniques permit analysts to study the behavior or performance of systems by creating a computer based model or imitation of a real system or process operating over a period of time. Simulation further allows for experimental studies and analyses in a hypothetical context that would be too expensive or too dangerous to carry out in an actual system.

Building a simulation is itself a complex process. After a problem is formulated and a mathematical or conceptual model built, data must be collected that typifies the actual environment in which the simulated system operates. Modeling the activities of the real system and generating random events that could occur in the real system are among the most critical aspects of simulation development.

Simulation would be an arduous and impractical analysis to perform manually; therefore, the process is automated by developing computer programs to perform the simulation. Steps must be taken to ensure that these programs are correct and appropriate for the study at hand. After simulation experiments are designed, the simulation study enters its production phase, during which the scenarios of interest are carried out via execution of the computer program. Analysts observe the computer simulation and gather statistics to compose a comprehensive picture of various aspects of the simulated system's performance.

By simulating a system, it is possible to make observations of the performance of an existing system, to determine the operating characteristics of a nonexistent system, or to project modifications to an existing system.

Key Terms

activities
attributes
collecting data
coding
discrete simulation
entities
event
experimental design
inverse transform method
problem formulation
production
pseudorandom numbers
simulation
simulation models
simulation–optimization
system state
validation
verification

Exercises

8.1 Select three appropriate applications of simulation analysis—one each from a business, engineering, or environmental setting. In each case, explain why analytical models might be inappropriate or infeasible; justify how simulation could successfully allow a useful and valid analysis of your chosen systems; and speculate on what might be learned from such a simulation study.

8.2 Consider simulating the operation of an emergency health clinic. Identify what issues should be studied, the questions to be investigated, uncontrollable characteristics and constraints within the clinic, controllable aspects of the operation of the clinic, and measures of performance of the clinic.

8.3 Suggest an appropriate method of gathering data for use in simulating the operation of the clinic described in the previous question.

8.4 Write a computer program that generates a sequence of random numbers that are Poisson distributed, with $\lambda = 10$.

8.5 Select a favorite bookstore or grocery store, and observe the pattern of customer arrivals at the checkout facility. Develop a simulation of the customer arrivals by writing a computer program that starts a software *clock* at time zero, then prints the times at which customers arrive over a four-hour period of time. Analyze the times, and determine the longest, shortest, and average interarrival times.

8.6 Select a traffic intersection that is convenient for you to observe. Identify the physical entities that characterize this intersection (such as lanes, directions of traffic flow, stoplights, pedestrian walks, and any obstructions). Observe the operation of the intersection and notice its operating characteristics (such as number of vehicles, patterns of arrival of vehicles at the intersection, speed of traffic, pedestrian or other types of arrivals). Design a model that could be used to simulate the activities of this intersection.

8.7 Simulation can be used to study and predict weather patterns. Using the transition probabilities given in Example 6.1, simulate the most likely daily weather conditions at a ski resort during a winter holiday season beginning December 20 and continuing through January 10, assuming that it was snowy on December 19.

8.8 Develop a computer simulation of a system in which cars arrive at a toll gate on a highway according to a Poisson distribution with mean rate of 90 miles per hour. The times for passing through the gate are exponentially distributed with average 38 seconds.

a. Make a chart that displays enough information so that you can analyze the waiting times experienced by the cars going through this facility.

b. How long must you run this simulation program to get reliable information about the queueing characteristics of your system?

c. Modify your simulation program so that it automatically gathers statistics, and reports the average number of cars waiting and the average waiting time of each car.

8.9   A computer center has one multi-user computer. The number of users in the cen-
      ter at any time is ten. For each user, the time for writing and entering a program
      is exponential with mean rate 0.5 per hour. Once a program is entered, it is sent
      directly to the ready queue for execution. The execution time per program is expo-
      nential with mean rate of six per hour. Assuming the mainframe computer is oper-
      ational on a full-time basis, and neglecting the possibility of down-time, develop a
      computer simulation that allows you to find:

      a. The probability that a program is not executed immediately upon arrival in the
         ready queue

      b. The average time until a submitted program completes execution

      c. The average number of programs in the ready queue

      State any assumptions that you made about the computer center or the multi-user
      computer in the system you have analyzed.

8.10 The mean time between failures of a computer disk drive is 3,000 hours, and fail-
      ures are exponentially distributed. Write a computer program that generates these
      failure events until 25 disk drive failures have occurred. Print out the number of
      hours separating successive failures that occur in your experiment.

8.11 Printer jobs are created in a computing system according to a Poisson distribution
      with a mean of 40 jobs per hour. Average print times are 65 seconds. Users com-
      plain of long delays in receiving their printouts, but the computing lab director
      will be willing to purchase a faster printer (twice as fast as the present one) only
      if it can be demonstrated that the current average queue length is four (or more)
      jobs, and only if the new printer would be idle for at most 20% of the time. Will the
      lab director be able to justify the acquisition of the new printer? You have already
      answered this question (in Exercise 7.4) using queueing formulas; now develop
      and run a simulation model to *test* your answer.

8.12 Computer programs are submitted for execution according to a Poisson distribution
      with mean arrival rate of 90 per hour. Execution times are exponentially distributed,
      with jobs requiring an average of 38 seconds. Users complain of long waiting times.
      Management is considering the purchase of a faster CPU that would decrease the
      average execution time to 30 seconds per job. This expense can be justified only if,
      under the current system, the average number of jobs waiting exceeds five. Also, if a
      new CPU is to be purchased, its percentage of idle time should not exceed 30%. Can
      the new CPU be justified? You made the necessary calculations to make a recommen-
      dation (in Exercise 7.5). Now develop a simulation of the aforementioned scenario
      that might provide an even more convincing explanation to users or to management.

8.13 Develop a simulation of the vehicle refueling system described in Exercise 7.9.
      Determine how long you must run your simulation to obtain performance mea-
      sures that are reasonably consistent with the ones you computed when you worked
      the problem using queueing analysis.

8.14 In Exercise 7.12, you were asked to select a system in your university, business, or
      community that involves queues, to develop a queueing model that describes that
      system, and to describe the performance characteristics of this system. Write a
      computer program to simulate the system you studied, and compare the statistics
      gathered by your simulation program to the analytical performance results that
      you computed with the formulas.

References and Suggested Readings

Abrams, M. 1993. Parallel discrete event simulation: Fact or fiction? *ORSA Journal of Computing* 5 (3): 231–233.

Adkins, G., and U. W. Pooch. 1977. Computer simulation: A tutorial. *Computer* 10 (4): 12–17.

Asmussen, S., and P. W. Glynn. 2007. *Stochastic Simulation*. New York: Springer.

Bagrodia, R. 1993. A survival guide for parallel simulation. *ORSA Journal of Computing* 5 (3): 234–235.

Banks, J., B. Burnette, J. D. Rose, and H. Kozloski. 1994. *SIMAN V and CINEMA V*. New York: John Wiley & Sons.

Banks, J., J. S. Carson, B. Nelson, and D. Nicol. 1984. *Discrete-Event System Simulation*. Englewood Cliffs, NJ: Prentice-Hall.

Banks, J., J. S. Carson, B. L. Nelson, and D. M. Nicol. 2005. *Discrete Event System Simulation*, 4th ed. Upper Saddle River, NJ: Prentice-Hall.

Belanger, R. 1993. *MODSIM II: The High-Level Object-Oriented Language*. La Jolla, CA: CACI Products Company.

Bell, P. C., D. C. Parker, and P. Kirkpatrick. 1984. Visual interactive problem solving—A new look at management problems. *Business Quarterly* 49 (1): 14–18.

Bratley, P., B. L. Fox, and L. E. Schrage. 1987. *A Guide to Simulation*. New York: Springer-Verlag.

Brown, J. J., and J. J. Kelly. 1968. Simulation of elevator systems for world's tallest buildings. *Transportation Science* 2 (1): 35–56.

Bulgren, W. 1982. *Discrete System Simulation*. Englewood Cliffs, NJ: Prentice-Hall.

Buxton, J. N. (Ed.) 1968. *Simulation Programming Languages*. Amsterdam, the Netherlands: North-Holland.

Carrie, A. 1988. *Simulation of Manufacturing Systems*. New York: John Wiley & Sons.

Cassandras, C. G. 1993. *Discrete Event Systems: Modeling and Performance Analysis*. Homewood, IL: R. D. Irwin and Aksen Associates.

Cates, G., M. Mollaghasemi, G. Rabadi, and M. Steele. 2001. Macro-level simulation model of space shuttle processing. *Military, Government and Aerospace Simulation Proceeding, Advanced Simulation Technologies Conference* 33 (4): 143–148.

Cates, G., M. Steele, M. Mollaghasemi, and G. Rabadi. 2002. Modeling the space shuttle. *Winter Simulation Conference*, San Diego, CA, pp. 754–762.

Chan, N. H., and H. Y. Wong. 2015. *Simulation Techniques in Financial Risk Management*, 2nd ed. Hoboken, NJ: John Wiley & Sons.

Chisman, J. A. 1996. *Industrial Cases in Simulation Modeling*. Belmont, CA: Duxbury Press.

Choi, B. K., and D. Kang. 2013. *Modeling and Simulation of Discrete Event Systems*. Hoboken, NJ: John Wiley & Sons.

Chorafas, D. N. 1965. *Systems and Simulation*. New York: Academic Press.

Christy, D. P., and H. J. Watson. 1983. The application of simulation: A survey of industry practice. *Interfaces* 13 (5): 47–52.

Clymer, J. 1988. *System Analysis Using Simulation and Markov Models*. Englewood Cliffs, NJ: Prentice-Hall.

Elizandro, D., and H. Taha. 2008. *Systems Simulation of Industrial: Discrete Event Simulation Using Excel/VBA*. New York: Taylor & Francis Group.

Evans, J. B. 1988. *Structures of Discrete Event Simulation: An Introduction to the Engagement Strategy*. Chichester, UK: Ellis Horwood.

Fishman, G. S. 2001. *Discrete-Event Simulation: Modeling, Programming, and Analysis*. New York: Springer.

Franta, W. R. 1977. *The Process View of Simulation*. New York: North-Holland.

Fujimoto, R. M. 1990. Parallel discrete event simulation. *Communications ACM* 33 (10): 30–53.

Fujimoto, R. M. 1993. Parallel discrete event simulation: Will the field survive? *ORSA Journal of Computing* 5 (3): 213–230.

Gantt, L. T., and H. M. Young. 2015. *Healthcare Simulation: A Guide for Operations Specialists*. Hoboken, NJ: John Wiley & Sons.

Godin, V. B. 1976. The dollars and sense of simulation. *Decision Sciences* 7 (2): 331–342.

Gordon, G. 1978. *System Simulation*, 2nd ed. Englewood Cliffs, NJ: Prentice-Hall.

Graybeal, W., and U. W. Pooch. 1980. *Simulation: Principles and Methods*. Cambridge, MA: Winthrop Publishers.

Hendriksen, J. 1993. SLX the successor to GPSS/H. *Proceedings of the 25th Conference on Winter Simulation*. Los Angeles, CA: ACM.

Hoover, S. V., and R. F. Perry. 1990. *Simulation: A Problem-Solving Approach*. Reading, MA: Addison-Wesley.

Isermann, R. 1980. Practical aspects of process identification. *Automatica* 16: 575–587.

Jain, S., K. Barber, and D. Osterfeld. 1990. Expert simulation for on-line scheduling. *Communications ACM* 33 (10): 55–60.

Kaylani, A., M. Mollaghasemi, D. Cope, S. Fayez, G. Rabadi, and M. Steele. 2008. A generic environment for modelling future launch operations—GEM-FLO: A success story in generic modelling. *Journal of the Operational Research Society* 59 (10): 1312–1320.

Kelton, W. D., R. P. Sadowski, and N. B. Zupick. 2014. *Simulation with Arena*, 6th ed. New York: McGraw-Hill Professional.

Kheir, N. A. (Ed.) 1996. *Systems Modeling and Computer Simulation*, 2nd ed. New York: Marcel Dekker.

Kirkerud, B. 1989. *Object-Oriented Programming with SIMULA*. Reading, MA: Addison-Wesley.

Knuth, D. E. 1981. The art of computer programming, 2nd ed., Vol. 2. *Seminumerical Algorithms*. Reading, MA: Addison-Wesley.

Kobayashi, H. 1978. *Modeling and Analysis: An Introduction to System Performance Evaluation Methodology*. Reading, MA: Addison-Wesley.

Kreutzer, W. 1986. *System Simulation: Programming Styles and Languages*. Reading, MA: Addison-Wesley.

L'Ecuyer, P. 1990. Random numbers for simulation. *Communications ACM* 33 (10): 85–97.

Law, A. M. 2007. *Simulation Modeling and Analysis*, 4th ed. New York: McGraw-Hill.

Lembersky, M. R., and U. H. Chi. 1984. Decision simulators speed implementation and improve operations. *Interfaces* 14: 1–15.

Maisel, H., and G. Gnugnoli. 1972. *Simulation of Discrete Stochastic Systems*. Chicago, IL: Science Research Associates.

Marsaglia, G. 2003. Seeds for random number generators. *Communications of the ACM* 46 (5): 90–93.

Marsaglia, G. 1985. A current view of random number generators. *Computer Science and Statistics, Sixteenth Symposium on the Interface*. Amsterdam, the Netherlands: Elsevier Science Publishers, North-Holland.

Mattila, V., K. Virtanen, and T. Raivio. 2008. Improving maintenance decision making in the Finnish air force through simulation. *Interfaces* 38 (3): 187–201.

Miller, D. J. 1990. Simulation of a semiconductor manufacturing line. *Communications ACM* 33 (10): 98–108.

Mollaghasemi, M., G. Rabadi, G. Cates, D. Correa, M. Steele, and D. Shelton. 2000. Simulation modeling and analysis of space shuttle flight hardware processing. *Proceeding of Harbour, Maritime & Multimodal Logistics Modeling and Simulation Workshop, a publication of the Society for Computer Simulation International (SCS)*, Portofino, Italy, pp. 59–62.

Neelamkavil, F. 1987. *Computer Simulation and Modeling*. New York: John Wiley & Sons.

Nelson, B. 2013. *Foundations and Methods of Stochastic Simulation: A First Course*. New York: Springer.

Park, S. K., and Miller K. W. 1988. Random number generators: Good ones are hard to find. *Communications ACM* 31 (10): 1192–1201.

Pawlikowski, K. 1990. Steady-state simulation of queueing processes: A survey of problems and solutions. *ACM Computing Surveys* 22 (2): 123–170.

Payne, J. A. 1982. *An Introduction to Simulation*. New York: McGraw-Hill.

Pegden, C. D. 1985. Introduction to SIMAN. *Proceedings of the 1985 Winter Simulation Conference*. San Francisco, CA: IEEE.

Pegden, C. D., R. P. Sadowski, and R. E. Shannon. 1991. *Introduction to Simulation Using SIMAN*. New York: McGraw-Hill.

Pidd, M. 1984. *Computer Simulation in Management Science*. New York: John Wiley & Sons.

Pooch, U. W., and J. A. Wall. 1992. *Discrete Event Simulation: A Practical Approach*. Boca Raton, FL: CRC Press.

Pritsker, A. A. B. 1974. *The GASP IV Simulation Language*. New York: John Wiley & Sons.

Pritsker, A. A. B., and C. D. Pegden. 1979. *Introduction to Simulation and SLAM*. New York: John Wiley & Sons.

Ravindran, A., D. T. Phillips, and J. J. Solberg. 1987. *Operations Research: Principles and Practice*. New York: John Wiley & Sons.

Reiser, M. 1976. Interactive modeling of computer systems. *IBM Systems Journal* 15 (4): 309–327.

Reitman, J. 1971. *Computer Simulation Applications*. New York: Wiley-Interscience.

Rice, S. V., A. Marjanski, H. M. Markowitz, and S. M. Bailey. 2005. The SIMSCRIPT III programming language for modular object-oriented simulation. *Proceedings of the 2005 Winter Simulation Conference*. San Diego, CA: CACI Products Company.

Ripley, B. D. 1988. Uses and abuses of statistical simulation. *Mathematical Programming* 42: 53–68.

Roberts, N., D. F. Andersen, R. M. Deal, M. S. Garet, and W. A. Shaffer. 1994. *Introduction to Computer Simulation: A System Dynamics Modeling Approach*. Portland, OR: Productivity Press.

Robinson, S., R. Brooks, K. Kotiadis, and D. J. Van Der Zee. 2010. *Conceptual Modeling of Discrete-Event Simulation*. Boca Raton, FL: CRC Press.

Ross, S. 1990. *A Course in Simulation*. New York: Macmillan.

Ross, S. 1996. *Simulation*. San Francisco, CA: Academic Press.

Rubenstein, R., B. Melamed, and A. Shapiro. 1998. *Modern Simulation and Modeling*. New York: Wiley.

Russell, E. C. 1983. *Building Simulation Models with SIMSCRIPT II.5*. Los Angeles, CA: CACI.

Russell, E. C. 1993. SIMSCRIPT II.5 and SIMGRAPHICS tutorial. *Proceedings of the 1993 Winter Simulation Conference*, San Diego, CA.

Salt, J. 1991. Tunnel vision. *OR/MS Today* 18 (1): 42–48.

Schmeiser, B. W. 1980. *Random Variate Generation: A Survey, Simulation with Discrete Models: A State of the Art View*. New York: IEEE.

Schriber, T. J. 1974. *Simulation Using GPSS*. New York: John Wiley & Sons.

Schriber, T. J. 1991. *An Introduction to Simulation*. New York: John Wiley & Sons.

Schriber, T. J. 1993. Perspectives on simulation using GPSS. *Proceedings of the 1993 Winter Simulation Conference*. Los Angeles, CA: IEEE.

Shannon, R. E. 1975. *Systems Simulation: The Art and Science*. Englewood Cliffs, NJ: Prentice-Hall.

Solomon, S. L. 1983. *Simulation of Waiting-Line Systems*. Englewood Cliffs, NJ: Prentice-Hall.

Swain, J. J. 1997. Simulation goes mainstream: 1997 simulation software survey. *OR/MS Today* 24 (5): 35–46.

Swain, J. J. 1999. Simulation survey. *OR/MS Today* 26 (1): 38–51.

Swain, J. J. 2001. Power tools for visualization and decision-making. *OR/MS Today* 28 (1): 52–63.

Swain, J. J. 2003. Simulation reloaded. *OR/MS Today* 30 (4): 46–49.

Swain, J. J. 2005. Gaming reality. *OR/MS Today* 32 (6): 44–55.

Swain, J. J. 2007. Software survey: New frontiers in simulation. *OR/MS Today* 34 (5): 32–43.

Swain, J. J. 2009. Simulation software boldly goes. *OR/MS Today* 36 (5): 50–61.

Swain, J. J. 2011. A brief history of discrete-event simulation and the state of simulation tools today. *OR/MS Today* 38 (5): 56–69.

Swain, J. J. 2013. Discrete event simulation software tools: A better reality. *OR/MS Today* 34 (5): 32–43.

Swain, J. J. 2015. Simulation software survey: Simulated worlds. *OR/MS Today* 42: 36–49.

Swain, J. J. 2017. Simulation software survey: Simulation takes over. *OR/MS Today* 44 (5): 38–49.

Taha, H. A. 1987. *Simulation Modeling and SIMNET*. Englewood Cliffs, NJ: Prentice-Hall.

9

Decision Analysis

9.1 The Decision-Making Process

Decision analysis is as much of an art as a science. Mathematical decision analysis must be considered in the context of an individual decision-maker. The techniques that have been developed in this area can be described as tools that encourage and assist people in making rational decisions. They are not intended as substitutes for the individual. Most of the techniques incorporate some interactive dialogue with the decision-maker to try to determine personal preferences and attitudes.

To truly appreciate this interaction, it is useful to try to imagine actually being faced with a particular problem. To illustrate this idea, consider the decision to buy a new car. We can easily develop a set of criteria that define a good car (price, mileage, maintenance, horsepower, etc.), and then we can devise a system of weights that measures the relative importance of each criterion. The car with the highest score is clearly the one to buy. Most people would agree that this sounds like a reasonable model. They might even be willing to recommend this selection to someone else. But imagine for a moment that you are making a decision concerning your own car. Would you be willing to accept the advice of this model without question? In fact, the majority of intelligent decision-makers tend to have reservations about accepting a strict mathematical interpretation and recommendation for their problem.

Decision analysis differs from the mathematical structure of many other areas of Operations Research in that it contains a high degree of uncertainty. The uncertainty is, in part, a by-product of any long range planning function. Traditional Operations Research problems in production planning and inventory analysis, for example, are concerned with a monthly sales forecast that may vary according to some probability distribution. In decision analysis, we may be deciding whether to develop and market a new product, build a new plant, or create a new government agency, or diversify our business interests. For example, the demand for an existing product next month is relatively predictable in most industries, but the demand for a new and unfamiliar product in five years' time is virtually impossible to estimate. Such issues as these can have a major impact, and an analysis of the effect of any current decision will not be fully appreciated for five or ten years into the future. The factors that must be considered in the decision process often involve a dramatic degree of uncertainty simply by virtue of the extended time frame.

Decision analysis can usually be expressed as a problem of selecting among a set of possible alternatives or courses of action. After making a choice, and at some future time, there will be a number of external, uncontrollable variables that will influence the final outcome. These external variables are often referred to as **states of nature** or **state variables**. An underlying assumption in decision analysis is that, if it were possible to predict accurately the result of these external variables, then the final outcome would also be predictable and the *correct* alternative would become obvious.

This section discusses a simple decision-making problem. Despite its simplicity, it illustrates many of the difficulties inherent in the decision-making process. Imagine yourself in the following situation. It is midnight on a Sunday night and you have just remembered that you were supposed to prepare a report for your boss for next week. Unfortunately, you cannot remember whether you were supposed to meet with him first thing on Monday morning, or if it was required for next Thursday. You are faced with a decision: should you stay up and work on the report for two or three hours, or should you take your chances and go to bed? This statement defines the alternative *courses of action* for the problem, which we will refer to as the **decision variables**.

The unknown external factor or state of nature is whether or not the report is due on Monday. If you knew that the report was not due until Thursday, you could go to bed and sleep peacefully. We will assume that, if it were known that the report was due tomorrow, the decision maker would feel obliged to stay up and work on it. Otherwise, there is no decision problem because the *preferred* action would be to go to bed independent of whether it is due on Monday or Thursday.

Having defined the alternatives (decision variables) and the external factors (state variables), the next aspect of decision analysis is to consider the possible **outcomes** or **payoffs** that would result from each possible combination of decision and state variables. In this example, as with many large practical problems, the outcome is not clearly defined. There may be a monetary component in the outcome (because the decision may affect future promotion potential and merit pay increases), but there are also a number of other less tangible consequences.

One method of concisely describing this type of problem is called a **payoff matrix**. The rows correspond to the possible states, the columns represent alternatives, and the entries in the matrix describe the outcomes associated with each possible combination of the problem variables. In traditional decision problems, an outcome is described by a single numerical value representing an associated profit, loss, or *value* of the result. For the moment, we address such problems using informal, verbal descriptions of the outcomes.

	Alternatives	
States	a_1: **Stay up, do it**	a_1: **Go to bed**
Θ_1: *Report due*	Tired, but happy • Lost some sleep • Guessed correctly	Miserable • Guessed wrong • The boss will be annoyed
Θ_2: *Report not due*	Depressed • Lost sleep for nothing	Relieved • Guessed right • Did you worry? • Sleep well?

This simple example illustrates some of the most difficult and frustrating aspects of decision making. Several observations can be made concerning the difficulties in quantifying the elements of decision-making:

Outcomes are often verbal descriptions. The problem of comparing outcomes is often complicated by the fact that the entries can be descriptive rather than numeric. In our example, is Depressed worse than Miserable? How much worse? Twice as bad? Is Tired but happy better than Relieved? Is the negative feeling of Depressed greater than the positive result of Relieved? The answers to these questions depend on the individual. For some people, the prospect of having to face the boss in the morning and admitting failure is unthinkable. Other people may do it regularly, presumably armed with a battery of excuses.

The outcomes often involve several conflicting criteria. The previous example illustrates the effect of multiple objectives that are commonly associated with practical decisions. The objectives of getting a good night's sleep and of maximizing one's credibility at the office are, in this case, conflicting goals. The same is true of corporate decision-making. Companies must distinguish between immediate profits and long-term advantages. For example, an investment today for upgrading present facilities will decrease this year's net profit, but may lead to increased future revenue. In addition, intangible costs and benefits such as worker attitudes, safety, environmental issues, legal liability, and customer satisfaction are difficult to quantify.

Even numeric outcomes are difficult to compare. Consider a decision problem in which all of the payoff matrix entries are described in simple terms of dollars of profit or loss. Most people do not consider a profit of $20,000 to be twice as good as a profit of $10,000. In economic theory, this principle is known as the Law of Diminishing Marginal Returns. The classic illustration of this concept says that three loaves of bread are not three times as valuable as one loaf of bread. If you had one loaf, you would eat it and satisfy your hunger. If you had three loaves, the third one would likely be unused.

The same logic applies to profits. People (perhaps unconsciously) normally employ some implicit ordering of the alternative ways of spending their money. The first dollar will be used for the most important item, while the last dollar may just go in the bank. The true value of the first dollar in terms of benefit or enjoyment is considerably greater than that of the last one.

This line of reasoning seems even more valid when comparing profits against potential losses. For most people, the *negative* feeling associated with losing $10,000 is much greater than the corresponding *positive* benefit of winning an equal amount. The profit would be very pleasant, but the loss would be terrible. Losses are generally viewed as being more dramatic consequences than gains. An important aspect of decision analysis concerns the determination of an individual's *attitude toward risk*. We introduce some approaches for dealing with these questions in Section 9.4 on Utility Theory.

The relative likelihood of the uncertain state variables must be considered. In the earlier example, suppose that you believed that the report is most likely due on Monday. In that case, you would be inclined to stay up and write it. However, the situation changes dramatically if you felt that the report was probably not expected until Thursday. If you trusted your judgment, you would go to bed. To make a choice, the decision-maker must try to associate a subjective probability value with each of the possible states. What is your best approximation of the likelihood of each uncertain event? We distinguish between three different approaches to defining probability.

Risk describes a situation for which an objective probability can be calculated. This includes most events that are repeated frequently as historical data is available. Based on past information, it is possible to compute a reasonably accurate probability assessment of the state variable. For example, there is a certain amount of risk associated with drilling oil wells, but using land form data and other inputs, the probability of success can be predicted and this information can then be used in drilling decisions.

Uncertainty normally applies to events for which there is limited historical or repetitive information. When attempting to estimate the probability of success of a new product, it is difficult to predict how the public will react. This is especially true when there have been no similar products introduced in the market. Although there is no data that allows precise computation of *objective* probabilities of success, an analyst may have some feeling or intuition or experience or limited history that allows at least a *subjective* assessment of the probability. In the decision example earlier, you might say that you are 60%–80% sure that the report is due on Monday. Note that the real distinction between *risk* and *uncertainty* is that *risk* is generally more precise. Under *uncertainty*, it may be possible to specify a range for the probability.

Complete ignorance describes a decision-maker who has no prior information of any kind with regard to the likelihood of a state variable. Such a person refuses to specify a subjective, intuitive probability range. Anything could happen and he would not be surprised. Many people feel uncomfortable about specifying subjective probabilities for state variables. Section 9.2 introduces the topic of Game Theory, and describes several methods that can be applied in the face of complete ignorance. It will become clear that the use of subjective probability assessments is often preferable.

Decision-makers are irrational. There is a rapidly growing literature describing the rather curious phenomenon of the irrational decision-maker in all of us. For the present, consider one simple example of this behavior: decision-makers will often lie about their true objectives. When middle managers are asked about their objectives in decisions, they will stress the importance of corporate profit and the overall benefit of the company. Their true objectives are often more selfish and reflect the desire that their own work centers look good. University students might claim that they are primarily interested in the quality of their education when, in fact, their main objective may be to get a diploma with the least amount of effort possible. A new car buyer will often rank safety as a high priority and then select the fastest and raciest sports model. We all have a tendency toward specifying objectives that we believe we should be using or that we think our boss would like to hear rather than being honest about them. We consider these and related issues when we discuss the psychology of decision-making in Section 9.5.

In summary, the foregoing example contains many of the underlying features that complicate decision analysis. In the remainder of this chapter, these features are presented in further detail. It should be mentioned from the outset that the amount of time and money invested in a decision should be a small fraction of the value of the potential outcome. As a general rule, one should spend about 1% of the potential value of a decision on the decision process itself. In the given example, the decision process should not take more than a few minutes. A detailed study of the options is unwarranted. However, in making the decision to buy a $200,000 house, it might be worth spending $2,000 in time and money on analyzing the alternatives and making a good selection.

9.2 An Introduction to Game Theory

Game theory addresses possible approaches to decision-making under the assumption of *complete ignorance*. It is described in terms of players, payoffs, and strategies. Consider a two-person game: the decision-maker (player one) selects an alternative and then nature (player two) selects a state. The payoff is given by the corresponding entry in a **payoff matrix**. Player two is assumed to be indifferent to the choices of player one (except when the decision-maker is slightly paranoid). Player one will make a selection based on some strategy intended to make the most of the opportunity. Throughout this discussion, we refer to the following payoff matrix:

	Alternatives					
States	a_1	a_2	a_3	a_4	a_5	a_6
Θ_1	5	3	0	3	2	3
Θ_2	5	3	8	6	7	3
Θ_3	0	3	0	1	2	2
Θ_4	4	3	0	2	2	1

Most decision-makers employ a process of elimination to reduce the number of alternatives. The simplest form of elimination is called **dominance**. An alternative a_k is said to **dominate** an alternative a_j if, for every possible state, Θ_i, alternative a_k is at least as good as alternative a_j. Alternative a_j can be eliminated from consideration.

In the example matrix, consider alternatives a_2 and a_6. Observe that, no matter which state eventually occurs, alternative a_2 is always at least as good as a_6. Therefore, alternative a_6 is dominated and can be eliminated from further consideration. By inspection, we can verify that no other alternatives are dominated.

A variety of strategies can be employed in making the selection of alternatives. We describe a few of the more common ones and, as they are introduced, we identify each one with a corresponding personality trait. The selection depends on the decision-maker's attitude toward risk. Because each choice has a different degree of risk associated with it, different people will make different selections. It is important to realize that there is no absolutely correct answer to this problem.

9.2.1 Maximin Strategy

For each alternative, a_j, pick the worst possible outcome (the minimum). Choose the alternative that has the maximum value of this minimum.

The Maximin strategy is associated with the *eternal pessimist*; the person who believes that, whatever they do will always turn out badly, and that nature is working directly against them. In the example, the worst outcomes for each of the first five alternatives are 0, 3, 0, 1, and 2, respectively. By choosing alternative a_2, the worst possible outcome is 3. The maximin player chooses this alternative to guarantee a payoff of at least 3 no matter what state occurs.

This strategy is characteristic of the conservative decision-maker. The given decision has the lowest risk. However, it also usually has the lowest variance. Not only will the decision-maker never make less than 3, they will never make more than 3 either. This strategy is commonly observed in people who invest all of their money into savings bonds with a guaranteed interest rate rather than participating in riskier forms of investment. They don't *really* believe that nature is out to get them; they just don't want to take any chances.

9.2.2 Maximax Strategy

For each alternative, a_j, pick the best possible outcome (the maximum). Choose the alternative which has the maximum value of this maximum.

The Maximax player represents the *eternal optimist*. Such people believe that anything they do will turn out right. They are gamblers by nature and are willing to take risks for a chance at the greatest possible prize. In the example, the maximum payoff of the first five strategies is given by 5, 3, 8, 6, and 7, respectively. The maximum possible outcome is 8, and the Maximax player will therefore select option a_3 and hope for state Θ_2.

This strategy is commonly identified with incurable gamblers who have an unrealistic or even unhealthy level of optimism. However, there is also a group of successful business people who regularly employ this strategy, but they do not rely on blind luck. These decision-makers will look for the best possible outcome, and determine the state(s) that must occur in order for the maximum profit to be realized. These people have great confidence in their ability to *make things happen*; they believe that they can influence and even control the state variables. There is often some truth in this when the decision involves the success of a new product, the potential market, and the ability of competition to react. Presumably, these people are using a modified form of Maximax in which they first eliminate any states that are unlikely or uncontrollable. They choose the maximum outcome corresponding to any state over which they believe they can exercise some influence.

9.2.3 Laplace Principle (Principle of Insufficient Reason)

Assume that every state is equally likely and calculate the *expected payoff* for each alternative. The alternative with the highest expected payoff is selected.

Because we have assumed *Complete Ignorance* with respect to the likelihood of each possible state, it is reasonable to assume that each state is equally likely. We have no reason to assume that any one state is more likely than any other. In our example, each state would be assigned a probability of 0.25 because there are four possible states.

The *expected payoff* for a given alternative is computed by taking each element in the corresponding column of the payoff matrix, and multiplying each payoff by the corresponding state probability. The expected payoff is the sum of these values. The expected payoff for each of the five alternatives in the example is given by 3.5, 3, 2, 3, and 3.25, respectively. Therefore, alternative a_1 has the highest expected payoff.

Observe that if each state really has equal probability, and we repeat the game a large number of times, the average payoff from selecting alternative a_1 will be 3.5. Unfortunately, in a real decision-making environment, we will be allowed to *play the game* only once. We will discuss expected value decision-making at greater length in subsequent sections.

9.2.4 Hurwicz Principle

Define $0 \leq \alpha \leq 1$ to be the Decision-maker's *Degree of Optimism* between the two extremes of Maximin ($\alpha = 0$) and Maximax ($\alpha = 1$). For each alternative, a_j, define the **Hurwicz measure**:

$$h_j = \alpha \left\{ \max_i p_{ij} \right\} + (1 - \alpha) \left\{ \min_i p_{ij} \right\}$$

where p_{ij} represents the payoff associated with alternative a_j and state Θ_i. Select the alternative with the highest value of h_j.

The **Hurwicz principle** is based on the assumption that the decision-maker is neither totally pessimistic (as with the Maximin strategy), nor totally optimistic (as with Maximax). Each individual decision-maker can select his own degree of optimism somewhere between these two extremes. Observe that when $\alpha = 1$, then h_j will simply be the maximum payoff for alternative, a_j, and Hurwicz will be equivalent to the optimistic Maximax rule. Similarly, when $\alpha = 0$, h_j will be the minimum payoff for alternative a_j. In this case, by selecting the largest h_j, the decision-maker is choosing the pessimistic, conservative Maximin alternative. However, when some intermediate value of a is chosen, we get an alternative that balances the risk against the potential gains. In the example, suppose we choose $\alpha = 0.6$. For alternative a_5, the maximum payoff is 7, the minimum payoff is 2, and the corresponding Hurwicz value is $h_5 = 0.6 * \{7\} + 0.4 * \{2\} = 5$. The values of the Hurwicz measure for the first five alternatives are 3, 3, 4.8, 4, and 5, respectively. By picking the maximum of these, we determine that the best strategy is to choose alternative a_5.

9.2.5 Savage Minimax Regret

Define the regret matrix by $r_{ij} = p_i^* - p_{ij}$ where p_i^* denotes the best outcome which could occur under state Θ_i. For each alternative, find the maximum regret. Select the alternative that minimizes this maximum regret.

This strategy is associated with *insecure* decision-makers. Such people are not primarily interested in making the highest profit; they are more concerned with how disappointed they are going to feel after the fact. To illustrate this, suppose that alternative a_3 is selected, and then state Θ_1 occurs. In hindsight, the decision-maker will wish he had chosen alternative a_1 for a payoff of 5 instead of the actual profit of 0. He will *regret* making the *wrong* choice, and the amount of this regret can measured by the difference between what he actually received and what he could have earned if he had known that state Θ_1 would result. He will experience a regret of $5 - 0 = 5$. The complete regret matrix for the example is:

States	Alternatives				
	a_1	a_2	a_3	a_4	a_5
Θ_1	0	2	5	2	3
Θ_2	3	5	0	2	1
Θ_3	3	3	3	2	1
Θ_4	0	1	4	2	2

The maximum possible regret for each of the five alternatives is given by 3, 5, 5, 2, and 3, respectively. The best strategy to minimize the possible regret is to select alternative a_4 with a maximum regret of 2. No matter which state occurs, the decision-maker will not regret his choice by more than 2. The decision-maker is protecting himself against the future prospect of someone coming along after the fact and telling him that he should have anticipated that the final state would result. Observe that, although this behavior does minimize regret, it also guarantees, at least for this example, that there will be some regret.

The strategies that have been described earlier are all based on logical and rational assumptions. Each of them proposes a different alternative as the *optimal* solution to the problem. Each of the alternatives is the correct choice for some decision-makers.

In practical problems, people have used all of these approaches in an attempt to reduce the number of original alternatives down to a small set of distinct options. For example, the USSR Siberian Power Institute was asked to make recommendations on the location of a new hydroelectric generation facility during the early 1970s (Bunn 1984). Three possible locations were being considered. The project would take many years to complete, and the potential impact on the economy and environment in the chosen area would be considerable, with a high degree of uncertainty. The committee developed 23 different possible scenarios concerning future energy supply and demand, potential investment, and operating costs. Under each scenario, and for each of the three possible sites, they calculated a net economic impact. The Institute identified the optimal actions using several criteria, including the **Maximax, Maximin**, and Regret techniques. Their report recommended, under each assumption of attitude toward risk, a different location. These results were then passed on to a higher political committee for final selection.

Although the methods have a natural and simple appeal, there are some definite problems having to do with their underlying assumptions. The **Laplace principle** provides an intuitively appealing method of dealing with complete ignorance. It seems logical to assume that each state is equally likely. Consider the example that we introduced in the previous section. Recall that the two states of nature were: Θ_1: *Report due* and Θ_2: *Report not due*. The Laplace principle asks us to assume that each has a 50% chance of occurring.

Suppose that we were to reformulate the problem, subdividing state Θ_1 into three different states:

> *Report due*—You are fired.
>
> *Report due*—But boss forgets to ask.
>
> *Report due*—But you get an extension.

This new decision problem now has four states instead of two. If we again apply the Laplace principle, we discover that the state *Report not due* has a probability of 25% and the aggregate states *Report due* have a probability of 75%. By changing the descriptions of the states, we can cause a change in the recommendations reported by the impartial method of Laplace.

The Maximin strategy also presents a problem in that it does not possess the property of **row linearity**. This property asserts that if we add a constant to each outcome in a row of the payoff matrix, this should not affect the chosen alternative. If that state occurs, all alternatives will be better by exactly the same amount, so this should not influence the choice. Consider the problem of deciding whether or not to take your umbrella with you in the morning. We assume that carrying an umbrella around all day is a nuisance; but if it rains, getting wet is a bigger nuisance. The following *payoff* matrix might represent the total amount of discomfort that you would experience under each condition.

	Bring Umbrella	Do Not Bring Umbrella
Rain	−4	−8
Sun	−6	0

The Maximin player will bring an umbrella to avoid the potential discomfort of getting wet (−8). Just before he is about to leave the house, the boss calls and says that if it rains today, she will be closing the office in the afternoon. Our decision-maker now has a more favorable attitude toward the prospect of rain and therefore increases all outcomes corresponding to the state *rain* by 4. The revised payoff matrix reflects this new attitude to inclement weather.

	Bring Umbrella	Do Not Bring Umbrella
Rain	0	−4
Sun	−6	0

The same player will now decide not to bring an umbrella because he will not mind getting wet quite as much. This is not really rational. The Laplace rule is the only strategy introduced here that maintains row linearity.

All of the rules, except Laplace, are concerned exclusively with the *extreme* outcomes (the best or the worst values), and ignore intermediate results. Consider the following payoff matrix:

	a_1	a_2
Θ_1	0	1
Θ_2	1	0
Θ_3	1	0
Θ_4	1	0
Θ_5	1	0
.	.	.
.	.	.
.	.	.
Θ_n	1	0

Under the rules, the two alternatives are equivalent: the maximum outcome is 1, the minimum outcome is 0, and the maximum regret is 1. It would be reasonable to assume that unless it were fairly certain that Θ_1 would occur, alternative a_1 is a much better option than alternative a_2.

The **Savage Minimax regret** strategy displays an additional, rather surprising, logical anomaly. It is possible to construct an example such that, if the decision-maker must choose between two alternatives, a_1 and a_2, he will choose a_2. But, upon adding a third, useless alternative a_3, he will now prefer alternative a_1. Suppose that this person, when choosing between a Ford or a Chevrolet, picks the Chevrolet. However, by offering him the additional option of a Volvo, he will now take the Ford.

This behavior is clearly irrational. The decision-maker who persistently applies the Savage regret method can be turned into a *Perpetual Money-Making Machine*. Consider the following example:

	a_1	a_2	a_3
Θ_1	1	9	5
Θ_2	9	5	1
Θ_3	5	1	9

Clearly, the three alternatives are identical. However, if this decision-maker were offered only alternatives a_1 and a_2, he would choose a_2. Moreover, if he currently has a_1, he might be willing to pay us \$1 to exchange a_1 for a_2. Similarly, when considering options a_2 and a_3, if he has a_2, he would be willing to pay \$1 to exchange it for a_3. Finally, now that he has a_3, he will pay \$1 to exchange it for a_1, and the cycle can continue indefinitely, or at least until our victim adopts a new strategy. (It is left as an exercise to construct and verify these pairwise regret matrices.)

In summary, game theory provides an interesting framework for classifying and analyzing general types of human behavior in the presence of uncertainty. It does not provide a very practical set of rules for solving decision-making problems. In particular, recall that our discussion of game theory has been based on the assumption of complete ignorance. The decision-maker was unwilling or unable to make any subjective probability assumptions. However, in reality, each of the approaches described in this section is equivalent to making very specific probability statements:

Maximin: The worst outcome for each alternative has probability 1.

Maximax: The best outcome for each alternative has probability 1.

Laplace: All states have equal probability.

Hurwicz: The best outcome has probability α and the worst outcome has probability $(1 - \alpha)$.

Savage: The highest regret outcome for each alternative has probability 1.

In every case, by claiming complete ignorance and then selecting a particular strategy, the decision-maker has implicitly assigned probabilities to the outcomes. Realizing this, the decision-maker would likely prefer to trust his or her own judgment about probabilities.

9.3 Decision Trees

Practical decision-making usually involves a sequence of simple decisions. For example, when corporate decision-makers consider developing a new product, they will normally first do a market survey and a feasibility study. If both of these are encouraging, they may decide to invest more time and money in the design and development of a prototype. If the model is successful, they will try limited production and possibly introduce the product in a test market. If the response in the test market is favorable, they might decide to proceed to full scale production and a national sales campaign. They will generally allow for a review of their progress after six months or a year to decide whether or not to continue. The simple payoff matrix methods introduced in the previous sections are inadequate for sequential decision-making.

Observe that even a simple personal decision such as buying a new car is really a sequential type of problem. The true cost of a new car depends on how long the owner decides to keep it, which in turn depends on the car's performance. The present decision is a function of future state variables and a sequence of future possible decisions.

Decision trees provide a method for representing sequential decisions and evaluating the alternatives. A **decision tree** is composed of the following basic building blocks:

1. **Decision fork.** A point in the tree where a decision-maker must choose one of several paths, or alternatives: represented by a square box in our diagrams.

2. **Chance fork.** A point in the tree where nature will choose a path according to some probability: represented by a circle.

3. **Gate or toll.** A branch of the tree where a cost will have to be paid if that path is selected: represented by a bar across a path.

Throughout our discussion on decision trees, we assume that the reader is familiar with basic probability theory.

Consider an example of the vice-president of sales for a medium sized manufacturer who must decide whether to market a potential new product. After consultation with people from the accounting and the marketing departments, she decides to consider three possible scenarios: high demand (1,000 sales per year), medium demand (500 sales per year), and low demand (100 sales per year). For each of these states, she estimates the expected annual net profit of $1 million, $200,000 and −$500,000 respectively.

The corresponding decision tree for this problem, shown in Figure 9.1, is organized chronologically from left to right. We begin at the extreme left and move along the path until a **fork** is encountered. At a decision fork, we must pick the best possible alternative according to some decision strategy; at a chance fork, a path is randomly selected for us, according to some probability function. Eventually, we arrive at some unique outcome at the extreme right-hand side of the decision tree.

Suppose that the decision-maker has determined subjective probabilities for each of the three possible states: high demand (0.2), medium demand (0.4), and low demand (0.4). Based on this assumption, we can calculate the **expected monetary value (EMV)** of the chance fork in the tree:

$$EMV = 0.2 \times (\$1,000,000) + 0.4 \times (\$200,000) + 0.4 \times (-\$500,000) = \$80,000$$

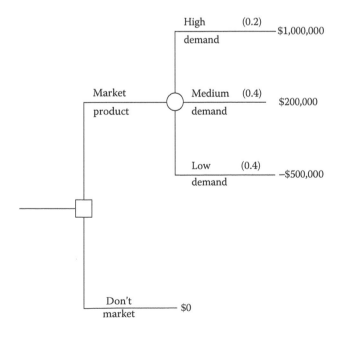

FIGURE 9.1
Decision tree.

The decision fork now becomes a choice between a chance fork with an expected value of $80,000 or a certain outcome of $0. We assume that the decision-maker will choose to market the product with an expected profit of $80,000.

This process is known as **folding back** the decision tree. Beginning at the extreme right-hand side, for each chance fork, we calculate the expected monetary value. For each decision fork, we select the branch with the highest EMV. The *value* of the decision fork is this maximum expected profit. Eventually, we arrive at the left-hand side of the tree. Each decision fork in the tree has a preferred branch. The set of preferred branches is called a **decision strategy**. In the example, the preferred strategy is to market the product with an expected profit of $80,000.

This approach, although intuitively appealing, is based on some implicit underlying assumptions that must be considered. A particular concern for most decision-makers is the issue of relying on expected monetary values. In the example, the suggested strategy involves a 40% chance of losing half a million dollars. This could have serious consequences on the future of the company, and many people would consider the risk too high when weighed against the potential gain. In the next section on utility theory, we illustrate how decision trees can be modified to incorporate attitudes toward risk.

Another issue involves the use of a *discrete* set of state variables. At the chance fork, we have assumed that the demand will either be high, medium, or low. In fact, the demand for the product is a continuous variable in our problem. The eventual outcome is drawn from a distribution anywhere between $1,000,000 and –$500,000. By limiting this range to three possible values, we have simplified the real problem. We have developed a model of the decision process that has lost some of the detailed structure of the original. At the same time, however, we can now ask the decision-maker to determine subjective probabilities and potential outcomes for a limited number of distinct possibilities. By adding more options, we could make the model more realistic, but the decision-maker would find it increasingly difficult to distinguish between the various scenarios. The model builder must be conscious of the delicate balance between model realism and the practical implications of too many subjective evaluations.

Now suppose that the decision-maker has the option of performing a market survey before making his final decision. The survey will cost $20,000 and will provide an estimate of the potential success of the product. We assume, for simplicity, that the survey results will be either *favorable* or *unfavorable*. This new problem can be represented by the decision tree shown in Figure 9.2. This example contains several additional interesting features. Observe that the decision to survey immediately costs $20,000 represented by a **gate**. When we fold back the tree, we must subtract $20,000 from the expected value of the survey chance fork to evaluate the decision fork.

The intended purpose of doing a survey is to improve our estimates of the probabilities of product demand. We would expect that a *favorable* survey result should increase the probability of *high demand*. To determine how these probabilities change, we must first know how much confidence we should have in the survey results. The company that does the surveys claims the following levels of accuracy based on its past experience:

	Favorable (F)	Unfavorable (UNF)
If high demand (HI)	70%	30%
If medium demand (MED)	60%	40%
If low demand (LOW)	30%	70%

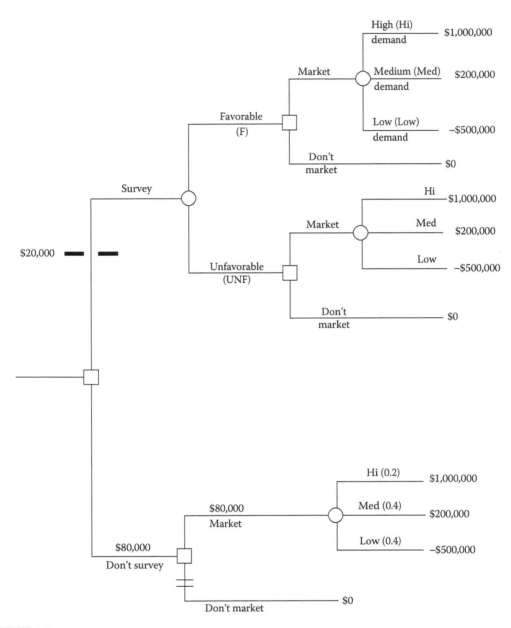

FIGURE 9.2
Decision tree with market survey.

Unfortunately, the marketing company has given us the probabilities in the reverse direction for our decision tree, saying that "The probability of 'favorable' response *given* 'high demand' is 0.70." This can be abbreviated as $Pr\{F|HI\} = 0.70$. In our decision tree, we need to know the probability of *high demand* given a *favorable* survey response. Recall that we have already assumed that $Pr\{HI\} = 0.20$. We can calculate the probability of getting a favorable survey response by adding up all of the favorable *conditional probabilities*:

$$Pr\{F\} = Pr\{F \mid HI\} \times Pr\{HI\} + Pr\{F \mid MED\} \times Pr\{MED\} + Pr\{F \mid LOW\} \times Pr\{LOW\}$$

$$= (0.70) \times (0.20) + (0.60) \times (0.40) + (0.30) \times (0.40)$$

$$= 0.14 + 0.24 + 0.12$$

$$= 0.50$$

Similarly, the probability of an unfavorable result is given by: $Pr\{UNF\} = 0.50$.

We can use this information to derive the required conditional probability using **Bayes Rule**:

$$Pr\{A \mid B\} = \left[Pr\{B \mid A\} \times Pr\{A\}\right] \div Pr\{B\}$$

This version of the formula is derived from a standard result in probability theory which states that

$$Pr\{A\&B\} = Pr\{A \mid B\} \times Pr\{B\}$$

and similarly,

$$Pr\{A\&B\} = Pr\{B \mid A\} \times Pr\{A\}$$

Equating the right-hand side of both expressions and dividing by $Pr\{B\}$ produces the desired result.

By applying Bayes rule, we can now derive the required conditional probabilities. For example:

$$Pr\{HI \mid F\} = \left[Pr\{F \mid HI\} \times Pr\{HI\}\right] \div Pr\{F\}$$

$$= \left[0.70 \times 0.20\right] \div 0.50$$

$$= 0.28$$

The complete table of conditional probabilities can be calculated in an analogous way:

	High Demand (HI)	Medium Demand (MED)	Low Demand (LOW)
If favorable (F)	28%	48%	24%
If unfavorable (UNF)	12%	32%	56%

The corresponding decision tree is shown in Figure 9.3.

With a favorable survey result, the expected value of the *Market* decision increases from $80,000 to $256,000. An unfavorable result decreases the value of the *Market* decision to a loss

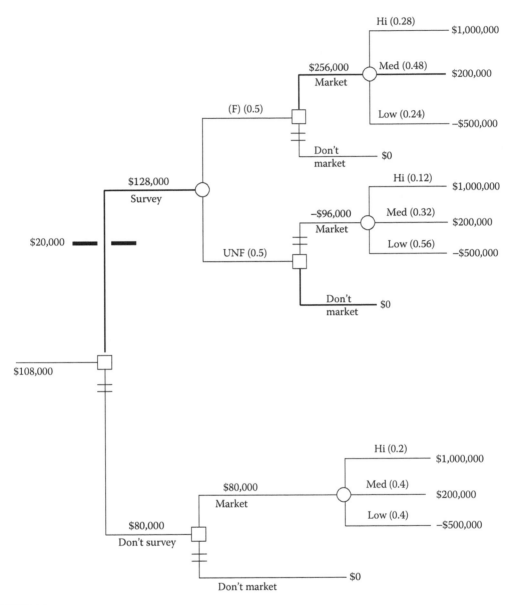

FIGURE 9.3
Completed decision tree with survey information.

of $96,000. In the latter case, the decision-maker would not market the product, and the $20,000 spent on the survey is written off as an inexpensive way to avoid the potential loss of $96,000.

The optimal strategy for this decision can be summarized as follows:

Do the survey;
   If *favorable*,
      Market the product (expected value: $256,000)
   If *unfavorable*,
      Do not Market (expected value $0)

This strategy is highlighted on the decision tree at each decision fork.

In a sense, the survey information is not very reliable. Even when the demand is low, we still have a fair chance of getting a favorable survey response. However, the adjusted probabilities are still sufficient to dramatically affect the expected profit. This leads us to consider the question of the value of survey or sample information. In the example, the survey increases the expected value of the market decision from $80,000 to $128,000. Therefore, we could say that the *Expected Value of the Sample Information* is $48,000. The decision-maker might be willing to pay up to $48,000 for the survey.

At any stage of a decision-making process, the decision-maker usually has the option of requesting more information. He could ask for a more detailed survey, or could try distributing the product in a small test market before making the final decision. One of the most important and difficult decisions is deciding when to stop collecting data.

A useful measure of the potential value of additional information assumes the existence of a source of perfect information. The **expected value of perfect information** (EVPI) is obtained from the decision tree by adding a chance fork at the beginning of the tree that tells us whether demand will be high, medium, or low. We then decide to market the product or not. This process is illustrated in Figure 9.4.

If there were a perfect survey that could accurately predict the true product demand, the expected value of our decision would change from $80,000 (with no information) to $280,000. The expected value of perfect information (EVPI) is $200,000. Sources of perfect information are rare, and they certainly are not free. However, the EVPI gives an indication of the potential value of looking for better surveys and tests.

Consider the position of the decision-maker in our example. Recall that the conditional probabilities for the survey results are presented as objective information. They are based on historical data from previous surveys and we have a reasonable degree of confidence in their accuracy. However, the estimates for the probabilities of the three levels of market demand are highly subjective. These are based on intuition, some past experience, and an

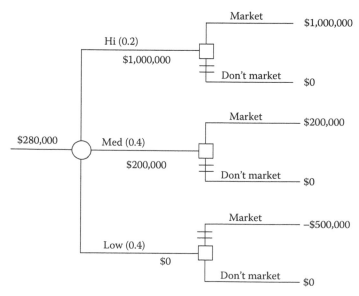

FIGURE 9.4
The expected value of perfect information.

educated guess. The decision-maker can have only as much confidence in the final strategy as he does in these estimates.

For this reason, it is important to perform some sensitivity analyses on the final decision. For example, the decision-maker in our example might be interested in knowing what the best strategy would be if the probability of high demand was only 10% and the chance of low demand increased to 50%. The revised decision tree is given in Figure 9.5. We discover that the expected value of the decision without the survey is now –$70,000. However, the value of the survey is $43,000. After subtracting the $20,000 cost of the survey, the expected profit is $23,000, and the optimal strategy still suggests marketing the product if the survey results are favorable. Because these new demand estimates are presumably pessimistic, our decision-maker's confidence in going ahead with the survey increases significantly.

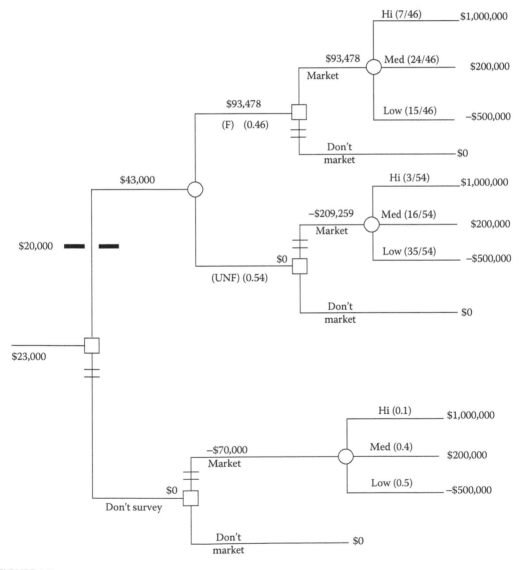

FIGURE 9.5
Decision tree with pessimistic estimation of demand.

The example given was of course deliberately simplified. Practical problems would have several decision forks and a large number of state variables with which to contend. The calculations required to evaluate even a moderate-sized decision tree can be very tedious. Fortunately, there are many software packages available that can handle large practical problems and relieve the decision-maker of considerable computational burden.

One of the most valuable uses of decision trees is simply for *organizing* and *modeling* decision problems. As a first stage, the decision tree can be drawn with only the decision forks and a few of the main chance forks. This preliminary tree is useful in determining the possible decision options in the decision process.

The decision-maker can then consider which of the possible environmental state variables could have a significant impact on the final outcome. In a practical setting, there will be a large number of state variables that can influence the final outcome. The *art* of decision analysis is deciding which of these variables are likely to change the optimal strategy. For example, in a production problem, the likelihood of a union strike would have a significant impact on expected profit. However, it may have no impact on the best decision selection if the strike reduces all outcomes proportionally.

An excellent example of the art of using decision trees is presented by Byrnes (1973). He describes an actual case study of a decision by a major soap manufacturer in England of whether to market a new deodorant bath soap in the 1960s, at a time when many companies were experimenting with the idea. The case is interesting because decision trees were used as a vehicle for understanding the problem. Although the final tree was used to predict expected profit, there was a sequence of decision trees that reflected the changes in the attitudes of management as they learned more about the decision at each stage. The case study describes each step, and, in particular, the mistakes and guesses that actually took place along the way.

9.4 Utility Theory

In Section 9.3, we made the assumption that people will choose the alternative that exhibits the highest expected value. Such people will be called *EMVers* for their use of **expected monetary value**. If a particular decision is to be repeated many times, then the EMV approach is perfectly sound. In the long run, the actual profit will be very close to the EMV sum of the individual decisions. Unfortunately, most practical decision processes apply to a single decision-making event.

For this reason, the vast majority of decision-makers do not rely solely on EMV, and will also make a subjective evaluation of the amount of risk involved in a decision. They will attempt to incorporate their attitude toward risk in a trade-off against the potential benefits of taking a chance.

As an experiment illustrating attitudes toward risk, we have tried the following game in our classes. We place $100 on the table at the front of the room. We tell the class that we are going to flip a coin with one student. If a head comes up, the student wins the $100; but if a tail occurs, we keep the money. We then ask the students what is the maximum amount that they would be willing to *pay* to play this game. (We will keep the money that they pay regardless of the flip.)

The EMV of this game is $50, and from a strictly mathematical point of view, people should be willing to play for any amount up to $50. However, students are generally not wealthy people. They begin to think of the consequences of the gamble. If they lose, they might not eat tonight, or they might not have enough money to pay the rent. If they win, they could take their friends out to dinner, but the value of winning does not compensate them for the risk of losing $50. Over the years, we have observed that the average amount that students are willing to risk for this gamble is around $20. (One student was willing to play for $75, but he was probably independently wealthy.)

People's willingness to use an EMV decision rule depends on their ability to absorb the potential loss. For relatively small values, they can afford to rely on EMV; but as the stakes increase, most people exhibit an aversion to risk. For example, few corporate decision-makers (in medium sized companies) would be willing to risk $400,000 for a 50–50 chance of earning $1,000,000. The prospect of such a substantial loss would be considered too risky.

It is important to distinguish between *gambling* and *decision-making*. The previous example with the students was clearly a gambling situation. It was a *game*, and the students had a *choice* of whether or not they wanted to play the game. However, in the real world of decision-making, the decision-maker is *forced* to pick one of several uncertain alternatives. Another distinction between gambling and decision-making is illustrated by the student who was willing to pay $75 to play the game. This student was a gambler, whereas the others were making a rational decision about their ability to pay versus the potential gains. In the quest for success, we cannot avoid taking some chances, but we can certainly avoid being foolish. In casino gambling, for example, the odds, in the long run, always favor the house.

Utility theory gives us a tool for characterizing an individual's attitude toward risk. It is based on the idea that people will associate an implicit value or utility with any given outcome that is not necessarily proportional to the associated dollar (monetary) value. For example, a particular individual may feel that the negative value associated with a loss of $100 is compensated by the positive value of a gain of $500. He would consider the utility or value of the two outcomes to be equal and opposite. A 50–50 chance of losing $100 or gaining $500 would be fair within his personal value system. Utility theory allows us to assign values to these outcomes which reflect this attitude.

9.4.1 The Axioms of Utility Theory

Utility theory depends on four basic assumptions or axioms. If we accept the validity of these axioms, then the subsequent material follows as a logical consequence. In the axioms, we use the term **lottery** to mean a single chance fork in a decision tree where one outcome is randomly chosen from several possible outcomes, each having a given probability. We first state the axioms, and then we discuss some of their more controversial aspects.

Axiom 1: *Every pair of outcomes can be compared.*

There is a preference ordering (possibly indifferent) associated with all outcomes. Moreover, this ordering is transitive: if outcome A is preferred to B, and B is preferred to C, then A is preferred to C. Similarly, if A is indifferent to B, and B is indifferent to C, then A is indifferent to C.

Axiom 2: *We can assign preferences to lotteries involving prizes in the same way that we assign preferences to outcomes.*

Consider a lottery L with probability p of an outcome A and probability (1 – p) of outcome B. This lottery itself has a value in our preference ordering, and we can decide whether or not we prefer lottery L to a third outcome C.

Axiom 3: *There is no intrinsic reward in lotteries.*

There is no fun in (or fear of) gambling.

Axiom 4: *Continuity Assumption.*

Given any three outcomes where A is preferred to B is preferred to C, then there exists some probability p such that we would be indifferent to getting outcome B for certain, or getting a lottery L with probability p of outcome A and probability (1 − p) of outcome C.

These assumptions are the subject of considerable controversy among decision theory authors. The first assumption implies that all outcomes can be measured by a single scalar value in order of preference. In decisions involving only dollar values, this appears reasonable. However, for decisions with multiple objectives, these axioms become less obvious. Consider the simple problem of choosing a new car. There are several conflicting attributes that define the best car. Utility theory assumes that we have some underlying value system that allows us to rank all possible car models in order of preference. The decision problem is reduced to one of explicitly determining this value structure.

Figure 9.6 illustrates the concept of assigning values to lotteries. In Figure 9.6a, a particular decision-maker might be indifferent to the decision fork alternatives when X = $230. In this case, the **Certain Monetary Equivalent** (CME) of the lottery (the chance fork) is $230. If X is greater than $230, he will take the certain cash. If X is less than $230, he will prefer the lottery. The lottery itself has a value equivalent to the utility of $230.

In Figure 9.6b, suppose that the same decision-maker is indifferent to the decision fork when Y = $220. Now, the lottery at the chance fork has a CME of $220. If this person were asked to choose between the two lotteries, he would select the first one, because it has a higher perceived value for him. Note that the EMVs of the two lotteries are $300 and $275, respectively. But, in both cases, the decision-maker puts a lower monetary value on the lotteries, because the cash values are certain, while the lotteries have an element of risk.

It is not always true that the CME values are in the same order as the corresponding EMVs. For example, our decision-maker could attach a very high value to having at least $150 when he is finished. The lottery in Figure 9.6a has some risk because he could finish with only $100. Figure 9.6b has very little risk because he can always be certain of earning $150. He might therefore be more inclined to use the EMV for the second lottery, and choose Y = $260. For this decision-maker, the CME of the second lottery is higher than that of the first, although the EMV is lower.

The third assumption—that there is *no fun in gambling*—refers to the attraction that some people have to the *thrill* of taking a chance. When people buy a lottery ticket, they get the chance of winning; but they also get the fun of just playing the game. They can watch the

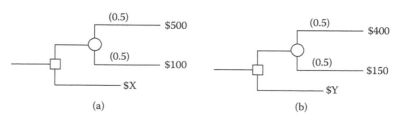

(a)   (b)

FIGURE 9.6
(a,b) Examples of the value of lotteries.

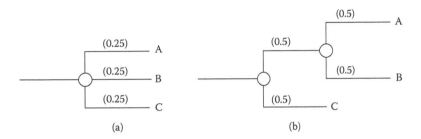

FIGURE 9.7
(a,b) Example of the effect of *fun in gambling*.

numbers being selected and cheer for their sequence. These people get an added positive value simply by being at a chance fork in a decision tree.

Consider the two simple lotteries depicted in Figure 9.7. Both of these lotteries are mathematically equivalent. However, the problem in Figure 9.7b has two chance forks, while the tree in Figure 9.7a has only one. A person who enjoys gambling might actually prefer the former because he would have the opportunity to gamble twice. The reverse is true for people who fear gambling. Utility theory assumes that people have neither an attraction nor an aversion to the opportunity of taking a chance.

Consider the following extreme example of the continuity assumption:

A	$1
B	$0
C	Death

The continuity axiom, when applied to these outcomes, states that we can find a value of p such that outcome B ($0) is equivalent to a lottery with a probability p of A ($1) and probability (1 − p) of C (Death). In other words, there exists some probability p such that you would be willing to risk death for a dollar. For example, suppose that you are walking along the street and you notice a 1-dollar bill on the opposite sidewalk. Many people would cross the street to pick up the bill although there is a remote chance of being killed on the way. The difficulty with the continuity axiom is not in the existence of a probability p, but rather in determining a value for it.

9.4.2 Utility Functions

If we accept the validity of these axioms, then it is possible to define a **preference function** or a **utility function**, u(A), with the properties that:

1. For any two outcomes, A and B, u(A) > u(B) if and only if outcome A is preferred to outcome B.
2. If an outcome C is indifferent to a lottery L with probability p of outcome A and probability (1 − p) of outcome B, then

$$u(C) = p \times u(A) + (1-p) \times u(B)$$

That is, we can *define* a utility function such that the utility of the lottery is equal to the mathematical expectation of the utilities of the prizes.

A utility function that satisfies these properties is invariant under linear scaling. If we add a constant to all utility values, or if we multiply all utilities by a constant, the new function will still satisfy both of the aforementioned properties. Therefore, we can assume any convenient scale for our function. In particular, we will assume that the best possible outcome has a utility, u(Best) = 1, and the worst possible outcome has a utility, u(Worst) = 0. Note that we could use any convenient scale (e.g., from 1 to 100, or from 0 to 10).

Consider the decision problem from the previous section which is displayed in Figure 9.8. We wish to repeat the analysis, but this time, we will incorporate the decision-maker's attitude toward risk using utility theory. Note that the *gate* associated with paying for the survey has been removed. Instead, the $20,000 cost of the survey has been subtracted from the final outcomes for all corresponding tree branches. This does not affect the EMV of the decision, but, in order to evaluate utilities, all outcomes must be expressed as a net effect of that complete branch.

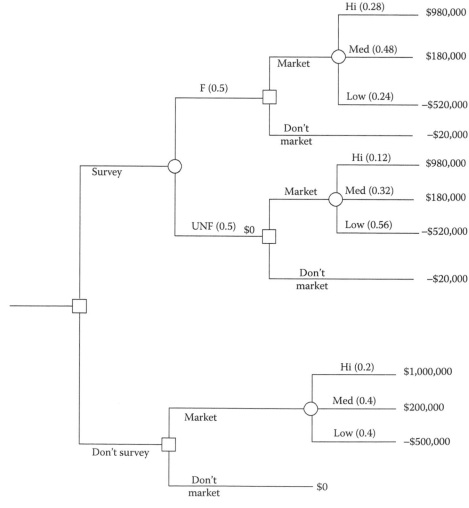

FIGURE 9.8
Marketing decision problem with survey information.

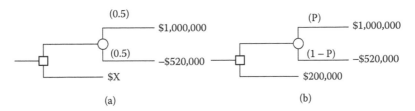

FIGURE 9.9
(a,b) Utility value assessment techniques.

In this decision problem, the best possible outcome is $1,000,000 and the worst possible outcome is –$520,000. Therefore, we can arbitrarily assign utilities:

$$u(\$1,000,000) = 1$$

and

$$u(-\$520,000) = 0$$

There are two commonly used methods for determining the utilities of the intermediate values. As seen in Figure 9.9, each will give us one new point for the utility function. In Figure 9.9a, the decision-maker is asked, "For what value of X are you indifferent to the alternatives at the decision fork?" Observe that the expected utility of the lottery is:

$$\left[(0.5) \times u(\$1,000,000) + (0.5) \times u(-\$520,000)\right] = 0.5$$

By the definition of a utility function, the utility of X must be u(X) = 0.5. Thus, the decision-maker is essentially saying that the utility of the lottery is equal to the utility of X.

In the approach illustrated in Figure 9.9b, the decision-maker is asked, "For what value of p are you indifferent to the options at the decision fork?" The expected utility of the lottery, in this case, is given by

$$(p) \times u(\$1,000,000) + (1-p) \times u(-\$520,000) = p$$

We conclude that u($200,000) = p, again relying on the definition. There are a variety of other assessment techniques, but the two approaches described here are the simplest, and the most common.

Suppose that we decide to use the first method, and our decision-maker selects a value of X = –$100,000. For this person, u(–$100,000) = 0.5. This decision-maker is very risk averse. Given a 50–50 chance of earning $1,000,000 or losing $520,000, he would prefer not to play. The chance of a loss of $520,000 is simply too great. In fact, he would prefer a certain loss of up to $100,000 to the lottery. Presumably, this decision maker feels that the smaller loss could be absorbed, while the potential large loss would be nearly intolerable. This rather dramatic behavior is not uncommon among actual decision-makers, and we will consider other examples later.

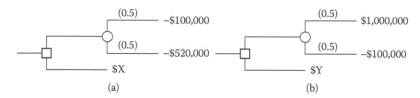

FIGURE 9.10
(a,b) Utility assessment.

Let us now repeat the utility assessment procedure using the two lotteries shown in Figure 9.10. Observe in Figure 9.10a that, when the decision-maker specifies a value for X at which he is indifferent, the utility of X is equal to the expected utility of the lottery:

$$u(X) = (0.5) \times u(-\$100,000) + (0.5) \times u(-\$520,000)$$

$$= (0.5) \times (0.5) + (0.5) \times (0) = 0.25$$

Similarly, in Figure 9.10b, when a value of Y is selected, we find that the u(Y) = 0.75.

Suppose that values of X = -\$350,000 and Y = \$250,000 are selected. We therefore have five sample points for the utility function. By plotting these points, the remaining values can be estimated by drawing a smooth curve through them, as shown in Figure 9.11.

Using this admittedly approximate utility function, we can now answer several lottery questions. For example, suppose he were faced with a lottery having a 50–50 chance of \$500,000 or –\$200,000. From the curve, u(\$500,000) ≈ 0.86 and u(-\$200,000) ≈ 0.41. The expected utility of this lottery is (0.5) × (0.86) + (0.5) × (0.41) = 0.635. Again using the **utility curve**, we find that the u(\$80,000) ≈ 0.635. Therefore, the lottery is approximately equivalent to a certain outcome of \$80,000. Hence, our decision-maker should be indifferent to a certain outcome of \$80,000 or a 50–50 lottery of either \$500,000 or –\$200,000.

Beginning with this simple function, we would then ask a variety of somewhat redundant lottery questions to validate the utility curve and adjust the shape at each iteration to

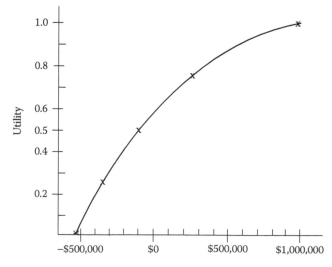

FIGURE 9.11
Utility function for marketing example.

best reflect the answers. We would then present the decision-maker with several examples of how the utility function would interpret his preferences for simple lotteries. In practice, this process is frequently implemented using an interactive dialogue between the decision-maker and a computer.

Finally, when the decision-maker is satisfied that the current estimate represents a reasonable approximation of the utility function, the function can be applied to the original decision problem. Each outcome is translated into its corresponding utility value. The expected utility of each chance fork in the decision tree represents the relative value of the lottery for the decision-maker. Averaging out and folding-back the decision tree in the usual manner produces a decision strategy that maximizes the expected utility. The marketing example is illustrated in Figure 9.12.

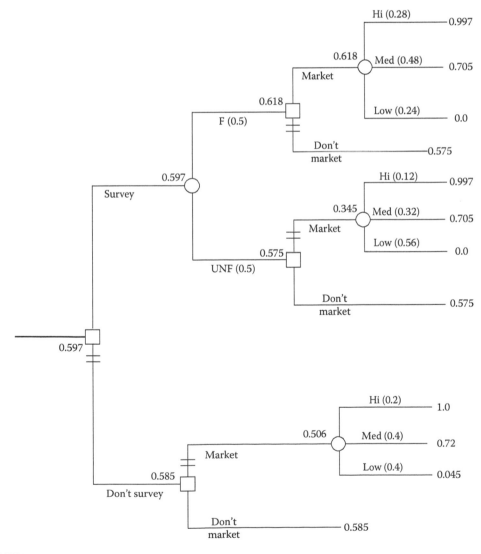

FIGURE 9.12
Marketing decision tree with utilities.

Observe that the use of a utility function did not change the optimal strategy. The decision-maker should still do the survey, and then market the product if the results are favorable. However, the expected utility of this decision is 0.597, while the expected utility of doing nothing is 0.585. This difference is very small, especially when we consider that the numbers are approximate. The decision-maker should probably consider these two options as being of equal value. Perhaps other criteria could be used to resolve the tie.

When we based our decision on expected monetary value in the previous section, the survey strategy produced an expected profit of $128,000. If we decided to market the product without a survey, the expected profit was $80,000. Both alternatives were clearly preferred to *Do Not Market*. When we consider the decision-maker's attitude toward risk, the survey strategy is only marginally preferred, while the alternative of marketing without a survey is definitely dominated by all other options.

9.4.3 The Shape of the Utility Curve

The primary types of utility curves are illustrated in Figure 9.13. For the EMVer, each additional dollar has the same utility value. The utility curve is a straight line indicating that the marginal value of each dollar is constant. The **risk averse** (RA) decision-maker has a curve with a *decreasing slope*, indicating a decreasing value of each additional dollar. Observe that this person derives 60% of the total utility from simply breaking even. Conversely, the **risk seeking** (RS) gambler has a curve with an *increasing* rate of change. The marginal value of each additional dollar is increasing. This individual is happy only when he is very close to the top of the scale. Breaking even has a very low utility. It is important to recognize that a person's attitude toward risk is reflected by the *rate of change* of the slope of the curve—not by the absolute value of the slope.

The gambler in a business environment is not the same personality that one would expect to find gambling in a casino. Such people are typically found at the head of a new venture. There is considerable risk associated with starting a new company, but these people have

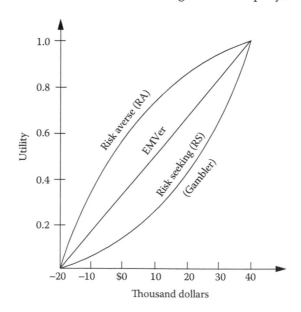

FIGURE 9.13
Basic utility curves.

enough confidence in their own abilities that they believe that they can succeed. They are willing to take chances for the large potential profits that they visualize in the near future. They are not depending on pure luck because they honestly believe that they can control and influence the final outcome.

The risk averse decision-maker is commonly associated with more established, conservative companies. These individuals have already developed a successful enterprise and they have no desire to jeopardize this position in a risky adventure.

There have been a number of psychological studies that suggest that people's degree of risk aversion is directly related to their personal feelings concerning luck and fate. People who approach life, friendship, business, and so on with their fingers crossed, hoping that they will be lucky, are often risk averse. They believe that external forces are controlling events and that the consequences are unpredictable. Risk seekers tend to believe that they have considerable control over their lives, and that their destinies are not completely controlled by external forces. Most people lie somewhere between these two extremes.

In reality, people are risk averse at certain levels and risk seeking at others. Consider the utility curve illustrated in Figure 9.14, which describes an attitude which is risk averse for values below $4,000, and above $25,000, but risk seeking for values between $4,000 and $25,000. This type of behavior is seen in people who have established a financial goal for themselves. For example, this person may have decided that, if he had $25,000, then he could open his own business, or buy some new equipment. The $25,000 figure has a very high utility, relative to say $10,000. As such individuals get close to their target, they are willing to take an extra risk to attain it. Outside of this perceived range, they are generally risk averse. A person with several financial goals could have a number of risk seeking segments in his utility curve.

Earlier in this chapter, it was stated that decision-makers are generally irrational. A prime example of this behavior can be found in the way that people assess their utility curve.

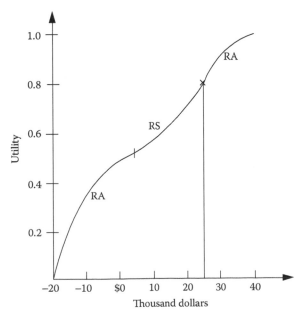

FIGURE 9.14
Utility curve with a $25,000 target level.

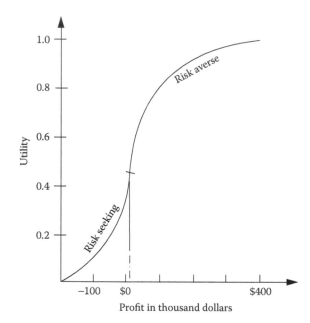

FIGURE 9.15
The *zero-illusion* utility curve.

A set of experiments was performed by Kahneman and Tversky (1979) in which subjects were asked to develop a utility function in the standard way. It was discovered that the typical curve of the majority of cases resembled the form shown in Figure 9.15. Most respondents exhibited RA behavior above zero, but they became gamblers when the values were negative. Apparently, people have a financial target of breaking even and are willing to take considerable risks to avoid a loss.

In decision analysis, this behavior is called *The Zero Illusion*. The problem is that the zero point is relative to the scale that has been defined in the choice of units. For example, if we use net profit as the outcome measure, zero represents the point at which the company makes or loses money on the product. If we use net assets to measure the effect of a marketing decision, we have not changed the problem, but zero now represents the point at which the company is in the black or in the red. Profit could be described in terms of this year's profit-and-loss statement. In each case, the method used to calculate outcome values has no effect on the real decision problem, but the scale has simply been shifted.

When a decision-maker produces a curve with this structure, he can usually be convinced to abandon the zero target by reformulating the questions in terms of a few different scales. He will soon adjust his answers to a more consistent behavior. Zero is indeed an imaginary target. We will come back to this notion when we discuss the *framing effect* in Section 9.5.5.

A classic example of apparently irrational behavior with respect to utility theory is known as the **Allais Paradox**. Consider the two decision problems shown in Figure 9.16. For the problem in Figure 9.16a, most people prefer to choose the certain outcome of $500,000 because the lottery branch looks too risky. Although there is a 10% chance of getting $1,000,000, there is also a 1% chance of getting nothing, so why take the risk?

In Figure 9.16b, the two lotteries look very similar, except that the second one has a payoff of $1,000,000 while the first gives only $500,000. In this case, the majority of people will

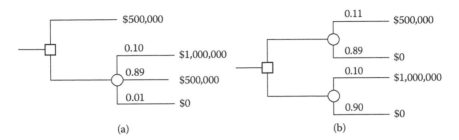

FIGURE 9.16
(a,b) The Allais Paradox.

choose the second lottery with a chance at $1,000,000. In both lotteries, the probability of a zero payoff is about the same.

Consider an individual who plays the aforementioned two lotteries as described for the majority. Not everyone will agree with this person, but the selections certainly appear to be logical. The paradox arises when we attempt to develop a utility function for this person: it cannot be done.

Without loss of generality, we can assume a utility value of u($0) = 0, u($1,000,000) = 1, and u($500,000) = p for some value of p. We wish to find a value of p such that the utilities of the two decision problems correspond with the choices of our decision-maker.

In the first problem, the utility of $500,000 certain is simply p. The utility of the lottery is

$$0.89 \times p + 0.10$$

Because our subject chose the certain branch, we conclude that

$$p > 0.89 \times p + 0.10$$

or

$$0.11 \times p > 0.10$$

Similarly, for the second problem, the utility of the first branch is given by 0.11 × p, while the utility of the second branch is 0.10. Because the second branch is preferred, we conclude that

$$0.11 \times p < 0.10$$

In other words, no matter how the utility function is defined, the decision-maker must either select the first branch in both problems, or the second branch in both problems.

There is no *rational* way to explain this dilemma. In the first problem, the decision-maker is frightened away from the lottery because there is a 0.01 chance of obtaining a zero. However, in the second problem, there is an *extra* 0.01 chance of getting zero in the second lottery. This additional risk apparently is not recognized. The decision-maker is effectively saying that a probability of 0.01 is significant enough to avoid the risk, but, at the same time, probabilities of 0.89 or 0.90 are essentially equivalent. A difference in probability of 0.01 should either deter a person in *both* cases or not at all. Even when we point this out, many people will stick with their original, irrational selections. We will discuss this and other aspects of irrational behavior in the following section.

Many practitioners believe that utility theory is the only solution to decision-making problems. Others argue that, although it is reasonable to assume that people have an implicit personal utility function, it is not a practical tool for decision analysis. The main objection, in addition to the problems already mentioned, is that the assessment procedures for determining a person's utility function are basically artificial. When people are asked to play lottery games, the prizes are not representative of real decision problems. Therefore, it is difficult for people to treat the answers seriously, especially as the number of questions increases. Despite these criticisms, utility theory has been used in a wide variety of practical decision situations, some of which are described and discussed in Raiffa (1968) and Keeney and Raiffa (1976).

9.5 The Psychology of Decision-Making

By now, it should be clear that decision analysis is an artful combination of mathematical logic and human intuition. Unfortunately, human decision-makers are prone to a number of misconceptions and idiosyncrasies that can severely limit their ability to make rational choices. We have already alluded to a few of the problems, and we will now expand on that theme in this section. Many of the examples are based on the research of Tversky and Kahneman (1982) and Kahneman (2013).

9.5.1 Misconceptions of Probability

Suppose that you are in Reno and that you have been casually watching people play roulette. You notice that red has come up 40 times in a row. Would you now bet everything you own on black? We know that getting red 41 times in a row is highly unlikely (a probability of approximately 4.5×10^{-14}), and therefore, for many people, black seems highly probable. But in fact, assuming that the wheel is fair, the probability of red on the forty-first spin, given that we already have 40 reds, is 0.50.

This assumption is known as the **Gambler's Fallacy**. The same behavior can be observed in more practical decision-making situations. When people observe a sequence of events with a high proportion of failures, they assume that the probability of success must be increasing, and they adjust their decisions and their attitude toward risk accordingly. As in the game of roulette, this is not rational when the individual observations are independent.

There is a popular lottery in which people pick six numbers between 1 and 49. Every week, six numbers are drawn at random, and anyone who matches all six wins the grand prize (usually in the millions of dollars). There is considerable speculation about "Which numbers are more likely?" Many people apparently believe that some combinations must surely be more likely than others. "You will never see the sequence 1,2,3,4,5,6" for example. If we try to explain that 1,2,3,4,5,6 is just as likely as any other, many people respond by telling us that we do not understand the basic laws of probability and true randomness. One can even purchase software to help us select six numbers that are truly random (and hopefully lucky).

Consider the following experiment: suppose that we ask people to select colored marbles from a large opaque jar, one at a time with replacement. The subjects are told that the jar contains two colors, red and white, and that two-thirds of the marbles are one color, while one third are the other color. The first individual draws six marbles and finds four red and two white. He concludes that the jar is two-thirds red. The second individual (drawing from a different jar) selects 20 marbles, of which 11 are red and 9 are white. She concludes that her jar is also two-thirds red. After making their draws, we ask the subjects how much confidence they have in their assessment.

Most people agree that the first person has a higher probability of guessing correctly. His draw corresponds precisely to the expected distribution if two-thirds of the marbles are red. The second subject found the colors to be almost equally divided and feels that the probability of guessing correctly is only slightly better than 50%. In fact, in both cases, the probability of two-thirds red is exactly 0.80. The larger number of draws in the second experiment greatly increases the accuracy of the conclusions. Generally, people do not appreciate the significance of sample size information. The same principle is true for market surveys and opinion polls. Assuming that the selection procedure is unbiased, even small samples can be very accurate predictors.

As already discussed with reference to the Allais Paradox in the previous section, people are inconsistent in their application of small probabilities. Probabilities with an obvious physical interpretation, such as a 50% chance of getting a head when tossing a coin, are easy to understand. However, probabilities of 0.48 and 0.52 are both considered close to 0.50, and we perform this substitution in our minds when we analyze a problem. The 4% difference is often essentially ignored.

At the same time, a probability of 0.01 is too small to visualize. Consequently, people have a tendency to either exaggerate the probability, or to decide that it is essentially zero. Down to a certain level, people will treat probabilities of 0.01 or 0.02 as if they were closer to 0.05 or 0.10. At some point, the associated probability is taken as being effectively zero. The same behavior is true for probabilities close to 1.0. The perceived probability is less than the actual probability up to some point at which people assume that the event is certain. This behavior, although understandable, is mathematically irrational.

Another common error in the appreciation of probability concerns the net effect of a series of conjunctive (or disjunctive) events. Consider a decision-maker who is responsible for a large project composed of a series of small components. The project could be the design and installation of a computer system, an office tower, or a nuclear reactor. We assume that each part must be successful in order for the project to succeed. This is a conjunctive event in that the probability of success of the project is the product of the probabilities of success of the components.

Let us suppose that the decision-maker and his staff investigate each component, and they determine that each has a 99% chance of success. They conclude that the success of the project is highly likely. In actual fact, if there are 1,000 components, the probability of a successful project is less than 0.00005. This problem is compounded by the fact that the people responsible for the individual components are not likely to estimate a 99% chance of success. At that level of certainty, they will usually say that they are *sure* that their part will work properly.

As a final example, consider the following problem based on the format of a popular television game show. Contestants are shown three doors and told that behind one door is a two week, all expense paid vacation to Hawaii, or something equally valuable. The other two doors conceal a consolation prize. Suppose the contestant selects door number 2, and then the host opens door 3 and shows the contestant that it contains one of the consolation prizes. (Doors 1 and 2 are not opened.) The host then asks the contestant if he/she wants to change his/her initial choice (from door 2 to door 1 in this case). Based on the probabilities, and on this *new* information, should he/she change doors? We will leave this question for the reader to ponder, and come back to it later in this chapter.

9.5.2 Availability

When decision-makers are asked to make subjective probability assessments of uncertain future events, their judgment depends on their personal available store of information. Unfortunately, the **availability** of information is often influenced by subjective external events. People make decisions based on the experiences related to them by a trusted friend, events they read about in the morning newspaper, or what they saw on the way to work.

To illustrate, suppose that we ask people to estimate the probability of an airline accident. Some people may actually go to the trouble of collecting statistics on flight accident rates over the past few years, but most would simply use their intuition. Their probability estimates would be strongly influenced by recently reading about an accident or by knowing someone who was involved in a crash. People who actually witness this kind of disaster often conclude that the risk is so high that they will decide never to fly again. Observe that none of these events reflects the true probability of an accident. People often make probability assessments based on very limited personal experiences.

The same logic applies in business decision-making. An executive who has previously been involved in a risky venture that failed will be very reluctant to try anything like it in the future. His own estimation of the probability of success has been greatly reduced. The availability and use of such highly subjective input can produce very irrational behavior.

9.5.3 Anchoring and Adjustment

When people make subjective assessments, they often begin with an initial estimate based on their previous experiences, or perhaps even based on ideas suggested by the wording of the question at hand. When they try to make a prediction, they can become **anchored** to their original estimate, even when they know it should not affect their decision. This produces insufficient or conservative **adjustment** in the direction of the new assessment.

Consider a rather dramatic example, described by Tversky and Kahneman (1982), of an experiment in which people were asked to estimate the percentage of African countries that are members of the United Nations. The experimenter would first spin a wheel of fortune in the presence of the subject. The wheel would randomly pick a number between 1 and 100. If the number was 10, the experimenter would ask, "Is it 10%?" The average response of subjects was, "No, it is closer to 25%." When the random number was 65, the experimenter would ask, "Is it 65%?" The average response was, "No, it is more like 45%." When people were given a number that they knew was irrelevant, they used it anyway. They were anchored to the initial wording of the question and then performed insufficient adjustment. Moreover, their performance did not improve when they were offered money for guessing correctly. Apparently, if people are given no information, they will use common sense, intuition, and/or statistical estimates. When people are given useless information, they will use it and ignore logic.

In one experiment, 32 *judges* were shown the case background for a patient. Eight of the judges were clinicians. The patient's file was divided into four sections and the judges were asked to give their opinion on the diagnosis after reading each section. The study showed that the accuracy of the diagnoses did not increase significantly with the amount of information. However, the judges' *confidence* in their diagnoses increased dramatically. Presumably, people became anchored to their initial impressions.

The same is true in management decision-making. When a manager has access to a great deal of data and reports, he will have a correspondingly high confidence in his ability to make decisions. This attitude does not necessarily depend on the quality of his information. People have a tendency to be influenced by the sheer *volume* of data available to them.

Expertise itself can be a source of the **anchoring** bias. Professionals, such as doctors, lawyers, managers, or stockbrokers, may develop a system of *standard operating procedures* based on years of training and experience. Expertise produces efficient responses to environmental signals and symptoms. When you describe your ailments to your family practitioner, he does not usually need to spend hours consulting his medical reference books. He will quickly identify a few possible diseases that match your symptoms and prescribe further tests or medication. The value of expertise is that we can get quality advice quickly.

Unfortunately, experts can become anchored to their own standard procedures. If some of the symptoms and signals are incompatible with their standard procedures, they tend to be ignored or re-interpreted by the expert to fit their existing models. Experts will put greater emphasis on information that is consistent with their own previous experience, and thus become anchored to their own expertise.

People can also be anchored to the mean of a distribution. Suppose that we asked a decision-maker to estimate the expected value of sales for a product next year, or to forecast the inflation rate. We then ask him to specify an upper and lower limit for the distribution, with a probability of 99%. In experiments with experts, people tend to specify a range that is accurate 70% of the time. They are conservative in their estimates of high and low values and are anchored to their initial estimate of the expected value.

9.5.4 Dissonance Reduction

Consider the decision to buy a new car. Most people will begin this exercise with total objectivity. They will develop a list of desirable features and decide on a budget limitation. After visiting several dealers, test driving the cars, talking to people and collecting brochures, they will compile a mental catalog of the possibilities, and start objectively removing certain alternatives that are too expensive, too slow, or too small.

As this process continues, the decision-maker reduces the set of options to some small group of items that are all, in some sense, equally acceptable. It becomes difficult to choose between them, and the decision-maker enters a phase called **dissonance**. A choice must be made; and at this point, the decision-maker will become very subjective, and simply pick one alternative. This is perfectly rational because all of the options have been judged to be of equal utility to him.

Having now made a choice, the majority of psychologically stable decision-makers then enter a completely irrational phase called **dissonance reduction**, in which they try to convince themselves that the alternative they selected was, in fact, the very best one by far. They will exaggerate favorable qualities and down-play the less attractive ones.

This type of justification after the fact is irrational, but it is also necessary in order to dispel the feeling of dissonance. People who do not enter this phase may spend the rest of their lives doubting themselves and worrying about whether they made the right decision, and they might never really be satisfied with their decision.

This behavior is important in decision analysis in a practical environment because business decision-makers will also subconsciously employ dissonance reduction. Once they have made their decision, they become increasingly stubborn about it. They will tend to discredit any new information that does not confirm the wisdom of their original choice. It may be very difficult to return to the initial objective context of decision-making after having mentally justified the choice that was made.

9.5.5 The Framing Effect

It has been observed that people sometimes change their answers when we simply alter the wording of the question. This **framing effect** is closely related to the idea of the *zero illusion* discussed earlier.

In one study, two groups of physicians were given the following decision problem. Suppose that a rare Asian flu is expected to hit the country next winter. If nothing is done, we expect that 600 people will die. The first group of physicians was told that there are two possible inoculation programs that could be used. Program A has been used in other countries and the results are highly predictable. Program B is a new, experimental treatment.

Program	Expected Result	Probability
A	200 people saved	1.00
B	600 people saved	0.333
	0 people saved	0.667

Observe that the two programs are equivalent in terms of the expected number of people who will be saved. The majority of the physicians preferred program A. They were being risk averse and preferred to save 200 lives for certain, rather than take a chance of saving all or none.

The second group of doctors was given the same problem, except that they were told that there are two possible inoculation programs, C and D.

Program	Expected Result	Probability
C	400 people die	1.00
D	600 people die	0.667
	0 people die	0.333

The majority of the subjects in this group preferred program D. Presumably, the thought of having 400 deaths on their conscience was too much, and they preferred to gamble.

In this experiment, both groups answered the same question, but changing the wording of the question changed the way they responded. The first group looked at the problem in terms of positive results (lives saved) and were risk averse, while the second group became more risk seeking for negative results (death). This is precisely the effect of the zero illusion.

In another experiment, subjects were asked to imagine that one of their friends had contracted a fatal, contagious disease. The disease has no symptoms that can be detected; people who have it will simply die in two weeks. There is a remote probability of 0.0001 that you have contracted the disease from your friend. Fortunately, there is an antidote that you can take now as a precautionary measure. What is the maximum amount that you would be willing to pay for this antidote? The average response was $200. If the drug cost more than $200, they would prefer to take their chances.

A second group of subjects was asked if they would be willing to volunteer for a medical research experiment. They were told that there was a remote chance (probability 0.0001) that they might contract a fatal disease. There is no antidote and, if they got the disease, they would suddenly and painlessly die in two weeks. What is the minimum amount that we would have to pay you to volunteer for this program? The average response was $10,000.

This is a dramatic example of the zero illusion. People are unwilling to pay more than $200 to avoid the risk of death. But these same people will not take less than $10,000 to face the same risk. Notice that the $200 is a loss, while the $10,000 is a gain. We can interpret this to mean that the positive utility of $10,000 is the same as the negative value of –$200. We can also assert that people are gamblers for losses and highly risk averse for profits. This attitude appears perfectly rational until people realize how easily we can move their zero.

Consider yourself in this simple situation. Someone sends you a card on your birthday with a $100 bill inside. A few days later, they come up to you, terribly embarrassed, and tell you that it was a mistake. They put the money in the wrong envelope, and could you please give it back—which you do reluctantly. Observe that, not counting the insult, the net effect of this pair of transactions is very negative. Receiving the $100 had a certain positive utility. However, once you had it in your pocket, and had already decided how to spend it, giving it back is a loss with a much higher negative utility. After you receive the $100, you move your zero.

Companies will often use the framing effect to their advantage in marketing strategies. Some years ago, credit card companies banned their affiliated stores from charging higher prices to credit card users. A bill was once presented to the U.S. Congress to outlaw this practice. Lobbyists for the credit card bloc realized that some bill would be passed, and they preferred that the new legislation call for a *discount for cash* rather than a credit card surcharge. The two options are identical because merchants simply add the surcharge to the cost of the merchandise. However, customers see the discount for cash as a positive gain (low utility), whereas the added cost of a surcharge would have much higher value, and many more people would pay cash.

A common marketing ploy is the "two week trial with a money back guarantee." People must make two decisions: one at the beginning and a second decision at the end of the two weeks. In the first decision, people will compare the value of a two-week trial against the transaction costs (pick-up, delivery, etc.). The cost of the item is not included because they can get it back. In the second decision, they compare the value of keeping the item to the utility of the *positive refund*. But, as we have seen, the utility of a refund is much smaller than the utility of a payment if we had bought the item outright in the first place. People are more likely to keep things that they would never have purchased otherwise. The mail order purchasing industry thrives on this principle.

There are examples of the framing effect that do not rely on the zero illusion. Consider the following two scenarios. Sam is waiting in line at a theater. When he gets to the window, the manager comes out and says, "Congratulations. You are our 100,000th customer, and you win two free tickets to the show!"

Sally is at a different theater. When the man in front of her gets to the window, the manager comes out and tells him, "Congratulations sir! You are our 1,000,000th customer, and you win $1,000." The manager then turns to Sally and gives her $100 as a consolation prize for being number 1,000,001.

Which of these two people had a better experience? Although Sam's net gain has a much smaller value (around $20), many people feel that Sally experienced a great loss at almost (but not) getting $1,000. By framing the question in terms of what could have happened, we can change the perceived value of Sally's $100 profit.

9.5.6 The Sunk Cost Fallacy

The sunk cost fallacy is really a specific variation of the framing effect. The relevant aspects are illustrated in Figure 9.17. Let us assume that, at some past time t_0, a decision was made to initiate a project. We are now at time t_1, and we must decide whether to continue the partially completed project or to quit now and cut our losses. We further assume that we have already invested some amount $S in the development. The question is: should the value of S, the sunk cost, be considered when the decision is made at time t_1?

An example of this issue occurs in a so-called *Continue/Discontinue* decision, where x represents the potential profit of successful completion with probability p, y represents the potential cost of failure with probability $(1 - p)$, and z denotes the expected cost of discontinuing the project. We will assume that x > z > y. The same decision tree structure occurs in an *Asset Disposal* problem. At time t_0, we purchased an asset for $S, and at time t_1, we must decide to either keep it with a risky future cost or dispose of it and take the current salvage value.

The question is: how does the value of S, the consequence of previous decisions, affect the current decision at time t_1? Authors in mathematical and economic theory refer to this question as the **sunk cost fallacy**. They argue that nothing can be done about S, and the decision at time t_1 should depend on the real options currently available.

Consider a man who joins a tennis club and pays a $300 annual membership fee. After two weeks, he develops tennis elbow, but continues to play (in pain) because he does not want to waste the $300. If the same man had been given a free annual membership, valued at $300, he would likely quit. The *sunk cost* directly affects future decisions.

Empirical studies have shown that the rate of betting on longshots increases during the course of a racing day. Presumably, people have not adapted to earlier losses and they are trying somewhat desperately to compensate for them.

This type of behavior is common in the business world. A manager initiates a project at time t_0 with an expected cost of $100,000, to be completed in one year, and a 50% probability of revenues of $500,000. At the end of the year, time t_1, he has already spent $150,000, and he estimates another six months and $50,000 to finish with a 10% chance of $200,000 revenue. There were unexpected delays and the market conditions have experienced an unfortunate decline. The project is no longer profitable. The decision tree is illustrated in Figure 9.18.

Figure 9.18b is the same as 9.18a except that the costs of both gates have been moved to the end nodes of the tree. In the first diagram, paying $50,000 for a 10% chance of making $200,000 is clearly a very risky proposition. However, when we consider Figure 9.18b, the prospects of losing either $200,000 or $150,000 are both considered almost equally bad outcomes by the risk seeking decision-maker. By defining *zero* in terms of time t_0 in

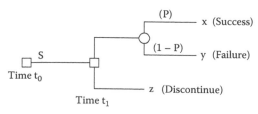

FIGURE 9.17
Decisions involving a sunk cost S.

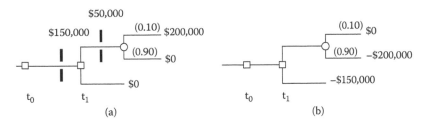

FIGURE 9.18
(a,b) Two views of the same *sunk cost* decision.

Figure 9.18b instead of time t_1 as in Figure 9.18a, we can change the decision-maker's utility function from risk averse to risk seeking.

This behavior is a form of **entrapment** in which a manager stays committed to his original course of action long after it ceases to be justified by any rational analysis. There is a strong psychological motivation to appear consistent in decisions made over time. People often feel that they cannot abandon a course of action without admitting that their previous decisions were wrong. In 1968, Lockheed started development of the L-1011 Tristar jumbo jet. At that time, they were predicting 10% annual increases in airline traffic. By 1971, actual increases were closer to 5%, and their expected sales were now well below the projected break-even point. Lockheed needed federal guarantees for $250 million of additional bank credit. (The banks did not consider Lockheed to be a good risk.) At the subsequent congressional hearings, Lockheed and the Treasury Secretary proposed that federal guarantees were in the national interest because it would be the height of folly to abandon a project on which close to $1 billion had already been spent. The continuation of the project was being justified on the basis of the sunk cost alone. In fact, the real question was whether the expected future income would be worth the additional expenditure of $250 million. The sunk cost argument is often used to explain many political decisions to support an apparent lost cause.

9.5.7 Irrational Human Behavior

Reconsider the game show problem described earlier. The contestant was asked to pick one of three doors and he chose door number 2. The host now opens door 3 to reveal a consolation prize and asks the contestant if he would like to switch his selection from door 2 to door 1. What is the probability that the grand prize is behind door number 1? Many people (and many actual contestants) believe that the probability has been reduced to a 50–50 proposition between the two remaining doors, and they will agonize over this new decision, often sticking to their original choice.

In actual fact, they should always switch. The probability that the door initially selected, door number 2, was the one that concealed the grand prize, was one third and it still is one third. Whether or not the initial choice was correct, it is certain that at least one of the other two doors contains a consolation prize. The host, who *knows* where the grand prize is, simply verified the fact that one of the two doors was wrong, and he has not really given any new probability information. The probability that the door initially selected was the right door is still only one third, and therefore, the probability that door number 1 is correct must be two-thirds. Consequently, the contestant should *always* switch. The reader who is still skeptical should try the experiment outlined in Exercise 9.1. As has been stated already, people often feel that additional information must improve the validity of their probability estimates. But, in fact, more information is not always helpful to an irrational human decision-maker.

9.5.7.1 What Can We Do about Irrational Behavior?

1. One of the most important tools for combating irrational behavior and biases is through proper **training**. Simply making people aware of the biases described earlier can improve their understanding of the decision process, and they can avoid making some of the common mistakes.

2. A **decision simulator** works on the same principle as the jet aircraft simulator for training pilots. The decision-maker is presented with a large number of different situations (one at a time) and asked to choose a course of action. The simulator immediately gives him the consequences of his decisions and, if possible, the results of the *optimal* decision. The cases used in the simulator can be actual historical problems with known outcomes. (e.g., Lembersky and Chi [1986] describe a computer simulator that helps decision-makers at Weyerhauser to decide more effectively how trees should be cut in order to maximize profit.)

3. A less-expensive form of training is **feedback**. Decision-makers will estimate probabilities of various market parameters in predicting the success of a product, but they seldom get any direct feedback on the quality of their intuition. If the product is successful, they must be doing something right. By comparing their original estimates with the actual results, it is possible to improve their future prediction skills.

4. Another method of reducing bias is by **automatic correction** procedures. For example, when we must forecast future sales, we could use an expert and a simple linear regression model, and then split the difference between the two. The assumption here is that the expert will have more information than the regression model, but that the human has a tendency to overreact.

5. A common approach to eliminating bias is to ask a number of **redundant questions**. In particular, we can reduce the effect of the zero illusion by rephrasing the same question in several different ways with the zero shifted to make people aware of the effect.

6. We must **recognize the limitations** of the human decision-maker as well as the strengths. Dawes (1982) compares human judgment with linear regression in a variety of selection processes and concludes that the linear models are generally superior to expert decision-makers. Human experts were often found to be much better at gathering and encoding information than they were at integrating it and using it effectively in decision-making.

9.6 Software for Decision Analysis

A large variety of software is available to support various aspects of decision-making. For a comprehensive survey of decision support software, refer to Oleson (2016). A few representative packages are mentioned here.

SAS/OR offers decision making support that allows users to create, analyze, and interactively modify decision tree models, incorporates utility functions and attitudes toward risk, and identifies optimal decision strategies.

Oracle's **Crystal Ball** and **Crystal Ball Decision Optimizer** provide advanced optimization capabilities to support predictive modeling, forecasting, optimization and simulation, with insights into critical factors affecting risk.

Palisade Corporation has acquired a reputation as a leading provider of fully integrated and comprehensive risk analysis and decision analysis software known as the **DecisionTool Suite** Risk and Decision Analysis software. The Suite includes a component called **@RISK** for risk analysis and another component called **Precision Tree** that supports visual decision analysis by building decision trees. Precision Tree incorporates the decision maker's attitude toward risk by creating a risk profile that illustrates the payoff and risks associated with different decision options, and also performs sensitivity analysis to track changes in the expected value of the decision tree when values of variables are modified.

Analytica is a spreadsheet-based visual software environment for building, exploring, and sharing quantitative decision models. It runs under Windows and has a web-based implementation for collaborative decision making. It has a Monte Carlo simulation capability to handle uncertainty and risk. It has also an automatic optimizer to find the best decisions. Decision trees can be developed and sensitivity analysis can be conducted. Graphical representations such as tornado diagrams, influence diagrams, and decision trees can be generated.

DPL has a family of decision analysis software packages that run under Windows, Mac and online with various capabilities to quantify uncertainties and enumerate options Influence diagrams, decision trees, tornado diagrams and rainbow diagrams can be produced. It has Monte Carlo simulation, risk tolerance, utility functions, and graphical sensitivity analysis capabilities. It has a component to compute the value of imperfect information.

TreePlan is among the early packages for decision analysis, and specifically for developing decisions trees. It is now an Excel Add-in that can generate decision trees for analyzing sequential decision problems under uncertainty and it runs under Windows and MacOS. TreePlan creates formulas for summing cash flows to obtain outcome values and for calculating rollback values to determine the optimal strategy. Although it lacks many of the decision analysis features that other packages include, TreePlan is included in many textbooks specifically for decision trees, which makes it quite well known among other packages.

Analytic solver platform is a suite of tools that are used for risk analysis, Monte Carlo simulation, optimization, forecasting and data mining. It works with Excel and it can handle uncertainty, risk tolerance, decision trees, and sensitivity analysis. It can also produce tornado diagrams and decision trees.

9.7 Illustrative Applications

9.7.1 Decision Support System for Minimizing Costs in the Maritime Industry (Diz et al. 2014)

International maritime transportation is responsible for shipping the majority of commercial goods in world trade transport. Because crude oil and its derivatives account for approximately one third of this total cargo, scheduling the global fleet of petroleum tanker ships is a critically important component in managing the cost of this huge transportation enterprise. The ship scheduling process involves allocating tankers from a given fleet to a set of cargoes, with the aim of minimizing costs. Significant academic research efforts over many years have resulted in the creation of new mathematical models, with details adequate to represent realistic situations, and new algorithms to optimize the ship scheduling models. However, the system described in this section is somewhat unique in that it is one of the few research studies which demonstrates a substantial and successful application of such analytical tools to control costs in actual industry ship scheduling problems.

In this application, we will see both the details and the broad scope of a successful decision support system. The development and implementation of this decision support system encompasses the difficulty of collecting reliable and critical data, the use and adaptation of powerful optimization techniques, the initial resistance of human experts to accept new methodologies, and the eventual inclusion and integration of the new decision support system into traditional corporate processes. The use of this decision support system provided a comprehensive approach for minimizing costs without compromising the service level standards set by the company. Moreover, the implementation of the new system also represents a turning point for this large corporation in the use of analytical support tools which ultimately fostered stronger cooperation among engineers, researchers, and managers.

The large multinational energy company Petrobras is a prominent player in global petroleum transport. Over a long period of time, Petrobras worked with a long-haul ship scheduling system to handle both importing and exporting crude oil for Brazil through numerous ports in five continents. In their system, the logistics department received information about cargoes when they were acquired or traded, and this data were constantly updated as cargoes were exchanged in open trade. The cargoes are defined by a narrow range of standard volumes of approximately a million barrels. Export cargoes must be loaded at ports on the Brazilian coast and transported to discharging ports in various parts of the world. The complicated pattern of importation and exportation involves transport both into and out of Brazil as dictated by Petrobras and customer supply demands.

The scheduling operation at this company assigns vessels to cargoes which are to be loaded, transported, and discharged. The size of the cargoes is consistent with the capacity of the Suez Max class of ships in their fleet. The assignment is subject to commercially negotiated time frames and operational constraints at ports and underway, and aims to minimize overall transportation costs. Petrobras maintains a fleet consisting of long-term chartered ships. For economic reasons, the capacity of this semi-permanent fleet is less than what is usually required for transporting all the cargoes. It is therefore routinely necessary to engage extra vessels on a per voyage charter basis in order to meet immediate demands for moving cargoes.

The scheduling process in use was practical, being based on the expertise gained through experience. For example, one heuristic rule used by the schedulers was to attempt to avoid idle time of long-term chartered ships; another strategy tried where possible to assure that a ship underway is loaded with cargo so that ships are used efficiently and not traveling empty unless necessary. The schedulers manually chose what they considered to be efficient routes. However, the methods employed merely generated several alternative feasible schedules, from which a few could be manually compared and selected based on low cost. The process did not produce an optimal minimum cost schedule.

Because of the very high and increasing costs anticipated in the maritime transport industry, the company undertook a thorough study of its long haul shipping operations for the transportation of crude oil. Until 2012, skilled and experienced schedulers manually carried out the ship scheduling processes, but the company hoped that a careful study would reveal cost savings that could be achieved by an automated decision support system (DSS) that would largely take over the scheduling process, and that would also evaluate and report the actual economic advantage of implementing the improved scheduler.

The study resulted in the creation of a DSS that provides a minimum cost schedule for transporting a set of cargoes on a given fleet of ships. The new system systematically verifies and confirms or updates all the original input data pertinent to a schedule, such as operational costs, unexpected restrictions, changes in vessel speed or capabilities, port circumstances, distances, freight rates, vessel availability and location, and cargo characteristics. Once the DSS determines a minimum cost schedule, any of the earlier unexpected changes can occur, and must be incorporated to produce a new feasible schedule.

The DSS operates as follows. First, an algorithm is applied to generate a complete set of feasible schedules based on the most recently updated cargo, fleet, and port data. The method used for this is to generate all possible routes for the cargoes that adhere to the specified loading and unloading times in port; then to match each route to a time chartered vessel, adding voyage chartered ships if needed. A cost is then computed for each route to vessel assignment; that is, the total cost for each voyage (which includes ship operating cost, port taxes and fees, and current market freight rates). Next, the DSS uses the previous set of voyages and costs as input parameters for an integer programming model, formulated to minimize the overall cost for the entire fleet. An efficient commercial solver based on the AIMMS optimization platform with CPLEX is then used to solve the integer programming problem, thus yielding the minimum cost fleet schedule.

The integer programming model in this case was a straightforward formulation, using binary decision variables to denote whether vessel v is assigned to route r; and other binary decision variables to flag an idle vessel or a cargo that will have to be assigned to a voyage charter vessel and thus not assigned to any time-chartered ship.

The DSS was thoroughly tested, then fully implemented, with positive outcomes including an average 7.5% savings representing hundreds of millions of dollars in costs. Furthermore, the new system solved more complex and complete problem models, and did so within measured computation times of approximately five seconds. With this new DSS, the importance of acquiring accurate data and the ability to deal with unforeseeable restrictions, mechanical degradations and weather conditions were recognized as being critical. In this respect, a key factor was the interaction between the user and the system that included the skillful and experienced manual schedulers to handle changes that could not always be dealt with during execution of the DSS. This interaction created a healthy environment of cooperation and promoted acceptance of the new system by Petrobras employees. The schedulers had, through experience, acquired a rich understanding of many subjective aspects of ship scheduling. Their ability and willingness to integrate their knowledge in dealing with last minute changes in the scheduling environment contributed significantly to the successful implementation of the DSS and to continued improvements in the economics of marine transportation.

9.7.2 Refinery Pricing under Uncertainty (Keefer 1995)

During the 1980s, when crude oil prices were fluctuating dramatically and refining overcapacity made the profitability of operating refineries unpredictable, an oil company shut down a large overseas refinery. Management's opinion of how best to dispose of this non-performing investment varied considerably: some thought it would not even be possible to give away a shut down refinery, while others hoped to sell the defunct facility for a substantial sum.

For purposes of price negotiations, management needed to develop some idea of what the refinery would be worth to a variety of types of potential buyers. Because the current owner had been unable to operate this refinery profitably (nor did it foresee being able to do so in the near future), presumably a likely buyer would not be a company just like the current owner. Instead of the buyer being a major international oil company, it was expected that the buyer would be a small, well-capitalized company, perhaps a newcomer to the industry, in the business of trading, refining, or marketing oil, and which would take a short term, entrepreneurial approach to using this refinery.

In the process of determining how to price the refinery for sale, the decision analyst drafted four operating scenarios describing how each of four categories of potential buyers could use the refinery profitably:

1. In the first scenario, the new owner would not actually operate as a refinery, but would instead use the facilities as a terminal, berthing ships, storing and trans-shipping crude oil, storing and blending certain products, and selling to local customers.

2. The second scenario consists of all the aforementioned activities plus operating the refinery itself opportunistically during periods of advantageous refining margins. (This posed some problems because positive refining margins in the near term were possible but very unpredictable.)

3. In a third scenario, the operation of the facilities as in the first scenario would be supplemented by refining under a so-called *netback* agreement with a crude oil producer. Under such an arrangement, the refinery agrees to buy and refine crude oil, at an agreed-upon steady supply rate for a prespecified refining margin. The predictability of throughput arising from this scenario is of considerable value, but this advantage is offset by the uncertainty of what netback margin could be negotiated with a crude oil producing country.

4. The fourth scenario is a combination of the first three: use of facilities for storing, transshipping, blending, a netback agreement to support a steady refining operation, and stepped-up refining activity during periods of positive refining margins.

The current owner of the refinery had traditionally analyzed and evaluated uncertainties using deterministic methods to calculate net present value, then applying sensitivity analysis. This simple approach turned out to be inadequate for pricing the refinery. Net present value (NPV) calculations were based on large and uncertain ranges for parameters such as margins and throughputs. These estimates led to discrepancies in NPVs that fluctuated too widely (over hundreds of millions of dollars) to give management much insight into how to price the refinery for purposes of negotiating a sale.

Rather than a deterministic model that allowed small changes in parameters, the analyst chose to utilize decision analysis techniques, treating the heretofore unwieldy parameters as random variables and basing much of the uncertain data on judgmental probability assessments.

The expected NPV was calculated using a decision structure known as a *probability tree*. This differs from a conventional decision tree in that the branches in the tree do not represent a timeline of sequentially made decisions in response to specific uncertainties. There are no decision nodes per se; rather, the branches associated with uncertainties denote the possible outcomes of the uncertainties. Working from the extremities of the tree back to the root of the tree, the probability-weighted average NPV was simple to calculate, and the entire model and computational process were easily understood by management.

Calculations for the fourth scenario showed the highest expected value of $53 million, while lower expectations were associated with all the other scenarios (as low as $23 million for the first scenario). The analyst's recommendation to management that it would be reasonable for certain prospective buyers to pay in the range of $23 million to $53 million for the shut down refinery constituted a hopeful alternative to the sad prospect of having trouble giving it away. As a result, a decision was made to wait for a reasonable offer, and indeed, a sale was eventually made for a price in excess of $50 million.

9.7.3 Decisions for Radioactive Waste Management (Perdue and Kumar 1999)

High-level radioactive waste resulting from spent nuclear fuel is sometimes dealt with by encapsulating the waste in glass, using a process known as *vitrification*. But just how much nuclear waste cleanup is necessary, desirable, and cost-effective. Determining the appropriate extent of this technologically difficult undertaking has proven to be quite a complex decision process. In a joint effort of the U.S. Department of Energy, the New York State Energy Research and Development Authority, and Westinghouse Electric Company, decision analysis techniques have been used to help analyze how this cleanup process should be properly accomplished.

The contaminated waste is contained in underground tanks. Waste is removed from the tanks, sealed in glass containers, and the tanks cleaned and rinsed. This process is repeated until the tanks are no longer classified as *high-level radioactive waste*, but the declassification criterion is not perfectly defined and includes safety issues, and technical capabilities, as well as social and economic considerations. Decision analysis tools were used to study alternative clean up processes based on expected monetary benefits and societal costs.

The different cleanup regimens studied range from one extreme in which only currently used technologies are employed, to the other extreme which assumes availability of all technologies under development. For each scenario, numerous levels of waste removal are considered, ranging up to 99.9% cleanup of the known initial radioactivity measured.

For each combination of technology and radiation removal level, the analysts develop projections of benefits and costs. Societal benefits are quantified by estimating the monetary value of an avoided radiation dose plus the value of not having to undertake construction of additional containers. Costs include operating expenses for the vitrification process, tank cleanup, and technology deployment. The decision model includes a time factor that addresses the time it takes to clean a tank, which would be important in case key equipment failures caused interruptions or delays at critical times during the cleaning process.

Sensitivity analysis was applied to determine the robustness of the projections and to reveal just which of several uncertainties are the ones that most critically affect the estimated outcomes. Results of this study are being examined by the U.S. Nuclear Regulatory Commission as it works toward establishing standards and requirements for nuclear waste management.

9.7.4 Investment Decisions and Risk in Petroleum Exploration (Walls et al. 1995)

The exploration division of Phillips Petroleum Company must routinely evaluate a broad range of exploration investments, determine an appropriate level of participation in each project available to the firm, and select the most advantageous mix of investments consistent with the division's budget. Petroleum exploration is an industry characterized by financial risk and uncertainty. There are often investment opportunities with high probabilities of small losses, and others with small probabilities of ruinous losses, not reflected in expected values. The expected value concept that had guided Phillips investment decisions in the past did not adequately address how sensitive managers are to exposure to the chance of substantial capital losses. There is a general perception in the petroleum industry that this exposure can be dealt with by entering into smaller capital allocations in more *different* projects, thereby *spreading the risk*. Yet Phillips Exploration had no formal way to quantify the value of such diversification. Their methods for controlling risk were often informal, and based strongly on the intuition of individual managers.

Attitudes toward risk interfered with traditional decision-making processes because managers at Phillips needed to look beyond expected values and consider downside risks as an integral up-front part of the investment picture. Management had evidently never realized how strongly risk averse they were (and in fact needed to be), and how poorly their decision-making framework had supported this position on risk.

A software package was developed to assist management in the process of deciding how to allocate investment capital across a set of possible exploration projects. Using some of the standard tools of decision analysis, this software not only provided a means of organizing the data associated with each investment opportunity, but it also offered a way of incorporating the company's attitudes toward risk and allowing decisions to reflect these attitudes.

The new decision software package met several of the company's needs. One requirement was to have a relatively consistent measure of risk to be used over the entire range of investment alternatives. Management needed to be able to compare the risk and upside potential of two projects; for example, one with an unlikely but large payoff versus one with a highly probable lower payoff, both of which may have equal expected values. The methodology incorporated into the software package facilitated this comparison between alternatives.

The package also allowed Phillips Exploration division to determine the optimal *level* of participation in each of many diverse projects having a desirable mix of risk characteristics. There are typically more investment opportunities than can be afforded with the scarce investment capital available; so rather than merely choosing projects to invest in, the company must also allocate and balance its investment capital.

In the exploration business, a *prospect* is a geological structure thought (or known) to contain petroleum potential, and a *play* is a collection of geologically similar prospects located in the same geographic locale. The decision software package assumes probabilistic independence among individual projects. But because prospects within the same play have, by definition, similar physical characteristics, they may not be independent at all. To deal with this interdependence, the package allows users to specify whether they wish to evaluate investment projects on a prospect basis or on a play basis.

Each new investment opportunity presents new alternatives to consider. And over time, there emerge decision patterns of which no one is really consciously aware. The decision support package measures the firm's risk tolerance by reviewing past decisions and encoding this information as a utility function. In so doing, the package thus captures the user's subjective (and perhaps unrecognized) perceptions about probabilities and risks associated with specific exploration outcomes. By creating a historical *risk personality* for the decision-maker, the system provides an integrated capability for ensuring a consistent risk attitude in evaluating and ranking projects for capital investment and determining participation levels in different prospects or plays that are consistent with attitudes toward risk.

This software package does not require the user to have any specialized technical knowledge of decision analysis, risk profiles, or utility theory. Instead, the user selects from several input formats and enters available data; then the software interprets the input and constructs a decision tree.

User reaction to this decision support system has been mixed. Management has displayed some initial reluctance to accept the utility functions generated by the software. And although the users acknowledge that they are not risk-neutral, there remains some hesitation on their part to quantify their risk aversion. Nevertheless, the use of this tool has raised awareness of the issues of risk tolerance and the importance of its role in capital investment allocation. Phillips has used this package to support companywide analysis of all exploration projects. This same software is also used by several other petroleum exploration firms, both to assist with small-scale individual decision-making and for comprehensive organizational decisions.

9.8 Summary

Decision analysis involves aspects of both mathematics and psychology. Because of the uncertainty that often surrounds decision-making, it is important to analyze the decision process as objectively as possible, and yet to realize the important role played by the human psyche.

Human attitudes toward risk and uncertainty often interfere with rational decision-making. Strategies in game theory help to identify and explain these attitudes, and several principles have been proposed that attempt to characterize human perspectives on risk. Utility theory gives us a mechanism for quantifying human attitudes toward risk.

Decision trees provide a framework for representing sequential decisions in which there is a response or some type of feedback at every stage in the decision process. Through the use of probabilistic information, optimal strategies can be identified and evaluated, using such measures as expected monetary value.

Decision-makers are prone to a variety of misconceptions and idiosyncratic behavior that can severely limit their ability to make rational choices. The availability of information can influence people in surprising ways. People are often unwilling to modify their decisions even when additional relevant information or evidence becomes available to them; or they may feel trapped by earlier decisions. Proper training and education can often help analysts develop an awareness of the psychological difficulties associated with decisions. Such an awareness, along with an understanding of the quantitative methods that are available to facilitate decision-making, can encourage and foster more rational approaches toward dealing with decisions.

Key Terms

Allais Paradox
adjustment
anchoring

automatic correction
availability
Bayes rule
certain monetary equivalent
chance fork
decision fork
decision simulator
decision strategy
decision tree
decision variables
dissonance
dissonance reduction
dominance
dominate
entrapment
expected monetary value
expected value of perfect information
feedback
folding back
framing effect
gambler's fallacy
gate
Hurwicz measure
Hurwicz principle
Laplace principle
lottery
maximax strategy
maximin strategy
outcomes
payoff matrix
payoffs
preference function
principle of insufficient reason
redundant questions
risk averse
risk seeking
row linearity
Savage Minimax regret
state variables
states of nature
sunk cost fallacy
toll
utility curve
utility function
utility theory

Exercises

9.1 Imagine that you are the contestant in the game show described in Section 9.5.1, choosing a door in hopes of getting the grand prize. If you were allowed to repeat the game 30 times, you would expect to pick the right door 10 times. And if you always switch when given the option, you should be right 20 times. Write a computer program to simulate this process.

9.2 An enterprising computer science student plans to provide computing services for clients, and is considering several alternatives. He can work all the problems given him by hand, which will cost him nothing; but he estimates his income in this case will only be $20,000 annually. He can buy an unknown brand desktop computer for $2,500. There is a 90% probability that this machine will be software compatible as advertised, but there is a 10% chance that our entrepreneur will have to spend $6,000 on software modifications to achieve a working system. In any case, he figures his income with this machine will be $100,000. His third alternative is to purchase a famous brand workstation computer that is certain to run the necessary software, and this system will cost $3,600 to purchase. With this system, he gets a maintenance contract but there is a 70% probability that hardware modifications and repairs will still cost him $1,000. His projected income from this system is $120,000. Draw a decision tree, and determine the course of action that yields the greatest expected net income for the entrepreneur.

9.3 A marketing strategist at the Complete Feet Shoe Company must decide whether to introduce a new product. At most, one type of new product will be introduced, either:

Product A (shearling lined vinyl thongs).

Product B (velcro closure ankle mufflers).

Product C (truck tread knee-highs).

If no new product is introduced, the company's public relations officer figures that the damage to the company's image as a dependable supplier of trendy footwear can be estimated at a value of $100,000. The cost of advertising any new product will be $150,000. Analysts predict the following probabilities of sales:

Product	Probability	Sales
A	0.80	$180,000
	0.10	$40,000
	0.10	$20,000
B	0.50	$100,000
	0.50	$200,000
C	0.60	$120.000
	0.40	$100,000

If product A is introduced, there is a 50–50 chance that the *Save-the-Sheep* Society will launch a smear campaign that will cause damages of $60,000 to the shoe company. If product B is introduced, there is a 50–50 chance that the inventor of velcro will sue the shoe company for *misapplication of technology* and such a lawsuit would cost the company $50,000. If product C is introduced, nobody will likely object. Draw the decision tree to display all of these options and the expected effects. Indicate what course of action should be recommended by the marketing strategist on the basis of the information given here, and state the expected loss or gain for your recommendation.

9.4   A long range planning committee is considering proposing that a new building be built on the campus of a university. The construction cost for the new building will be $30,000,000. If the new building is built, there is a 25% probability that publicity associated with the new facility will cause increased enrollment, which will result in $2,500,000 in revenues for the university. If the new building is not built, there is a 75% probability that some students will choose to attend another university, resulting in $10,000,000 in lost revenues. Even if there is no loss in enrollment, the overcrowded conditions will be such that there is a 50% chance of faculty rebellion, which can be quieted only by increased employee benefits, costing the university $2,000,000. Draw the decision tree to display all these options and the expected effects. Indicate what course of action should be taken, and state the expected loss or gain from this decision.

9.5   Recall from Section 9.5.1 the experiment involving selecting colored marbles from a large opaque jar. Both subjects conclude, based on different experiences drawing marbles from their jar, that two-thirds of the marbles in their jar are red. Prove, in both cases, just what is the probability that two-thirds of the marbles are red.

9.6   Suppose that your Operations Research mid-term exam will consist of one question worth 10 points, and you have only three hours to study for it. You are told that the instructor will pick the question from one of three possible topics: decision trees (D), utility theory (U), or game theory (G). If you spend your three hours studying one topic, and that question occurs on the exam, you will probably get 10 out of 10 points. For two hours studying, you expect to get 8; for one hour, you would get 5; and if you do not study the correct topic at all, you will get 2 points. By taking a quick look at past exams, you discover the following frequency of each topic:

Topic	Number of Times
Decision trees (D)	8
Utility theory (U)	7
Game theory (G)	5

a. Use decision tree analysis to determine your best study strategy. How many hours should you spend on each topic, and what is your expected grade on the exam?

b. Your friend Steve says he has inside information that he will sell to you for $5. Steve's hot tips have not been very accurate in the past and you estimate the conditional probability of his information being correct as follows:

		Given: Actual Question		
		D	U	G
Steve	D	0.8	0.2	0.3
says	U	0.1	0.7	0.2
	G	0.1	0.1	0.5

You decide that you need the points, so you pay him and he tells you that the exam question will be a game theory question. How does this influence your study strategy and what is your new expected grade on the exam?

c. Suppose that you find the idea of failing your exam particularly unattractive, so you decide to do an analysis of your utility for points:

   i. You would consider a grade of 5 to be the same as a 50–50 chance between getting 2 or 10.

   ii. You are indifferent between a grade of 4 for certain and a 50–50 chance of either 2 or 5.

   iii. You are indifferent between a grade of 7 for sure and a 50–50 chance of either 5 or 10.

Based on this information, how would you reevaluate your decision in part (a)? Forgetting about Steve for now; what is your optimal strategy and what is your expected utility?

9.7 Suppose that you are in the position of having to buy a used car, and you have narrowed down your choices to two possible models: one car is a private sale and the other is from a dealer. You must now choose between them. The cars are similar, and the only criterion is to minimize expected cost. The dealer car is more expensive but it comes with a one year warranty. You decide that if the car will last for one year, you can sell it again and recover a large part of your investment. If it falls apart, it will not be worth fixing. After test driving both cars and checking for obvious flaws, you make the following evaluation of probable resale value:

Car	Purchase Price	Probability of Lasting One Year	Estimated Resale Price
A: Private	$800	0.3	$600
B: Dealer	$1,500	0.9	$1,000

Which car would you buy? What is the value of perfect information?

Suppose you have the opportunity to take car A to an independent mechanic, who will charge you $50 to do a complete inspection and offer you an opinion as to whether the car will last one year. For various subjective reasons, you assign the following probabilities to the accuracy of the mechanic's opinion:

	The Mechanic Will Say:	
Given:	Yes	No
A car that will last one year	70%	30%
A car that will not last one year	10%	90%

Assuming that you must buy one of these two cars, formulate this problem as a decision tree problem. What is the true value of the mechanic's advice? Is it worth asking for the mechanic's opinion? What is your optimal decision strategy? (*Note:* It is not necessary to ask for advice on car B because its problems could be repaired under the warranty.)

9.8   Give two examples of the *framing effect*.

9.9   Consider the following payoff matrix:

	Actions					
States	a_1	a_2	a_3	a_4	a_5	a_6
Θ_1	2	6	4	4	5	7
Θ_2	8	2	5	2	4	2
Θ_3	0	5	2	4	3	3
Θ_4	3	5	2	5	3	2

a.   Suppose that the decision-maker claims *complete ignorance* of the probabilities of occurrence of the four states. Can any alternatives be eliminated? What is the optimal action under each of the strategies: Laplace, Maximin, Maximax, Savage Minimax regret? What types of decision-makers should use each of these strategies?

b.   Under the Hurwicz principle, the decision-maker is assumed to have some level of optimism α between 0 and 1. Characterize the optimal decision for the range of all possible values of α. At what values of α does the optimal solution change?

9.10   The product manager of a large firm is faced with the decision of whether to proceed with a national marketing campaign for a new product. The monetary return from sales generated by the campaign will depend on prevailing market conditions. The manager believes there is a 40% chance of *good* market conditions and a 60% chance of *bad* conditions. The monetary returns (in thousands of dollars) for each condition are summarized in the following:

	Good Conditions	Bad Conditions
Market	$800	−$400
Do nothing	$0	$0

The manager may decide to purchase the services of a marketing firm that will do a survey for $75,000. The firm claims that their results are 75% reliable. (That is, when conditions are *good*, they correctly identify it 75% of the time, and similarly for *bad* conditions.)

a. The manager must decide whether to accept the survey and whether to proceed with national marketing. Construct the corresponding decision tree and compute the optimal strategy and expected payoff.

b. What is the *expected value of perfect information* for this problem? How do you interpret this value?

9.11 Consider the following apparent paradox.

a. The average person is risk averse at all levels of money.

b. The average person will insure his house for $5 per week, which is risk averse because the insurance company is making a profit.

c. The average person may buy a lottery ticket for $5 per week, which is a gamble.

If we let X be the insured value of the house and Y be the prize in the lottery, then the two situations can be described as shown in Figure 9.19. Let p and q denote the small probabilities of losing the house and winning the lottery, respectively. The outcomes depicted are expressed in terms of net change to assets in a given week. Is this *normal* or *average* person irrational? Can this behavior be described by a reasonable utility function? Discuss the possible motivations or perspectives of this person.

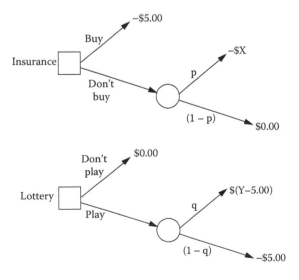

FIGURE 9.19
Apparent paradox.

References and Suggested Readings

Allais, M. 1953. Le comportement de l'homme rationnel devant le risque; critique des postulats et axiomes de l'école americaine. *Econometrica* 21: 503–546.

Arkes, H. R., and C. Blumer. 1985. The psychology of sunk cost. *Organizational Behavior and Human Decision Processes* 35 (1): 124–140.

Bunn, D. W. 1984. *Applied Decision Analysis.* New York: McGraw-Hill.

Byrnes, W. G. 1973. *Decision Strategies and New Ventures: Modern Tools for Top Management.* London, UK: George, Allen and Unwin Ltd.

Clemen, R., and T. Reilly. 1996. *Making Hard Decisions: An Introduction to Decision Analysis,* 2nd ed. Pacific Grove, CA: Duxbury Press.

Corner, J. L., and C. W. Kirkwood. 1991. Decision analysis applications in the operations research literature. *Operations Research* 39 (2): 206–218.

Davis, M. 1983. *Game Theory: An Introduction.* New York: Basic Books.

Davis, M. 1986. *The Art of Decision-Making.* New York: Springer-Verlag.

Dawes, R. M. 1982. The robust beauty of improper linear models in decision-making. In D. Kahneman, P. Slovic, and A. Tversky (Eds.), *Judgement Under Uncertainty: Heuristics and Biases.* Cambridge, UK: Cambridge University Press.

Diz, G., L. F. Scavarda, R. Rocha, and S. Hamacher. 2014. Decision support system for Petrobras ship scheduling. *Interfaces* 44 (6): 555–566.

Eapen, G. 2009. *Decision Options: The Art and Science of Making Decisions.* Boca Raton, FL: CRC Press.

Eden, C., and J. Harris. 1975. *Management Decisions and Decision Analysis.* New York: John Wiley & Sons.

Ehrgott, M. (Ed.). 1999. *Decision-Making Using Optimization Software.* Aachen, Germany: Shaker Verlag.

House, W. C. 1983. *Decision Support Systems: A Data-based, Model-Oriented, User-Developed Discipline.* New York: Petrocelli Books.

Howard, R. A. 1988. Decision analysis: Practice and promise. *Management Science* 34 (6): 679–695.

Ignizio, J. P., and J. N. D. Gupta. 1975. *Operations Research in Decision-Making.* New York: Crane, Russak and Company.

Kahneman, D. 2013. *Thinking, Fast and Slow.* Toronto, Canada: Anchor.

Kahneman, D., and A. Tversky. 1979. Prospect theory: An analysis of decision under risk. *Econometrica* 47 (2): 263–291.

Keefer, D. L. 1995. Facilities evaluation under uncertainty: Pricing a refinery. *Interfaces* 25 (6): 57–66.

Keen, P. G. W., and M. S. S. Morton. 1978. *Decision Support Systems: An Organizational Perspective.* Reading, MA: Addison-Wesley.

Keeney, R. L., and H. Raiffa. 1976. *Decisions with Multiple Objectives: Preferences and Value Tradeoffs.* New York: John Wiley & Sons.

Lembersky, M. L., and U. H. Chi. 1986. Weyerhauser decision simulator improves timber profits. *Interfaces* 16 (1): 6–15.

Marakas, G. M. 1999. *Decision Support Systems in the Twenty First Century.* Upper Saddle River, NJ: Prentice-Hall.

Maxwell, D. T. 2006. Software survey: Decision analysis. *OR/MS Today* 33 (6): 51–61.

Mendelson, E. 2005. *Introducing Game Theory and Its Applications.* Boca Raton, FL: Chapman and Hall/CRC Press.

Merkhofer, M. W. 1987. Quantifying judgmental uncertainty: Methodology, experiences, and insights. *IEEE Transactions on Systems, Man, and Cybernetics* SMC-17 (5): 741–752.

Meyerson, R. 1991. *Game Theory: Analysis of Conflict.* Cambridge, MA: Harvard University Press.

O'Keefe, R. M., and T. McEachern. 1998. Web-based customer decision support systems. *Communications ACM* 41 (3): 71–78.

Oleson, S. 2016. Decision analysis software survey. *OR/MS Today* 43 (5): 36–45.

Ozcan, Y. A. 2017. *Analytics and Decision Support in Health Care Operations Management*. 3rd ed. San Fransico, CA: Wiley.

Perdue, R. K., and S. Kumar. 1999. Decision analysis of high-level radioactive waste clean-up end points at the West Valley Demonstration Project waste tank farm. *Interfaces* 29 (4): 96–98.

Pratt, J., H. Raiffa, and R. Schlaiffer. 1965. *Introduction to Statistical Decision Theory*. New York: McGraw-Hill.

Raiffa, H. 1968. *Decision Analysis: Introductory Lectures on Choices under Uncertainty*. Reading, MA: Addison-Wesley.

Saaty, T. L. 1980. *The Analytic Hierarchy Process: Planning, Priority Setting, Resource Allocation*. New York: McGraw-Hill.

Saaty, T. L. 1994. *Fundamentals of Decision Making*. Pittsburgh, PA: RWS Publications.

Sauter, V. L. 1997. *Decision Support Systems*. New York: John Wiley & Sons.

Sauter, V. L. 1999. Intuitive decision-making. *Communications ACM* 42 (6): 109–115.

Sharda, R., and D. M. Steiger. 1996. Inductive model analysis systems: Enhancing model analysis in decision support systems. *Information Systems Research* 7 (3): 328–341.

Smith, J. E., and R. L. Winkler. 2006. The optimizer's curse: Skepticism and post-decision surprise in decision analysis. *Management Science* 52 (3): 311–322.

Tversky, A., and D. Kahneman. 1982. *Judgment under Uncertainty: Heuristics and Biases*. Cambridge, UK: Cambridge University Press.

Von Furstenberg, G. M. (Ed.). 1990. *Acting under Uncertainty: Multidisciplinary Conceptions*. Hingham, MA: Kluwer Academic.

Wagner, H. M. 1969. *Principles of Operations Research with Applications to Managerial Decisions*. Englewood Cliffs, NJ: Prentice-Hall.

Walls, M. R., G. T. Morahan, and J. S. Dyer. 1995. Decision analysis of exploration opportunities in the onshore U.S. at Phillips Petroleum Company. *Interfaces* 25 (6): 39–56.

Webb, J. N. 2007. *Game Theory: Decision, Interaction, and Evolution*. New York: Springer.

Winkler, R. L. 1990. Decision modeling and rational choice: AHP and utility theory. *Management Science* 36 (3): 247–248.

Winterfeldt, D., and W. Edwards. 1986. *Decision Analysis and Behavioral Research*. Cambridge, UK: Cambridge Press.

10

Heuristic and Metaheuristic Techniques for Optimization

Combinatorial optimization involves determining how best to arrange (or group, sequence, or assign) the controllable elements in large complex systems to achieve a specified objective or goal. Combinatorial optimization models have been used to describe problems as diverse as vehicle routing, workforce scheduling, manufacturing plant layout, portfolio selection, production scheduling, supply chain problems, aircraft scheduling, computer CPU job scheduling among many others. Combinatorial problems are ubiquitous, arising commonly in engineering, financial, industrial, computing, and social and human services applications.

Many combinatorial optimization problems are remarkably simple to state and intuitively easy to understand, requiring little mathematical sophistication. As an example, there is a famous problem popularly known as the **knapsack problem** in which a hiker considers which of n objects to pack into a knapsack. Each object has a weight and a value. The goal is to select a subset of the objects that have the greatest combined value and whose total weight does not exceed the capacity of the knapsack.

The knapsack model could be applied to as obvious a problem as packing suitcases for a trip without exceeding the baggage weight limitations imposed by airline regulations. Or the model could be used to select experiments and instrumentation packages to include in a deep space probe. Each candidate package has a potential value (technical payoff or social merit), but each package also requires certain resources such as electricity, cooling, oxygen for the mice, carbon dioxide for the soybean sprouts, space (volume) needs, weight, or waste disposal. For these requirements, one might imagine a multidimensional knapsack capacity which can supply only a limited amount of each of the resources (electricity, air, heat dissipation, space, and weight).

The simple knapsack problem can be formulated using n decision variables, x_i, where $x_i = 1$ if object i is to be included in the knapsack and $x_i = 0$ if not. Knapsack capacity is denoted as c. For each object, there is an associated weight w_i and a value v_i. Then, to select the most valuable feasible subset of objects, it is necessary to find the values of the variables to

$$\text{maximize} \quad \sum_{i=1}^{n} v_i x_i$$

$$\text{subject to} \quad \sum_{i=1}^{n} w_i x_i \leq c$$

$$x_i = 1 \text{ or } 0$$

Another famous combinatorial optimization problem, known as the **traveling salesman problem**, seeks to find the least costly route for a salesman who must visit n cities each exactly once, returning finally to his city of origin. Assume that the distances between cities are recorded in an n × n matrix D, where d_{ij} is the distance (or cost) to travel from city i to city j. Let decision variables $x_{ij} = 1$ if the route contains the road from i to j, and $x_{ij} = 0$ if not. If the salesman enters and leaves every city exactly once, it appears that his tour would be a feasible one, and the optimal tour can be determined by finding values of the variables to:

$$\text{minimize} \qquad \sum_{i=1}^{n}\sum_{j=1}^{n}d_{ij}x_{ij}$$

$$\text{subject to} \qquad \sum_{i=1}^{n}x_{ij} = 1 \qquad \text{for every } j, \text{ salesman enters city } j \text{ exactly once}$$

$$\sum_{j=1}^{n}x_{ij} = 1 \qquad \text{for every } i, \text{ salesman leaves city } i \text{ exactly once}$$

$$x_{ij} = 1 \text{ or } 0$$

At first glance, this familiar formulation (which is precisely that of the assignment problem discussed in Section 3.3.2) might tempt us to try to solve the traveling salesman problem using the Hungarian method for the assignment problem. Indeed, if the solution found by the Hungarian method really were a feasible tour, then it would be an optimal tour for the salesman. However, the solution obtained in this way may fail to represent the kind of tour needed by the salesman, although he enters and leaves each city exactly once. For example, suppose the salesman begins at city 1, and must visit cities 2, 3, and 4 in any order, and finally return to city 1. Then, all of the following tours are feasible:

$$1 - 2 - 3 - 4 - 1$$
$$1 - 2 - 4 - 3 - 1$$
$$1 - 3 - 2 - 4 - 1$$
$$1 - 3 - 4 - 2 - 1$$
$$1 - 4 - 2 - 3 - 1$$
$$1 - 4 - 3 - 2 - 1$$

However, notice that the constraints written in the previous formulation would permit decision variable values that describe not only those six feasible tours but also *sub-tours* (round trips that do not visit every city) such as:

$$1 - 2 - 1 \text{ and } 3 - 4 - 3$$
$$1 - 3 - 1 \text{ and } 4 - 2 - 4$$
$$1 - 4 - 1 \text{ and } 2 - 3 - 2$$

These latter solutions do not meet the salesman's requirements. Thus, the Hungarian Method cannot be relied upon to yield feasible traveling salesman problem solutions; additional constraints must be imposed in our formulation so that subtours are excluded.

As simple and easy to understand as these two famous combinatorial problems are, it is surprising that no efficient algorithms have been developed that are guaranteed to find optimal feasible solutions. In fact, both the hard problem and the traveling salesman problem belong to the set of NP-hard problems, and in that set, they are in good company with hundreds of other important practical problems.

For many practical problems in science, engineering, and management, the only way to be sure of finding an optimal solution is to search completely through the whole set of possible solutions. If there are infinitely many possible solutions, we know right away that this approach is unsatisfactory. But if there is a very large but *finite* number of possible solutions, the idea of a complete search is tempting, and often is quite easy to express as an algorithm and to implement in software. The difficulty is of course that the *time required to carry out* such an exhaustive search is, although finite, far greater than most mortals can afford. (Look again at Table 1.1 in Chapter 1 to be reminded just how many centuries such a computation might take. Clearly, technological advances, such as increasing CPU chip speed by several orders of magnitude, do not provide adequate computational tools against these formidable computational demands.)

The question then is to try to find short cuts that will allow us to organize the search process so that it is no longer a complete search over *all* possible solutions, but rather it becomes an affordable search that is *likely* to discover a good, or near-optimal, solution. Such methods are called **heuristic methods**. They are most often applied to the computationally intractable NP problems, simply because otherwise the best (most efficient) methods we know of for solving these problems exactly (or optimally) can take an exponential amount of computation time. Heuristic methods are usually rather problem specific, and often are based on simple common-sense ideas inspired by, or tailored to, the type of problem being solved. They are, however, vulnerable to falling into local optima (i.e., suboptimal solutions). As a result, metaheuristics emerged as more intelligent search techniques that can help heuristics escape such solutions. This chapter examines some heuristic and metaheuristic methods that are currently popular, effective, and practical.

10.1 Greedy Heuristics

Greedy heuristics are probably the simplest type of heuristics in which a partial solution is constructed step by step towards a complete solution based on basic known information of a problem instance. This can be accomplished by adding elements based on certain attribute(s) and in some cases based on the best contribution an element makes to the objective function at the point at which the element is selected. They also must make sure that the constraints on the problems are not violated. In their most basic form, greedy heuristics do not account for long term consequences of the decision made, but they rather consider the immediate impact in the short term; hence the term used is *greedy*.

As an example, consider again the traveling salesman problem discussed earlier. A greedy way of constructing a complete tour is to select the *closest unvisited city next* until all cities are visited. Suppose the distance matrix for a four city example is as given in Table 10.1. A greedy solution using this heuristic would be 1–3–2–4–1 with a total distance of 52.

TABLE 10.1

Traveling Salesman Problem Example

City	1	2	3	4
1		20	4	25
2	20		8	15
3	4	8		11
4	25	15	11	

Note that even for a small example like this, the greedy heuristic obtained a reasonable solution; yet it is suboptimal as the optimal tour for this instance is 1–2–4–3–1 with a total distance of 50.

It is very common to use greedy heuristics as a starting solution followed by a local search heuristic in which a better solution is sought by iteratively attempting to improve the greedy solution. This concept is discussed next.

10.2 Local Improvement Heuristics

Local iterative improvement techniques begin by placing the system being optimized in a known configuration; usually, any simple to obtain greedy or arbitrary configuration will do. Then some simple rearrangement or reorganization of the problem elements is performed repeatedly to various local parts of the system until a configuration is discovered whose objective function value is better than that of the previous configuration. When this occurs, the better configuration becomes the *current* configuration, and the process is repeated until no better configuration can be found by means of simple local rearrangements.

Because at each iteration, only *simple* changes involving neighboring elements are considered, the method is often referred to as a **local search** procedure. From any given configuration, only *nearby* configurations are considered, that is, configurations that differ from the current one by minor modification to the problem elements (variables). As might be guessed from this, local search heuristics can easily, and typically do, get stuck in (or converge to) a local but not global optimum. Therefore it is customary, and not terribly time-consuming, to carry out the entire procedure several times, beginning each time with a different arbitrarily chosen initial configuration. Having repeated the process many times and therefore likely having found many *different solutions*, the problem-solver would use the best result that was ever discovered during any of the searches.

To illustrate the kinds of rearrangements of problem elements that have been found to be effective, we look at a few classical combinatorial optimization problems. In the traveling salesman problem, a solution is any sequence of cities that includes each city exactly once, in the order visited. A very effective local improvement mechanism for generating a new configuration, known as 2-opt, is to select a pair of (directed) edges (i,j) and (m,n) and replace them with the *crossing* edges, (i,m) and (j,n) in such a way that the result is a new tour. For example, the tour shown on the left in Figure 10.1 is represented by the sequence 1–2–3–4–5–6–1. If we select the edges $(1,2)$ and $(4,5)$, the resulting sequence 1–4–3–2–5–6–1 represents the tour shown on the right in the figure. The quality of these two tours could

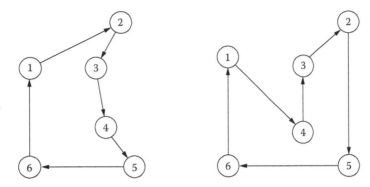

FIGURE 10.1
Sub-sequence reversal.

be compared, and the better one selected by the heuristic. Clearly, if the length of the two edges added is less than the two removed, then the new tour is an improvement. This simple method was introduced by Lin and Kernighan (1973), and extended as a k-opt method. Remove any k edges and replace them with (the best possible) crossing edges. It is easy to implement, and usually executes in a very reasonable and affordable amount of computation time. The 2-opt requires $O(n^2)$ operations each iteration, and 3-opt requires $O(n^3)$. Solutions for k = 3 are typically very good for practical applications.

Many combinatorial problems can be described as *placement* problems. For example, the placement of electrical components on a circuit board can be designed with the goal of minimizing the length of wiring required. The placement of equipment in a manufacturing plant would likely be done to facilitate the flow of manufactured products through the various pieces of equipment. Or the placement of data files in a computer network might be based on the amount of memory space available at the various workstations as well as the cost of transmitting files from one workstation to another. Also, another application is the placement of facilities in different locations in supply chains to optimize some objective(s) such as minimizing the transportation cost. In any of these applications, a **local improvement heuristic** would begin with any arbitrary feasible placement of the elements, then repeatedly consider the effects of exchanging any two elements: any two electrical components, any two manufacturing machines, any two files or two facilities. These are often called **local exchange heuristics**.

A minor modification to the exchange or *swap* idea is to arbitrarily select three objects and consider various ways to move, shift, or rotate the three objects around to different places in the system, continuing until no advantageous local rearrangement can be found. This approach belongs to a class of methods that have been termed *k-opt* heuristics (in this case, k = 3). These methods have been shown to give somewhat better results than just moving two objects at a time (Carter and Price 1988), and they do not take appreciably more computation time than simple swaps or exchanges.

Local iterative improvement heuristics are generally conceptually simple, easy to program, efficient to execute, and give reasonably good results. However, like greedy heuristics, they are susceptible to reaching suboptimal solutions as they search within their local search region. Therefore, another class of search methods known as **metaheuristics** is used to guide heuristics out of local optima. Among the most common methods are **simulated annealing, genetic algorithms**, and **tabu search**, which are discussed later in this chapter. Figure 10.2 demonstrates the concepts of local versus global minima where a greedy and local search algorithms may reach a local minimum and stop there as they

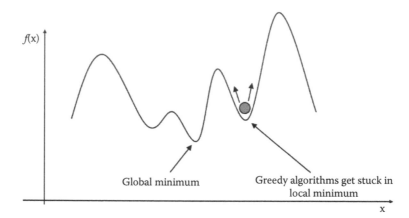

FIGURE 10.2
Escaping local optima.

cannot see any further improvement beyond their local neighborhood; while a meta-heuristic can guide them out by occasionally accepting worse solutions in the hope of escaping the local optimum.

10.3 Simulated Annealing

Simple local improvement heuristic techniques for optimization typically suffer from a tendency to converge to a local optimum that may not be a global optimum. This phenomenon is a natural consequence of a computational process that moves monotonically in an improving direction, from an arbitrary starting point, as was discussed in Chapter 5 on nonlinear optimization. Simulated annealing is a local improvement mechanism with a probabilistic twist, in which *non-improving* moves are occasionally made, and therefore offers chances to escape from local optima, in the hope of arriving at a global optimum. This metaheuristic is based loosely on concepts from thermodynamics, which deal with how a liquid substance is slowly cooled into a solid to produce a stronger, more stable (less brittle) final product. The use of simulated annealing as an optimization tool is due to the work of several researchers who were actually working in different disciplines at different times.

In the field of statistical mechanics, methods were developed in the 1950s to model the evolution of a physical system through a series of slowly decreasing temperatures (an *annealing* process) into a state of high order and low energy. During the annealing process, the temperature is reduced slowly to maintain system equilibrium with respect to temperature. Both positive and negative energy fluctuations are allowed, in contrast to a rapid quenching that would result in a disordered or unstable system.

About 30 years later, researchers interested in mathematical optimization had the breadth of scope and keen insight to perceive an analogy between the behavior of a physical substance in low energy states and the nature of the iterative improvement that can be made in a large and complex mathematical system that is in a nearly optimal configuration. States of low energy in the physical system are viewed as being analogous to the nearly optimum configuration (as measured by a very low objective function value) in a minimization process.

The analogy with combinatorial optimization is really just a variation on conventional iterative improvement methods that begin with an initial feasible solution, repeatedly generate and consider changes in the current configuration, and accept only those that improve the objective function. To avoid the characteristic convergence to a local optimum that typifies deterministic local heuristic methods, simulated annealing methods probabilistically accept configurations that temporarily deteriorate the quality of the system being optimized. An acceptance probability is computed, based on the change in the objective function and a *temperature* parameter. As the temperature is appropriately reduced (this is called an *annealing schedule* or a *cooling schedule*), fewer non-improving moves are accepted; thus, a coarse global search evolves into a fine local search for optimality, and the probabilistic *jumps* provide avenues to avoid sinking into non-global optima.

Let us now look more carefully at simulated annealing as it applies to statistical mechanics, and then we will investigate more precisely how to make use of the analogy to combinatorial optimization. All physical systems are composed of large numbers of atoms, and only the most probable behavior of the system is observed when the system is in thermal equilibrium at a constant temperature. This behavior is characterized by the average small fluctuations of the atoms or molecules about their mean positions within the substance. To observe different behaviors of a substance (or system), atoms are allowed to change their atomic positions by altering the temperature and then letting the system attain thermal equilibrium again. The most stable state of a system is the state associated with the lowest energy level. Under the assumption that atoms with configurations close to ground states dominate the properties of the system at low temperature, the temperature of the system is lowered in search of the ground state.

The process of lowering temperature slowly so that thermal equilibrium is always maintained is called an **annealing process**. A mathematical model has been developed to describe a system in a stable state, that is, the most probable state with respect to temperature. Each possible configuration of the system is defined by the Boltzmann probability factor

$$P(r_i) = e^{-\left(\frac{E(r_i)}{kT}\right)}$$

where each configuration r_i belongs to the set of all possible atomic configurations, and

   $P(r_i)$ is the probability of a configuration r_i
   $E(r_i)$ is the energy (in joules) of the system in configuration r_i
   k is the Boltzmann constant (in joules per degree Kelvin)
   T is the temperature in degrees Kelvin

As is shown by the nature of the curve in Figure 10.3, when the temperature approaches a very low value, the probability of the occurrence of a new configuration approaches zero because the system is already in a nearly stable state. At low temperatures (i.e., when the system is in either liquid or solid state), the exponent becomes very large and negative, and hence $P(r_i)$ approaches zero. On the other hand, at higher temperatures, there is more atomic movement within the substance, hence more different configurations occur, and therefore the probability of *occurrence* of any given r_i becomes greater as temperature increases.

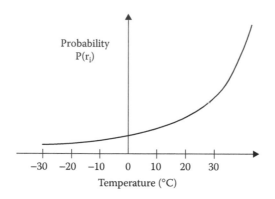

FIGURE 10.3
Boltzmann probability.

In the 1950s Metropolis et al. (1953) developed an algorithm known as **simulated anneal-ing,** which is used as a computational tool for efficiently simulating a collection of atoms in equilibrium at a given temperature. In each step of the algorithm, an atom is hypotheti-cally given a small displacement. Before a displacement is admitted, initial energy E_i of the system is noted, and final energy E_f is measured after the displacement. The difference between these two energy states is calculated as

$$\Delta E = E_f - E_i$$

If E_f is less than E_i, then the system has moved from a high energy level (state) to one at a lower energy level that is more stable than the previous one, and hence this displacement is *accepted*. (The system now assumes this new configuration.) In short, the new configura-tion is unconditionally accepted when $\Delta E \leq 0$.

But if ΔE is positive, the new configuration *may* be rejected and the current (more stable) configuration maintained. The acceptance criterion is based on the Boltzmann distribu-tion, thus, the probability that the new configuration is accepted is

$$P(\Delta E) = e^{-\left(\frac{\Delta E}{kT}\right)}$$

At any given temperature, the simulation must continue long enough for the system to reach a steady state. In other words, at a given temperature, the system at equilibrium is characterized by a certain distribution of configurations, and the precise distribution emerges as the simulation takes place.

In case it seems that we have wandered afar from the business of combinatorial optimi-zation, let us now restate the simulated annealing procedure, using terminology that is applicable to optimization, as Kirkpatrick et al. (1983) so ingeniously did in the early 1980s. In this context, a *configuration* means some assignment of values to the decision variables. The temperature is indicated by a simpler parameter which we will call θ (theta), because physical temperature has no absolute meaning in the optimization scenario. We will gener-ate a sequence of classes of configurations. Within each class, a parameter θ determines the magnitude of objective function value fluctuations that occur within that class. Each class is asymptotically distributed as a Boltzmann distribution, and the process of determining this distribution for any given value of θ is called **equilibration**. The optimization process is actually comprised of a *series* of equilibrations; each equilibration is associated with a temperature parameter θ, and equilibrations are done at successively lower temperatures.

Each equilibration begins with the system in some initial configuration, and carries out the following process until a stable distribution of configurations has been generated:

Generate a new configuration arbitrarily.

Calculate ΔF = new objective function value—current objective function value.

If $\Delta F \leq 0$, then accept new configuration unconditionally.

If $\Delta F > 0$, then accept new configuration only with probability $e^{-(\Delta F/\theta)}$.

When it is reasonably obvious that further iterations of this process will have no significant effect on the distribution of configurations, the equilibration at the current temperature is complete. The most frequently occurring configuration is chosen as the initial configuration for the next equilibration process that will take place using a lower temperature parameter θ.

Equilibrations are carried out until it is observed that practically no configurations are being generated (and accepted) that have a better (lower) objective function value than the current configuration (i.e., until the *acceptance ratio* or *probability of acceptance* is essentially zero). At this point, the heuristic optimization process is complete, and the best configuration seen so far is taken as the result. The entire process is illustrated in Figure 10.4.

Thus, in the same way that physical substances are cooled in a controlled manner (perhaps to attain a crystalline structure instead of an amorphous glass structure), so can combinatorial systems be first *stirred up* and then slowly sloshed around until they congeal into an orderly (perhaps nearly optimal) configuration having a low objective function value. Conceptually, the simulated annealing process can be presented as shown in Figure 10.5.

Simulated annealing is a technique that can be quite easy to implement. Specific details of an implementation often depend on the type of problem being solved.

- The annealing schedule (or cooling schedule) is usually determined by trial and error, or dynamically through real-time observation during the process itself. The practitioner must choose the initial value of the temperature parameter θ and the amount by which θ is to be decreased at each equilibration.

- It must be decided how to generate new random configurations, what decision variables to change, and whether to check feasibility of each new configuration. And if infeasible configurations are allowed, a means must be invented to measure the objective function (quality) of an infeasible configuration.

- How many new configurations should be generated and considered during each equilibration? It may be some fixed number of new configurations, or until the configurations that occur have appeared some specified number of times. Perhaps every entity (decision variable) should be changed, or at least have had a chance to be changed at least once. This issue has a strong impact on the computation time required for the simulated annealing process to execute.

- Implementation of the probabilistic decision of whether to accept a *bad* move is simple, and usually done in the following way. Generate a random number r in the interval (0, 1); if r is less than $e^{-(\Delta F/\theta)}$, then make the change; otherwise maintain the current configuration.

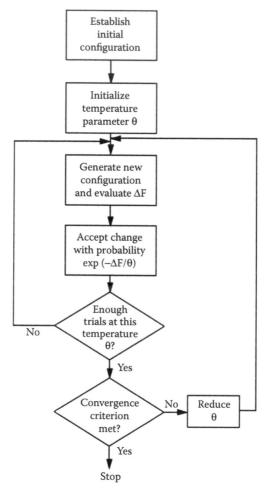

FIGURE 10.4
Simulated annealing algorithm.

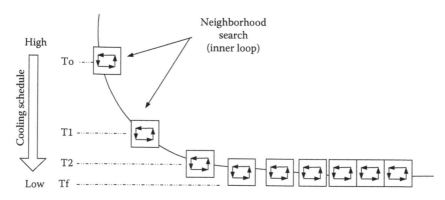

FIGURE 10.5
The simulated annealing process.

- At the end of each equilibration, some implementations choose the *best* configuration seen, rather than the *most frequently occurring* one, to use as the initial configuration in the next equilibration. Although this practice may seem to accelerate the convergence process, it can also be argued that it tends to drive the process more rapidly toward a non-global optimum.

Although there are some theoretical results that describe the performance and convergence properties of optimization by simulated annealing (Kirkpatrick 1984, Anily and Federgruen 1985, Lundy and Mees 1986), the most valuable guidelines for the analyst are gained through experience and observations of empirical results on specific application problems. Simulated annealing generally takes somewhat longer (more computation time) than simple local-improvement heuristics, but there is typically some *performance advantage* that results from the structured randomness of simulated annealing.

Example 10.3

Let's consider the **Capital Budgeting** problem in which we have a limited budget to execute projects from a set of possible projects as given in Table 10.2. Each project contributes a certain value v_i and the objective is to select the set of projects that will maximize the total value.

This problem is in essence a knapsack problem similar to what was discussed earlier in this chapter. Although this small instance can be modeled and solved as an integer program as in Section 4.3.2, we will use it to demonstrate how simulated annealing works. We first define x_i to be 1 if a project is selected and 0 otherwise. The energy function becomes the objective function and the only constraint is not to exceed the total budget of \$450 million.

Suppose that a neighbor (or a move) can be generated by switching one project selection from 0 to 1 or from 1 to 0. If we start with an initial solution S_0 based on a greedy rule that selects the project with the highest value first (without violating the budget constraint), S_0 would be represented by 01100100 with an objective function value of 950 and total expense of 437, which is feasible because it is below the maximum budget (see Table 10.3). If we randomly select a binary digit to switch from 0 to 1 or from 1 to 0 and it happened to be the second digit, then the new neighbor to evaluate would be 00100100 for which the total value would be 750.

The next step is to decide whether to move to this neighbor or not by computing the difference in the objective function ΔF, which is in this case a reduction of 200 in the total value. In local search, such a solution would be immediately rejected, but not in simulated annealing as we must first evaluate the probability of acceptance $Pa = e^{-(\Delta F / \theta)}$. Here, the temperature θ is assumed to be 180, and therefore $Pa = 0.329$. To decide whether to accept or reject this move, we generate a random number $r \sim U(0, 1)$, which

TABLE 10.2

Capital Budgeting Problem Example

Project	1	2	3	4	5	6	7	8	Budget
Expense (\$M)	50	92	144	22	67	201	88	112	450
Value v_i	120	200	300	84	150	450	180	220	

happened to be 0.251. Since r < Pa, we accept this move and S_1 becomes the current solution. In another iteration, the neighbor S_2 is generated by randomly selecting a digit to switch, which happened to be the first digit resulting in the neighbor 10100100 with F(s) = 870. Since this is an improvement of 120 over S_1, it is accepted without computing Pa, and S_2 becomes the current solution. Similarly, S_3 becomes the next neighbor to move to without checking Pa.

However, when S_4 is generated by unselecting project 6, the objective function value drops by 450 to 504, making the probability of accepting such a solution quite small (Pa = 0.082) and that move is rejected, which means the search algorithm stays at S_3 and another neighbor (S_5 in this example) is generated from S_3. Note that although S_5 also has a lower objective function, r happened to be less than Pa; hence this worse solution is accepted. Continuing this way, solutions S_6 and S_7 are generated with better objective function values. This process is summarized in Table 10.3.

It turns out that S_7 is actually optimal and using this neighborhood generation scheme (switching 0 to 1 or a 1 to 0), the simulated annealing could not have reached this solution from where it had started at S_0 without accepting worse solutions such as S_1 and S_5. It is important to recognize that the simulated annealing cannot tell that it has reached the optimal, but after running many more iterations, it should stop trying when no further improvement is achieved. Also, after attempting new neighbors at the same temperature θ with no improvement (or if the maximum number of iterations is reached), the temperature at iteration i is reduced (usually via a decay function such as $\theta_i = \alpha*\theta_{i-1}$) where $0 < \alpha < 1$. We did not do this in this basic example, but one can easily see that Pa depends on the amount of deterioration in the objective function and the current temperature where the larger the value of ΔF, and/or the lower the temperature, the smaller the probability to accept a new worse neighbor solution. Therefore, at higher temperatures, the simulated annealing accepts neighbors more frequently and as the temperature drops, it becomes more selective like a greedy algorithm.

It is noteworthy that the greedy rule used in this example is very simple and was used to demonstrate the simulated annealing process rather than solving the knapsack problem efficiently. A better rule would be to select the project with the maximum ratio of value to expense first until no more projects can be selected due to exceeding the allocated budget. If this rule were used instead, the optimal solution would have been found in one iteration. In fact, this rule is optimal if we allow the last unit selected to be fractional.

TABLE 10.3

Simulated Annealing Iterations with $\theta = 180$

Solution	Neighbor	Evaluation F(s)	Expense	Δ	r	Pa	Outcome
S0	01100100	950	437				
S1	00100100	750	395	−200	0.251	0.329	Accept
S2	10100100	870	345	120			Accept
S3	10110100	954	417	84			Accept
S4	**10110000**	504	216	−450	0.813	0.082	Reject
S5	10010100	654	273	−300	0.157	0.189	Accept
S6	11010100	854	365	350			Accept
S7	11011100	1004	432	350			Accept

10.4 Parallel Annealing

The good quality solutions obtained by simulated annealing heuristic methods are often paid for with substantial computational effort. Although the staged cooling regimen seems to be an inherently *sequential* process, recent research has been aimed at the development of models to reduce computation time through *parallel processing*.

In conventional simulated annealing, each new random configuration is typically generated by changing the value of *one* (or a very few) decision variables at a time. But imagine instead a multiple-processor computer in which there is a processing unit associated with every decision variable in the problem being solved. Then the processing units could independently and asynchronously consider changing the values of their individual associated decision variables, each applying a simulated annealing process to evaluate the merit of such a change.

As long as processing elements consider their changes only one at a time, asymptotic convergence to a global optimum is guaranteed (Aarts and Korst 1989). Unfortunately, processing units operating in parallel are basing their simulated annealing decisions on information that is unstable, because other variables may be simultaneously undergoing changes that are not currently recorded in any centrally accessible location.

If some element of centralized control were introduced into this asynchronous system, then statistical convergence guarantees could be preserved. Examine Figure 10.6, in which it is assumed that there are N processing units, one for every decision variable, each individually carrying out a simulated annealing process, but unaware of decisions being made by any other processing unit. In the figure, the portion of the computation that could be performed by parallel processors is outlined in dashed lines. After all processing units have either accepted or rejected their proposed changes (based on a first level temperature parameter θ_1), a centralized control component then assimilates the individual changes and constructs a new global configuration. This new configuration now must pass through a global filter, which is another simulated annealing acceptance test based on a global temperature parameter θ_2. In this way, the computational power of many free-wheeling asynchronous processors is checked at intervals by the centralized control, which ensures eventual convergence (Lucas and Price 1992).

Parallel annealing systems such as just described have been given the name **Boltzmann machines**. Boltzmann machines have taken many forms, depending on the problem at hand and the analyst's viewpoint, goals, and experience. In most cases, although there is parallelism, an element of sequentiality has been maintained because of the inherent requirement for monotonic cooling, and hence monotonic reduction of temperature parameters. More recent research has revealed that collapsing the timeline to a point, and randomly activating processing units at different temperatures (acceptance parameters) also works remarkably well, while alleviating any need for centralized control over the synchrony of the processing units. Cascading Boltzmann machines together in this way, with data-sharing among corresponding processing units at different temperatures, has proven to be an effective means of overcoming the time dimension through the use of multiple processors (Coughlin and Baran 1995, Price and Wahsheh 1999). Through this mechanism, spatial complexity is employed to compensate for temporal complexity—a common trade-off in the world of parallel computing that may serve us well in the realm of combinatorial optimization. Examples of parallel simulated annealing applications can be found in Wang et al. (2015), Ferreiro et al. (2013), and Santé et al. (2016).

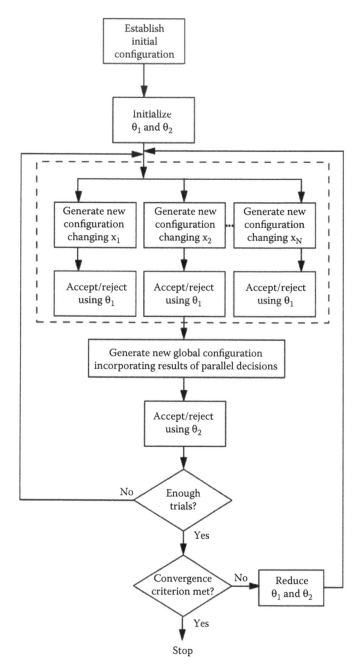

FIGURE 10.6
Parallel annealing.

10.5 Genetic Algorithms

Analogies between computational processes and natural phenomena seem to be quite appealing to problem-solvers, and simulated annealing is but one such analogy that has been effective and therefore popular. Biological analogies are particularly fascinating, and over the past 50 years have sparked many debates over whether machines can think or reason, and what techniques could and should be used to make machines compute in *clever* ways. **Genetic algorithms** (GA) are a type of search algorithm for finding optimal solutions to computationally difficult problems, and are based on analogies to biological reproductive processes. Computers and biological genes are similar to each other in the sense that both are able to record, copy, and disperse information. Genetic algorithms operate *iteratively*, over many *generations*, in such a way that only the *fittest* solutions survive, and thus these algorithms function as mechanisms for optimization.

The basic ideas for these methods were developed by Holland (1975), Goldberg and Holland (1988) during their investigations on how to build computing machines that are capable of learning. Inspired by the flexibility and adaptability that he observed in biological systems, he contended that rather than using and refining a *single* learning strategy, it was more advantageous for a machine to use a *breeding* of *multiple* strategies. The term *genetic algorithm* was popularized in a 1975 publication of Holland's work. Immediately thereafter, genetic algorithms began to be used successfully in scores of applications, which now include job-shop scheduling, pipeline systems, vehicle routing, keyboard design, and variations of the traveling salesman problem, to mention just a few. More important, these successes have prompted active research into the study of how various biological analogies can influence computing, as well as how computational models can give insight into the workings of biological systems.

Genetic algorithms operate by maintaining a *population* of feasible solutions to a problem. Each solution is evaluated (for example by using its associated objective function value). The best solutions are selected for reproduction and are grouped into pairs. Solutions that are less fit tend to not be selected and therefore die off and get replaced by other solutions. Then, within each pair of solutions, genetic modifications take place, which are described in terms of mutations and crossovers, resulting in a new breeding population that can repeat the process. The goal of optimization is served by selecting the best solutions for breeding, and introducing possible improvements through genetic crossovers, while mutations are introduced occasionally to prevent rapid convergence to a local non-global optimum.

Biological terminology abounds, although the adaptation of terminology is not always completely consistent with the corresponding biological meaning. Within the breeding population, individual solutions (encoded as strings) are referred to as **chromosomes**; the individual features in each chromosome are called **genes**; and the value of a feature in a given chromosome is called an **allele**. Using this terminology, we can now describe the entire process in greater detail.

First, a method is devised for mapping each feasible problem solution into a string (usually a binary string). The encoding mechanism depends entirely on the type of problem being solved, but usually it involves the values of the decision variables. Then it must be decided how many of these *chromosomes* to include in the breeding pool; a large pool increases diversity, but will have the effect of slowing the operation of the algorithm. An initial population is typically chosen arbitrarily, although other ways exist.

Next, the fitness of each string (chromosome) is evaluated, based on the objective function value corresponding to the encoded solution, and possibly also on problem constraints. For uniformity, the fitness values are typically normalized into the range of 0–1.

The selection of chromosomes (solutions) that will participate in reproduction is inspired by Darwin's (1859) *survival-of-the-fittest* theme. A proportional selection scheme favors a larger number of fit solutions, and allows fitter solutions to be chosen more than once, and weaker solutions to be possibly excluded entirely. A *roulette wheel* model provides a simple mechanism for this. Each string is associated with a sector on the wheel whose angle is proportional to the string's fitness. A random number is generated and assigned a point on the wheel. If the point falls within a particular string's sector, then that string is selected.

After selection, pairs of chromosomes are formed at random and are subjected to certain genetic manipulations; that is, modifications to the genes in the parent chromosomes. A process called **crossover** swaps a part of the genetic information contained in two chromosomes. Typically, a substring position in the chromosome is randomly chosen and the genes (string elements) within that substring are exchanged, forming two new offspring to replace the parents. The exact nature of crossovers is application specific, and must be done in such a way that resulting strings correspond to meaningful and feasible problem solutions. The recombination process can introduce improved genetic building blocks but will, on occasion, inadvertently disrupt favorable genetic structures. This (together with the selection of the fittest) may have the effect of driving the evolutionary process toward a local optimum. To overcome this, mutations are allowed to occur.

A **mutation** is simply a random reversal of one or more bits in a chromosome. Mutations are infrequent, but have the effect of reintroducing bits into the string that may be essential for an optimal solution and that may be currently absent in the breeding population. A higher probability of mutation tends to make the genetic search more broadly random, which can slow the convergence of the algorithmic process.

The offspring strings produced through these genetic manipulations may either replace the entire previous population (generation replacement method) or just the less fit members of the population (steady-state replacement method). In either case, the cycle of creation, evaluation, selection, and manipulation is repeated until a stopping criterion is met such as a specified number of generations have passed or until acceptable problem results are achieved.

Example 10.4

Consider the knapsack problem presented earlier in Table 10.2 and suppose that an initial population with six chromosomes (i.e., population size = 6) is randomly generated as shown in Table 10.4. The fitness of each solution is considered the same as the objective function in this case. To compute the probability of selection for each string, its fitness is divided by the population's total fitness. Strings with higher fitness would have higher chance of being selected. For example, String 2 with selection probability P_{select} of 0.27 is about twice as likely to be selected as String 1 which has a P_{select} of 0.151. The probabilities are shown in the table and are reflected in Figure 10.7. To select parent strings for the mating pool, imagine spinning a biased Roulette wheel like the one shown in the figure. It is more likely to land on strings with higher fitness proportionate to the area they occupy on the wheel. To select strings using this roulette wheel process in a systematic way, the cumulative probability is computed as shown in Table 10.4. A random number $r \sim U(0,1)$ is then generated, and depending on where it falls on the cumulative probability spectrum, the corresponding string will be selected. Suppose for example that six random values for r are generated as follows: 0.39, 0.68, 0.21, 0.64, 0.04, and 0.97. This will result in selecting strings 2 and 4 twice each, strings 1 and 6 once each and strings 3 and 5 not selected at all as shown in Table 10.5. Of course, if six other

TABLE 10.4

GA Initial Population

String										Fitness	P_{select}	Cumulative Probability
1	1	0	1	1	0	0	0	0		504	0.151	0.15
2	1	0	1	1	0	0	1	1		904	0.270	0.42
3	0	1	0	0	0	0	0	1		420	0.126	0.55
4	0	0	1	0	1	0	1	0		630	0.188	0.73
5	1	0	1	1	0	0	1	0		684	0.204	0.94
6	1	0	0	1	0	0	0	0		204	0.061	1.00
							Total			3,346	1.0	
							Avg fitness			557.6657		

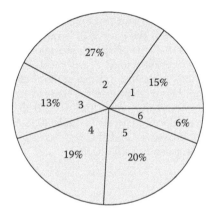

FIGURE 10.7
Roulette wheel selection.

TABLE 10.5

GA Mating Pool

String		Mating pool							Fitness
2	1	0	1	1	0	0	1	1	904
4	0	0	1	0	1	0	1	0	630
2	1	0	1	1	0	0	1	1	904
4	0	0	1	0	1	0	1	0	630
1	1	0	1	1	0	0	0	0	504
6	1	0	0	1	0	0	0	0	204

random numbers were generated, the outcome will likely be different, but we can easily see that strings with higher fitness will have a higher chance of being selected.

The next step is to randomly select parents from the mating pool for reproduction to create a new generation of solutions. Suppose that we randomly select the pairs of parents (2, 4), (4, 6), and again (2, 4). Note that a parent can be selected multiple times as the process is random. We then need to decide whether to perform a crossover operation or simply pass the parents as they are. This is done in a probabilistic way via the probability of crossover (P_c) which usually ranges between 0.5 and 1.0. A random number U~ (0,1) is generated and if $u \leq P_c$, crossover is performed; otherwise, exact copies of the parents are passed to the new generation. In this example, assuming that the pair (2, 4) undergoes a **single-point crossover** at a randomly selected location like gene 5 (along the line in Table 10.5), the second part of parent 2 (the last 3 genes) is swapped with second part of parent 4 resulting in two new strings as shown in Table 10.6 for strings 7 and 8. Similarly, if the pair (4, 6) is selected for single-point crossover at gene 3, along the line, the strings 9 and 10 will result as given in Table 10.6. Assuming that the last two parents 2 and 4 were again selected for crossover but failed the crossover test, they will then be passed to the new generation as they are. The new 4 strings (7, 8, 9, and 10) and copies of strings 2 and 4 in Table 10.6 represent Generation 1 with their fitness values and probabilities of selection for the next iteration.

It is important to note that any string included in any generation along the process must be feasible. In this example, a solution that violates the knapsack constraint should be precluded from competing. Another approach is to heavily penalize the fitness of infeasible solutions to make them very unlikely to be selected for future iterations.

Following the same Roulette wheel selection process, suppose that the values of the randomly generated number r were 0.01, 0.41, 0.60, 0.70, 0.79, and 0.63. This means that out of Generation 1, strings 7, 9, 10 and three copies of string 2 will be selected for mating, respectively (Table 10.7). If the pair (9, 2) is randomly selected for crossover at the location indicated in Table 10.7, the offspring strings 13 and 14 in Table 10.8 will be produced. Similarly crossing over the pair (7, 2) will produce the offspring strings 15 and 16 as shown in Generation 2 of Table 10.8. It is assumed that the parent strings 10 and 2 failed the crossover test and they were passed unchanged to Generation 2.

Randomly generating six new values for r (0.37, 0.89, 0.25, 0.96, 0.84, 0.50) will result in a mating pool from Generation 2 as given in Table 10.9. Assuming that only strings 15 and 14 were selected for crossover and the rest were passed as copies of string 2, Generation 3 would result as given in Table 10.10.

TABLE 10.6

GA Generation 1

Parents	String									Fitness	P_{select}	Cumulative Probability
2 and 4	7	1	0	1	1	0	0	1	0	684	0.175	0.18
2 and 4	8	0	0	1	0	1	0	1	1	850	0.218	0.39
4 and 6	9	0	0	1	1	0	0	0	0	384	0.098	0.49
4 and 6	10	1	0	0	0	1	0	1	0	450	0.115	0.61
2 (No crossover)	2	1	0	1	1	0	0	1	1	904	0.232	0.84
4 (No crossover)	4	0	0	1	0	1	0	1	0	630	0.16	1.00
								Total		3,902		
						Generation 1		Avg fitness		650.33		

TABLE 10.7

GA Mating Pool from Generation 1

String			Mating pool						Fitness
7	1	0	1	1	0	0	1	0	684
9	0	0	1	1	0	0	0	0	384
10	1	0	0	0	1	0	1	0	450
2	1	0	1	1	0	0	1	1	904
2	1	0	1	1	0	0	1	1	904
2	1	0	1	1	0	0	1	1	904

TABLE 10.8

GA Generation 2

Parents	String									Fitness	P_{select}	Cumulative Probability
9 and 2	13	0	0	1	1	0	0	1	1	784	0.185	0.19
9 and 2	14	1	0	1	1	0	0	0	0	504	0.119	0.30
7 and 2	15	1	0	1	1	0	0	1	1	904	0.214	0.52
7 and 2	16	1	0	1	1	0	0	1	0	684	0.162	0.68
10 (No crossover)	10	1	0	0	0	1	0	1	0	450	0.106	0.79
2 (No crossover)	2	1	0	1	1	0	0	1	1	904	0.21	1.00
							Total			4,230		
					Generation 2		Avg fitness			705		

TABLE 10.9

GA Mating Pool from Generation 2

String	Mating pool								Fitness
15	1	0	1	1	0	0	1	1	904
2	1	0	1	1	0	0	1	1	904
14	1	0	1	1	0	0	0	0	504
2	1	0	1	1	0	0	1	1	904
2	1	0	1	1	0	0	1	1	904
15	1	0	1	1	0	0	1	1	904

TABLE 10.10

GA Generation 3

Parents	String									Fitness	P_{select}	Cumulative Probability
15 and 14	19	1	0	1	1	0	0	1	0	684	0.136	0.14
16 and 14	20	1	0	1	1	0	0	0	1	724	0.144	0.28
2 (No crossover)	2	1	0	1	1	0	0	1	1	904	0.180	0.46
2 (No crossover)	2	1	0	1	1	0	0	1	1	904	0.180	0.64
2 (No crossover)	2	1	0	1	1	0	0	1	1	904	0.180	0.82
2 (No crossover)	2	1	0	1	1	0	0	1	1	904	0.180	1.00
							Total			5,024		
					Generation 3		Avg fitness			837.33		

This process continues for hundreds or thousands of generations until the genetic algorithm converges. Even with three iterations only (with a small population of six individuals), note how almost all individuals converged to identical copies of string 2—an indication that the genetic algorithm is about to converge on this solution. The optimal solution for this problem can be obtained by solving via integer programming, which turns out to be 1 1 0 1 1 1 0 0 with an objective function value of 1004. Although the genetic algorithm did not obtain this optimal solution, note how the average fitness value for the population has steadily increased from one generation to the next and can potentially reach to the optimum. However, in this simple example, it is not very likely that the optimal solution will be obtained even if it continues for more iterations due to the population size and possibly the initial population used. In fact, the only hope one would have in this case is for a mutation to make a change to one of the genes to kick the genetic algorithm out of its premature convergence. Since the probability of mutation P_m is typically small, after many iterations one of the genes can randomly be selected and switched from 0 to 1 or vice versa. In this example, mutation is the only way for item 2 to be selected since all strings in Generations 1, 2, and 3 had unselected this item, and a crossover operation would not have reversed this choice.

Note that although this example implied a trivial encoding of the eight binary decision variable values into a chromosome string, other more elaborate problem solution encodings may be necessary for different problems. For example, a mathematical programming problem with continuous decision variables may require many bits to encode a binary representation of each of many decision variables, resulting in long strings of thousands of bits for each solution (chromosome).

Genetic algorithms often seem to work quite well, no matter how they are designed. Yet, during the 1980s, genetic algorithms were recognized as having certain shortcomings that rendered them suspect as optimization tools. Practitioners introduced a number of modifications that improved the performance of genetic algorithms while preserving the attractive image of the concept of evolution by combination. Variations such as combining more than two parents simultaneously, using multiple point crossovers, and generating local *improvements* (rather than merely random mutations) in the breeding population, all seem to challenge the integrity of the biological model but do contribute to the quality of optimization results. One easily gets the impression that our experience with genetic algorithms is entirely empirical and unfounded in theory. However, error bounds have been developed, indicating some recent theoretical progress (Goldberg 1989, Goldberg et al. 1992, Goldberg 1994). A good overview of recent work on genetic algorithms is found in Reeves (1997) and Sivanandam and Deepa (2007).

Genetic algorithms lend themselves readily to computational parallelism at several levels. Because optimization is typically being performed within a large search space, different processors could be used to search different neighborhoods simultaneously. Alternatively, different processors could operate on different breeding populations over the entire search space at the same time. At a lower level, once pairs are selected, genetic manipulations are independent of each other, so multiple processors could perform crossovers and mutations simultaneously. Then, offspring would migrate across the network into either centralized or distributed selection processes in the next generation.

10.6 Tabu Search

Tabu search (TS) is a metaheuristic that utilizes a memory capability in escaping local optimal search regions by forbidding previously made moves from being revisited for a certain number of iterations. Tabu search was introduced by Glover (1989) and since then many extensions and variants of the method have been published in various areas of applications. The core concept of tabu search is similar to local search where it starts with a single solution and, using some neighborhood generation scheme, identifies a candidate list of moves (neighbors) whose contributions to the objective function are evaluated. The move with the greater contributions is selected and the reverse of the move (put it back) is placed on the *tabu* list to prohibit returning to the previous solution for a certain number of iterations as imposed by the size of the tabu list. Once a move is off the list, it becomes permissible to be revisited in future iterations. An exception to the rule is when a move passes the *aspiration criterion* which is commonly when a tabu move can identify a solution that is better than the *best-found-so-far*. The rationale in temporarily forbidding moves that had been tried is to allow other moves to be visited even if they are worse than the best local solution. This lets the search algorithm escape local optimal search regions to potential global search regions.

Example 10.5

Consider again the knapsack problem presented earlier in Table 10.2 and suppose that a new move or neighbor is created by switching a 0 to 1 or 1 to 0 representing the selection or unselection of an item respectively. The neighborhood size is therefore $|n|$ and each move must comply with the maximum allowable budget of 450 for all items; otherwise it will be considered infeasible. Assuming a tabu list size of 3, Table 10.11 includes an initial solution (S1) followed by ten tabu search iterations. The initial solution can in general be selected randomly or via some other method or heuristic. In this example, S1 was created based on the simple heuristic of *item with the largest value is selected first*. Z is the objective function value and ΔZ is the change in the objective function if a move or neighbor is selected. X means that a move produces an infeasible solution due to exceeding the maximum allowable budget, and the tabu list (TL) represents the number of future iterations for which a move is prohibited (unless it passes the aspiration criterion).

Starting with S1, the only possible moves are to unselect items 2, 3, or 6, which will reduce the objective function value by 200, 300, or 450 respectively.

Although all three moves will worsen the current value of Z, the algorithm must pick one as it may be the path to a better solution in future iterations; otherwise, the algorithm will be stuck in its current solution. The least damaging move is to unselect item 2 which will deteriorate Z by 200 as identified by the box around it. This move leads to S2 with Z = 750 and puts item 2 on the tabu list for the coming three iterations. In S2, the best move is to put item 2 back into the solution as it adds the maximum value of 200 to Z; however, this move is tabu and therefore the next best move is to select item 7 that adds 180 to the objective, resulting in a Z value of 930 as shown for S3. Note that item 7 is now tabu for the coming three iterations, while item 2 will remain prohibited for the coming two. The process continues in this fashion until a stopping criterion is met. For this small example, our stopping criterion was to run it for 10 iterations after the initial solution, which happened to produce the optimal solution in S11. An important point to note is that as long as a move is on the tabu list, it cannot be taken unless it passes the aspiration criterion, which did not occur in this example. After a certain number of moves defined by the tabu list length (3 in this case) a move is off the list and it can be selected again. In this example, item 2 could not be selected in S1, S2, and S3 as it was on the tabu list. But after that it was off the list and was indeed selected again in S7 at which time it became tabu once again for three more iterations. Of course, the best found solution will be updated as better solutions are obtained and will be reported at the end.

The tabu list is used to explore new search areas where larger list size would force the search algorithm to move to new neighborhoods that might not otherwise be explored. The tabu list is a form of short-memory that is used to explore the space based on search *recency* and its length or tenure can be either static (e.g., 3) or dynamic where it can be randomly selected between two values (e.g., between 3 and 7). Tabu search can be extended to include *intensification*. The algorithm can use *intermediate term memory* to keep track of the frequency that certain solution components have occurred. During intensification, the method will fix some of those components that have occurred frequently during past iterations, and then do a more intense search, for example, by expanding the size of the search neighborhood. This tends to accelerate the movement toward a local optimum. The method may also incorporate *diversification*. Tabu search

TABLE 10.11

Tabu Search Iterations for the Knapsack Problem

Solution									Z
S1	0	1	1	0	0	1	0	0	950
Neighbor	1	0	0	1	1	0	1	1	
ΔZ	X	[−200]	−300	X	X	−450	X	X	
TL		3							
S2	0	0	1	0	0	1	0	0	750
Neighbor	1	1	0	1	1	0	1	1	
ΔZ	120	200	−300	84	150	−450	[180]	X	
TL		2					3		
S3	0	0	1	0	0	1	1	0	930
Neighbor	1	1	0	1	1	0	0	1	
ΔZ	X	X	[−300]	X	X	−450	−180	X	
TL		1	3				2		
S4	0	0	0	0	0	1	1	0	630
Neighbor	1	1	1	1	1	0	0	1	
ΔZ	120	200	300	84	150	−450	−180	[220]	
TL			2				1	3	
S5	0	0	0	0	0	1	1	1	850
Neighbor	1	1	1	1	1	0	0	0	
ΔZ	X	X	X	[84]	X	−450	−180	−220	
TL			1	3				2	
S6	0	0	0	1	0	1	1	1	934
Neighbor	1	1	1	0	1	0	0	0	
ΔZ	X	X	X	−84	X	−450	[−180]	−220	
TL				2			3	1	
S7	0	0	0	1	0	1	0	1	754
Neighbor	1	1	1	0	1	0	1	0	
ΔZ	120	[200]	X	84	150	−450	180	−220	
TL		3		1			2		
S8	0	1	0	1	0	1	0	1	954
Neighbor	1	0	1	0	1	0	1	0	
ΔZ	X	−200	X	−84	X	−450	X	[−220]	
TL		2					1	3	
S9	0	1	0	1	0	1	0	0	734
Neighbor	1	0	1	0	1	0	1	1	
ΔZ	120	−200	X	−84	[150]	−450	180	220	
TL		1			3			2	
S10	0	1	0	1	1	1	0	0	884
Neighbor	1	0	1	0	0	0	1	1	
ΔZ	[120]	−200	X	−84	−150	−450	X	X	
TL	3				2			1	
S11	1	1	0	1	1	1	0	0	1004
Neighbor	0	0	1	0	0	0	1	1	
ΔZ	−120	[−200]	X	−84	−150	−450	X	X	
TL	2	3			1				

is intended to enable the search to escape local solutions. Under diversification, we use *long term memory* to track solution components that have *not* been explored. We begin a new search by incorporating some of these components, thereby forcing the algorithm to cover the space more broadly. Further detail and nuance of tabu search is available in Glover and Laguna (1997), Aarts and Lenstra (1997), Du (2010), and Taillard (2016).

10.7 Constraint Programming and Local Search

Constraint programming assumes that the problem can be described as a set of variables and a set of constraints that restrict the feasible solutions for a problem. Each variable has a domain: a set of feasible values. The constraints can be logical relations (for example, only one of a set of variables can be true), mathematical constraints (e.g., $x \leq 5$), integer, Boolean, real valued, and so on. Constraint programming defines the problem independent of the solution method.

Disjunctive constraints: When the domain consists of discrete values, a disjunctive constraint states that only one of a set of conditions can be true. When the variables are real valued ranges, the constraint states that the ranges must not overlap.

Conjunctive constraints: These are similar to disjunctive constraints, but in this case, there is a limit on the number of conditions that can be true.

Temporal constraints: In scheduling applications, these constraints specify that one activity must precede another activity by at least some amount of time. For example, job B cannot start until job A has ended.

To illustrate the concept, consider the popular Sudoku puzzle. The game consists of a 9×9 square. Each row and column must have the numbers from 1 to 9 exactly once. The board is also divided into nine 3×3 squares, each of which must also contain the digits from 1 to 9 exactly once. The initial puzzle has some of the squares filled in, and the challenge is to complete the design with a unique solution.

Consider the following example in Figure 10.8:

a		1				3		
	3		8					
	9	7		b	4	c	1	
				9			7	6
y		9	6	2	8	5	x	
6	8		7					
	7		3			1	4	
					2		6	
		6				7		

FIGURE 10.8
Sudoku puzzle example.

For each square in the puzzle, we can create a variable and the initial domain is the set [1, 2, ..., 9]. For example, the square marked "a" begins with 9 elements. It is involved in three sets of disjunctive constraints for the row, the column and the 3 × 3 square. The first row already has a 1 and a 3; the first column already has a 6; and the 3 × 3 square already has 1, 3, 7, and 9. Therefore, we can reduce the feasible domain of "a" to the set [2, 4, 5, 8]. This process is referred to as **constraint propagation**. If any square only has one possible entry, we can fill in the square. For example, the square marked "x" can only be 3.

Constraint propagation can also be used to identify variables that must be set. When you look at the middle row, the square marked "y" is the only one that can be 7. By repeated application of these two rules, you can solve the puzzle. Square "b" is the only one in row 3 that can be 3. Then, square "c" is the only cell in row 3 that can be 6.

One of the original applications of constraint based programming was for scheduling problems. Given a set of n jobs, where job i has processing time p_i, ready time, r_i, a set of precedence constraints (where job i must finish before job j begins), and perhaps resource constraints (where job i requires resource R_k that has limited capacity).

Typically, the methods for solving these problems, after the problem has been reduced as much as possible using constraint propagation, involve some form of local search heuristic. As the heuristic proceeds, further applications of propagation can be used to speed up convergence.

Note that constraint based methods are searching for a feasible solution as contrasted with an optimal one. Many implementations of constraint based methods have been proposed to create a minimization procedure. For example, once we have identified a feasible solution, we can add a new constraint that we are only interested in better solutions. For a detailed description of the topic, readers may refer to Hentenryck and Michel (2009).

10.8 Other Metaheuristics

In the last few decades, the area of heuristics exploded with algorithms especially with the advances in computer technology and programming languages. Du and Swamy (2010) classified the most common metaheuristics into four approaches:

1. Evolution-based methods such as:
   - Genetic algorithms
   - Genetic programming
   - Evolutionary strategies
   - Differential evolution
2. Swarm-based methods:
   - Particle swarm optimization
   - Artificial immune systems
   - Ant colony optimization
   - Bee metaheuristics
   - Swarm intelligence

3. Sciences-based methods:
   - Simulated annealing
   - Biomolecular computing
   - Quantum computing
4. Human-based methods
   - Memetic Algorithms
   - Tabu search
   - Scatter search

Many metaheuristics are based on the idea of introducing randomness into the search process as a mechanism to escape local optima where random solutions are sometimes selected over greedy solutions. Examples of such methods include GRASP (Greedy Randomized Adaptive Search Procedure) (Resende and Ribeiro 2003) and Meta-RaPS (Metaheuristic for Randomized Priority Search) (Rabadi et al. 2006, Garcia and Rabadi 2011, Kaplan and Rabadi 2013, Moraga 2016).

Finally, there could be many different variants of metaheuristics that fall under a class of metaheuristics. For example, Swarm Intelligence algorithms include firefly, frog, bat, monkey, fish, cuckoo search algorithms among some other ones. More elaborate material on metaheuristics and their types is available in Glover and Kochenberger (2006), Du (2010), Burke and Kendall (2014), Gendreau and Potvin (2010), and Siarry (2016).

10.9 Software for Metaheuristics

Software implementation of greedy, local, and metaheuristic algorithms is commonly developed in general-purpose programming languages such as C, C++, Java, Python, and so on. Appropriate data structures can easily be chosen that represent not only a current problem configuration, but also proposed modifications to the current configuration. Standard library functions for generating random numbers are convenient for effecting the probabilistic acceptance of such modifications, as required in simulated annealing and genetic algorithms. Because of the ease of developing such programs, and because the details of the implementation are often very application specific, commercial software is not typically needed for these heuristic techniques. Furthermore, even when (meta)heuristic algorithms are implemented in software systems for specific applications or industry, the design details tend to be hidden from the user for proprietary and competitive reasons.

Among the limited offerings of metaheuristic software tools is Evolver, spreadsheet-based product from Palisade Corporation which works as an Add-in to Microsoft Excel in which models are implemented in a spreadsheet, and solved as constrained optimization problems using genetic algorithms. Microsoft Excel itself comes with an Evolutionary Solver (a simpler form of genetic algorithms) to solve models that are implemented as spreadsheets.

MATLAB from MathWorks offers a Global Optimization Toolbox that includes genetic algorithms, simulated annealing, and particle swarm solvers. While using programming languages to implement metaheuristics might be computationally more efficient,

MATLAB's toolbox can reduce the implementation effort significantly with the tool boxes and functions it offers. It also gives the user some control over algorithm design. For example, the user can select the type of crossover and mutation to use with the genetic algorithms, and in simulated annealing, the user can decide on the temperature schedule and acceptance criteria among other things. Furthermore, developers can use MATLAB as a platform to develop software environments and tools including optimization environments. For example, **TOMLAB**, a general purpose development and modeling environment, implements a real-coded genetic algorithm called **TOMLAB/GENO** that can be used with various optimization problems.

Metaheuristic algorithms are typically used for problems that are computationally complex and messy to model and solve using structured modeling approaches such as mathematical programming. Hence, they often need to be tailored to the problem at hand. Therefore, it is no surprise that there are not many canned metaheuristic software systems as they need to be customized to specific problems. Nevertheless, the internet is full of codes and binaries in different computer languages that can be utilized in software development, the vast majority of which are freely available for download. COIN-OR (Computational Infrastructure for Operations Research), for example, includes some open source libraries and frameworks for metaheuristic development. Similarly, Google offers a suite of portable software called Google Optimization Tools for solving combinatorial problems. It is almost impossible to list all software sources that pertain to various metaheuristics and the readers are encouraged to refer to the book's website and conduct their own online search. The website GitHub.com contains open source code for several algorithms of interest.

10.10 Illustrative Applications

10.10.1 FedEx Flight Management Using Simulated Annealing (Campbell et al. 1997)

Federal Express (FedEx) is one of the world's largest express transportation company. Handling 3.4 million packages in over 220 countries every working day, with 650 aircraft and over 4,700 pilots, it is not surprising that the company must rely on a variety of analytical tools for scheduling and coordinating its activities.

In 1993, during negotiations involving pay rates and work rules with the Air Line Pilots Association, the 20-year-old company recognized the need to be able to evaluate alternatives to its traditional methods for scheduling work for its pilots. In particular, they needed a way to automatically build individual trips (flight legs) into lines of work (called *bid lines*). The method needed to be sufficiently fast and efficient that many alternatives could be generated, compared, and considered during, as well as after, negotiations with the pilots' association.

The scheduling questions demanded the use of a so-called *bid-line generator*, software that could compose units of work for pilots to bid on. The goal is to maximize the amount of flying assigned to bid lines and minimize the number of bid lines. Pilots submit bids by listing their preferred sequences of flights, and work assignments are made according to the pilot's seniority.

The number of inputs and constraints for generating the bid lines make the problem almost overwhelming. Considerations include

- Aircraft type.
- Crew size and requirements.
- Origin and destination cities.
- Layover cities.
- Number of trips in a line.
- Scheduled times and days.
- FAA regulations governing flight periods and rest periods.
- FAA day off and maximum duty length regulations.
- Crew turnarounds.
- International/domestic mixtures (generally undesirable).
- Week on/week off mixtures (generally desirable).

The bid line generator should generate bid lines that not only meet the hard constraints but that maximize line value (desirability to the pilots and productivity for FedEx) and minimize cost over all bid lines.

Details were kept to a minimum, but so many factors contribute to the composition of bid lines that the 0-1 integer programming model, with all its constraints, quickly became unwieldy—even to formulate, and much more so to actually solve. Simulated annealing proved to be the solution method of choice for this problem. Implementation was in C++ on a Unix workstation. The random changes to a current configuration involved arbitrarily selecting two bid lines, then in each, selecting a trip (flight leg) and exchanging them. The exchanges were accepted according to the usual probabilistic threshold until, as temperature parameters were lowered, there were no new changes accepted.

It is not especially surprising that FedEx analysts chose simulated annealing as their optimization heuristic, nor that simulated annealing eventually served their needs successfully. The real lessons to be learned here are first to notice how very awkward the analysts found this real-world problem to be. The sheer number of constraints from federal agencies, labor organizations, company resources, and normal crew preferences, were a serious challenge that had not been adequately faced throughout the previous 20-year history of FedEx. Second, although simulated annealing appears on the surface to be a relatively straightforward heuristic, the practical implementation presented several hitches.

Some of the drawbacks of simulated annealing were anticipated. Performance is very sensitive to the control parameters and the annealing (cooling) schedule. Extensive experimentation was done to fine-tune the system, and the maximum number of equilibrations was finally set to 300. Also, the heuristic can be fairly time-consuming to execute and there is no guarantee of optimal solutions. And because it generates potential changes randomly, it does not easily incorporate strategies for directed search. Nevertheless, despite these obstacles, some of the analysts had prior experience in using simulated annealing to solve problems in aircraft container loading and personnel and task scheduling, and they had great confidence in this heuristic method. Yet unanticipated difficulties followed.

The heuristic tended to produce too few valid lines and too much unassigned open time. This was remedied by tacking on a greedy algorithm (as a second pass after simulated annealing) to distribute open time into new lines (without modifying the high quality lines built during the annealing phase).

It was discovered that the initial heuristic did not give proper consideration to coordination of morning and afternoon trips, an important element in the minds of the pilots. The introduction of weighting factors addressed this problem satisfactorily.

One surprising observation was the critical importance of the initial solution in the behavior of the simulated annealing algorithm, which had previously been thought to be irrelevant and arbitrary. It became necessary to jumpstart the process by concocting initial lines by putting trips to the same first layover city on the same line, and making fewer lines.

There were other problems as well. The bid line generator was first built for the FedEx Boeing 727 fleet of aircraft. When initial implementations seemed stable, additional fleets were introduced, but the process then immediately yielded poor results. The problem was studied, and analysts found that the difficulty lay in the fact that different types of aircraft flew different length trips. When the process was tuned in favor of shorter trip aircraft such as Boeing 727 and DC10, the longer Boeing 747 flight legs became problematic. The solution to this issue involved some fundamental changes to the simulated annealing process based on categorizing the fleet according to average trip lengths.

It was also recognized along the way that the system needed additional data about its trips and lines that simply were not readily available. And some data files were found to be erroneous. A time consuming effort to upgrade the underlying databases proved necessary and beneficial, and taught the analysts to be extremely cautious about blindly assuming that input data files are complete and free of errors.

Finally, in this implementation, the simulated annealing process did not always converge at all. The *churning* behavior resulted when proposed changes having a net cost of zero were accepted, and the phenomenon was worst when a large proportion of the proposed changes had no impact on the objective function value but nevertheless involved complicated changes to the bid lines being constructed. No direct solution to this difficulty ever materialized, and the analysts viewed this as evidence of the limitations of any heuristic method in solving very complex real-world combinatorial problems.

Run times for the simulated annealing heuristic vary with fleet size, requiring 30 minutes for the smaller fleets and up to 10 hours of SPARCstation time for the largest (Memphis based) fleet. Churning can affect all of these run-times.

FedEx generally considers this system to be a valuable and practical analytical tool, which can automatically produce bid lines of a quality comparable to those produced laboriously by other methods. As is typical of many heuristic methods, simulated annealing clearly cannot build a tidy solution out of a messy problem, but it does appear to be a practical tool for effectively handling problems that heretofore could not be dealt with at all.

10.10.2 Ecosystem Management Using Genetic Algorithm Heuristics (Hughell and Roise 1995)

Managing a forest with the aim of profitable timber production and wildlife preservation is a good example of a multi-objective problem, in an environment of uncertainty, for which no single conventional optimization technique is adequate. A decision support system developed for ecosystem management in a North Carolina pine forest couples a wildlife behavior simulation model with an integer programming model that is solved using a genetic algorithm.

Foresters in the Croatan National Forest needed to address the question of how best to manage a 3,000-hectare region to sustain a dependable flow of timber while not destroying the foraging territories and nesting sites of the endangered red-cockaded woodpecker species. Conventional management schemes are typically based on *optimal* activity schedules; but in this case, the planning horizon covered 20-year harvesting cycles over a period of up to 200 years, during which there would be considerable environmental uncertainty as well as normal periodic re-evaluation. A strictly optimal harvesting schedule could easily become infeasible over time. What was needed was a decision support system that permits flexibility and presents a selection of *good* harvest schedules that could still be implemented in the face of environmental changes.

Stochastic wildlife group behavior simulation models have become valuable tools in the study of wildlife species viability. The red-cockaded woodpecker (RCW) model involves groups of individual birds having given attributes and foraging and breeding characteristics in five year cycles. The complex behavioral activities of RCW groups are abstracted down to fit into a lattice of 4-hectare forest landscape stands. Nesting and foraging suitabilities are calculated at the beginning of the cycle, and then simulations are carried out to determine the probabilities of various eventualities, including:

- Migration or mortality of RCW groups with inadequate or unsuitable foraging and nesting resources.
- Sharing of landscape by multiple groups of RCW.
- RCW group splits.
- Successful breeding and nesting.
- RCW extinction.

Simulation results are stored for subsequent incorporation into the larger decision process.

The timber stand model covers successive 20-year cycles of harvesting and regeneration. Details of the model include appropriate intermediate cuts, understory management through controlled burns, and primary stand harvests (which leave around 15 trees per hectare, 6 trees per acre). The overall management decision is the selection of a harvest schedule that maximizes the minimum timber volume harvested in any one management period and that supports the RCW proximity constraints. It is known that the optimal stand age for timber production is around 60 years, while the optimal stand age for woodpecker foraging is over 100 years. To represent this apparent mismatch, buffers are defined around each RCW nesting group, and parameters are introduced into the model to specify the minimum harvest age inside the buffer and outside the buffer. Through these constraints, the harvest schedule can respond to changes in the location of RCW groups; and herein lies the multi-objectivity of the optimization problem and the need for a feedback management policy.

The most obvious way to solve a two objective problem is to perform a series of single-objective optimizations with one objective fixed and the other optimized. Because neither timber nor endangered wildlife are to be treated as fixed constraints in this ecosystem, this traditional approach is not appropriate. Instead, varying the parameters in the RCW proximity constraints permits the development of management policies that balance the benefits for both timber production and woodpecker viability.

The resulting optimization model takes the form of an integer linear programming problem. The problem was solved with a conventional branch-and-bound algorithm but it was recognized that in the natural world of uncertainty and changing assumptions, the concept of optimality may itself be problematic. The harvesting schedule deemed to be optimal at one time may turn out to be infeasible in the long run.

To achieve the flexibility needed to make the decision support system workable, the forest managers turned to the use of a genetic algorithm. The genetic algorithm heuristic starts with a random set of feasible harvest schedules (a population of solutions) that, based on their quality, are copied into the next generation. Genetic operations of crossover and mutation take place, and then the process repeats. (Here, the quality of a solution is the minimum one-period wood volume harvested, which is to be maximized.)

In this evolutionary algorithm, each feasible harvest schedule (i.e., each configuration of decision variable values) is a *chromosome*, which is comprised of *genes* (decision variables associated with individual stands), each of which is assigned an *allele* (a set of decision variable values prescribing a harvest schedule for the stand). After initially random schedules are created, chromosomes are copied into the next generation in such a way that those of superior quality contribute multiple copies at the expense of under-representation by those chromosomes with inferior quality. Randomly chosen chromosomes are paired for crossover; and for each pair, a certain percentage of the genes are selected and their alleles switched. Mutations occur as a certain percentage of chromosomes are chosen and in each a randomly selected gene (stand) is assigned an arbitrary feasible set of decision variable values (stand harvest schedule).

By allowing this genetic process to repeat over many generations, a population of *good* harvest schedules is generated in a small fraction of the time that it takes a branch-and-bound algorithm to generate a single optimal solution. Croatan National Forest managers are convinced that a set of good choices, for a system fraught with uncertainty, is much more valuable than one *optimal* solution whose feasibility may become suspect in a changing ecological environment. In this context, the set of stands chosen for harvesting in the current management period is that set of stands represented in the largest number of *good* harvest schedules in the evolved solution population.

The decision support system that incorporates the wildlife behavior model, the stand characteristics, and the RCW proximity constraints together with the genetic heuristic search process, identifies the best solutions and displays the *critical* solutions for which an improvement in one objective (timber or woodpeckers) is gained only at the expense of the other. As had been expected, those schedules specifying longer rotations support larger populations of woodpeckers, while shorter rotations increase timber production. Feedback at 20-year cycles allows for the selection of a harvest plan, followed by adjustments to the RCW simulation model, followed by another timber harvesting decision, repeated throughout the 200-year horizon.

The set of options produced by this system allows forest managers to dynamically achieve a sustainable flow of timber production throughout the long planning horizon, which can be modified in response to the requirements for successful co-existence with wildlife. The system was developed in C++ with object oriented programming techniques, and run on a PC prior to being ported to a workstation platform. The ORSYS Operations Research System was used to obtain the branch-and-bound solutions.

10.10.3 Efficient Routing and Delivery of Meals on Wheels (Manikas et al. 2016)

Meals on Wheels America is an organization dedicated to combatting hunger and poverty by delivering around a million prepared meals every day to individuals in need. Apart from food preparation activities, the major challenge each day is to efficiently route delivery vehicles to approximately 30 destinations per vehicle, deliver the appropriate meals to each recipient, and return to the point of origin to return coolers and heaters for use on the following day. Finding near optimal delivery routes is a complex problem that is often addressed by using mathematical optimization tools such as CPLEX. But in a low-budget humanitarian organization staffed in large part by volunteers, it is unrealistic to incur the high cost of such tools, and to engage skilled and experienced analysts to develop routing solutions.

Other humanitarian operations had previously created tools and efficient solutions such as for scheduling and routing in home healthcare delivery programs. Transportation is an expensive and critical aspect of humanitarian logistics and operations in general, and indeed Meals on Wheels discovered its own similarities to other relief operations. These organizations typically rely on volunteers who have various levels of abilities and qualifications, must operate with limited time and financial resources, are expected to provide time-critical delivery of goods and services, and in practice often have only limited access to technological support.

Meals on Wheels deliveries generally originate at an institutional kitchen. Specific locations for delivery are pre-determined and are roughly clustered according to neighborhood proximity with at most 30 delivery points in a cluster for a given driver (the limitation being due to space in the volunteer's car and the need to get the fresh meals delivered in a timely manner). Manually routing vehicles and preparing delivery instructions for individual drivers is an extremely time-consuming process that must be completed for each day's unique pattern of deliveries. The vehicle routing problem encountered in delivering meals has been recognized by operations researchers as a computationally difficult problem for which no simple solution is known; dealing with such a challenge manually is far too time-consuming.

A local branch of Meals on Wheels in Boise, Idaho, wisely undertook to develop a much more practical approach. An affordable solution was found through the use of a Microsoft Excel spreadsheet and an Application Programming Interface (API) that connected older personal computers at no cost to an existing Internet connection providing access to mapping services from MapQuest and Google. Researchers familiar with the old manual system of delivering meals analyzed this system and made some recommendations for improving efficiency of meal delivery. Excel spreadsheets were already available; Excel's Visual Basic for Applications was used to write programs to access mapping data and build a *travel matrix* to store information about travel times and distances in the meal delivery area.

However, for the routing process itself, Excel's Solver would have required customized optimization models for each route. Instead the system was built so that a VBA procedure could use publically accessible time and distance information from MapQuest and Google. Researchers and analysts then developed customized code for a solver to provide all the vehicle routing processes needed for meal delivery.

In the system developed, users are able to input the addresses for each day's deliveries. Then in preparation for vehicle routing, the system constructs an accurate and up-to-date *travel matrix* that contains the travel time (or distance) between each possible pair of locations in the system. With the information from this matrix, the system then creates driving information for all routes, that is, for all possible pairings of delivery stops. Finally, the system applies a genetic algorithm to select for each vehicle the optimal or near optimal route for that vehicle's deliveries for the day.

The researchers created this genetic algorithm, customized to meet the requirements of the meal delivery application. The algorithm was initialized with a population of *chromosomes*, encodings of possible routes consisting of a series of stops in the order to be visited. Then through *crossovers* and *mutations*, alternative routes are chosen and evaluated with the aim of identifying the route with the least driving time. A crossover is accomplished by considering the next stop in the route represented by one parent and the corresponding next stop in another parent, and then randomly (50% chance) choosing the next stop from either parent. In this way a new *offspring* chromosome (route) is constructed with representative components from each parent. Mutation occurs by selecting a chromosome and exchanging two stops in the route sequence, thereby creating a slightly different chromosome representing a slightly different route driving time.

The best performance within the Meals on Wheels routing application was achieved by conducting multiple runs of the GA process, each run starting with unique random initial population of routes. Multiple runs of course require more computation time, but users agreed that the increased wait time was reasonable in order to obtain a faster route for delivery of meals.

At this Meals on Wheels site, a centralized planning location coordinates kitchen operations, special dietary needs of recipients and delivery logistics. Route coordinators found this system simple and convenient to use, and route drivers found the driving instructions to be accurate and easy to follow. With the application of this vehicle routing system, route driving time reductions range from 2% to 27% for each route. Therefore vehicle operating costs were substantially reduced, and volunteer driving commitments were met more easily.

This application is organized into a comprehensive package for ease and convenience of use. The delivery scheduling and routing system reviews the list of customers requesting meals each day, and accordingly updates the list of stops to be made and the specific meals to be delivered at each stop. It makes updates to mapping data related to delivery stop locations, optimizes the route using the GA algorithm, and finally automatically prints driving directions to be given to each driver.

For a long route with 30 stops, the route sequencing and instruction generation takes about 15 minutes which is a welcome improvement over the manual system in which route coordinators spent over an hour every day planning a single route. The system continues in use, and plans are underway for several enhancements, as well as sharing the system with Meals on Wheels organizations in other localities.

10.11 Summary

Heuristic and metaheuristic techniques are efficient and practical methods that can be used to find good (but not necessarily optimal) solutions to a wide variety of difficult combinatorial problems. Such techniques are employed to find acceptable solutions to problems, when otherwise the best-known algorithms for finding optimal solutions take far too much computation time to be usable in practice or when problems are too complex to model using the traditional methods.

The simplest of these heuristic methods operates by making local improvements to a feasible solution, merely by rearranging randomly a few elements in the solution, to achieve a slightly better feasible solution. While there are seldom any guarantees of reaching an optimal solution in this way, remarkably good results can be obtained quickly with minimal computational effort.

Heuristic techniques guided entirely by opportunities for improvement often converge rapidly to a local optimal solution. To broaden the search in hopes of finding a global optimum, metaheuristic techniques such as simulated annealing, genetic algorithms, and tabu search rearrange the entities in the solution so that not only better solutions but occasionally also worse ones are admitted. The algorithms for doing this bear a resemblance to science based processes such as the process of annealing in physical substances, or are inspired by nature such as the biological processes as we have seen in genetic algorithms. Many more promising metaheuristics have emerged in the last a few decades and have demonstrated effectiveness at solving challenging problems.

Key Terms

annealing process

Boltzmann machine

capital budgeting

chromosome

constraint programming

constraint propagation

crossover

equilibration

genetic algorithm

greedy heuristic

heuristic methods

knapsack problem

local exchange heuristic

local improvement heuristic

local search

metaheuristics

mutation

parallel annealing

simulated annealing

sub-sequence reversal

tabu search

traveling salesman problem

Exercises

10.1 One of the recurring themes in Operations Research is how best to explore a range of possible actions in pursuit of well-defined goals. The use of heuristic search methods has been suggested. Define the term *heuristic search.* and indicate why such methods are attractive.

10.2 Develop a local improvement technique for the knapsack problem described at the beginning of this chapter.

a. Design several possible methods for creating initial feasible solutions.

b. Develop a method for computing the objective function value for a current solution.

c. Design an exchange or swap technique, using a random number generator to select the items to be swapped. For each proposed exchange, compute the new objective function to determine whether to accept the change. Decide how many iterations of this exchange step you think would be necessary for a knapsack problem with n objects.

10.3 Implement your design in Exercise 10.2 by developing a computer program.

a. Demonstrate the results by running your program on a problem instance with n = 17 objects to be considered for a knapsack having a capacity of 3,876. The weights and values of the objects are shown in Table 10.12.

b. Try different numbers of iterations of the exchange process, such as 100, 1,000, and 10,000. Chart the improvements in the objective functions that take place throughout the execution of your algorithm, and determine how many

TABLE 10.12

Knapsack Data

Object Number	Object Description	Weight	Value
1	Life raft	800	900
2	Shark knife	050	550
3	Sun shades	010	475
4	Reef runners	240	850
5	Canteen	080	600
6	Iodine pills	350	350
7	OR book	738	900
8	Gnat spray	548	290
9	Nylon cord	310	500
10	Carrot cake	200	010
11	Firewood	300	800
12	Solar blanket	850	215
13	Dried apricots	490	285
14	Parachute	500	630
15	Space suit	300	320
16	Alien bane	480	850
17	Dry matches	150	400

iterations is a reasonable number. Would 1,000,000 iterations improve the quality of solution obtained by your algorithm?

10.4 Reconsider Exercise 10.2b. Is it necessary to recompute the objective function at each local improvement step? Refine your program so that objective function re-evaluations are as simple as possible.

10.5 Design and implement an algorithm that exhaustively enumerates all feasible packings of n objects in a knapsack having a given capacity. Use the results obtained from this algorithm as a benchmark to gauge the quality of the solutions generated by your exchange heuristic.

10.6 Design a simulated annealing heuristic algorithm for the knapsack problem. Use your local improvement exchange heuristic, and modify it so that it probabilistically accepts bad exchanges.

a. Design a cooling schedule for your algorithm. What should be the initial temperature parameter? By what amount should this parameter be reduced after each equilibration? At what temperature should the annealing process cease?

b. Apply your algorithm to the knapsack problem data shown in Table 10.12.

c. How many exchanges actually take place at each temperature? How many exchanges take place at the coolest temperature?

10.7 Compare the local improvement heuristic and the simulated annealing heuristic on the basis of the computation time required for each method to execute and the quality of the solutions obtained by each method.

10.8 Design a local improvement heuristic technique for solving the traveling salesman problem described at the beginning of this chapter. For purposes of this exercise, assume that we wish to find the least costly tour from city 1 through all the other cities and back to city 1.

a. Design a method for establishing an initial tour.

b. Develop a method for computing the objective function for a given tour.

c. Design a swap or exchange mechanism for local improvements, involving just two cities. After each proposed exchange, compute the change in the objective function value.

10.9 Implement your traveling salesman problem heuristic, and apply it to the problem instance with n = 10 cities, in which the cost of traveling from city i to city j is shown as the entry in the i-th row and j-th column of the following cost (or distance) matrix.

0	99	45	55	10	15	86	90	33	41
97	0	10	15	18	93	56	23	84	75
88	22	0	35	46	57	68	79	99	90
75	64	53	0	14	63	74	77	54	20
32	53	64	86	0	97	94	91	90	10
24	35	46	57	68	0	98	96	95	99
55	79	26	10	96	65	0	35	49	22
30	50	80	50	86	53	81	0	28	65
35	57	26	11	14	76	25	89	0	30
40	50	60	23	41	11	18	90	47	0

10.10 Modify your traveling salesman heuristic, replacing the exchange mechanism by a subtour reversal mechanism. Use a random number generator to select the endpoints of a subtour of cities, and then create a new tour with that subtour reversed. Compute the objective function value associated with this new tour, and accept the new tour if it is an improvement over the previous one.

10.11 Extend your algorithms from Exercises 10.9 and 10.10 to include possible acceptance of a new tour having a worse objective function value than that of the previous tour.

   a. Design a cooling schedule for this simulated annealing method.

   b. Determine the other operational parameters necessary to complete an implementation of simulated annealing for the traveling salesman problem.

10.12 Write a computer program that exhaustively enumerates all feasible traveling salesman tours.

   a. Apply this algorithm to the ten-city problem data given in Exercise 10.9. Compare the quality of solutions and the computation time performance characteristics of your exchange heuristic, your implementation of simulated annealing, and the complete enumeration algorithm.

   b. Estimate the amount of time your exhaustive enumeration method would require to find the optimal tour among 100 cities.

10.13 Collect or create data for a large routing problem that involves approximately 100 locations. For example, consider routing delivery trucks, ordering the pickups and deliveries in a campus mail or courier service, or sequencing the safety inspection sites in a large complex of buildings. Construct the 100×100 matrix of distances. This problem is significantly larger than the 10-city problem addressed in previous exercises. Experiment with your local improvement and simulated annealing programs to determine how effectively and efficiently they solve this larger problem.

References and Suggested Readings

Aarts, E., and J. Korst. 1989. *Simulated Annealing and Boltzmann Machines: A Stochastic Approach to Combinatorial Optimization and Neural Computing.* New York: John Wiley & Sons.

Aarts, E., and J. K. Lenstra. 1997. *Local Search in Combinatorial Optimization.* Chichester, UK: John Wiley & Sons.

Anily, S., and A. Federgruen. 1985. Probabilistic analysis of simulated annealing methods. Preprint. Technical Report, Graduate School of Business. New York: Columbia University, pp. 289–304.

Burke, E. K, and G. Kendall. 2014. *Search methodologies: Introductory Tutorials in Optimization and Decision Support Techniques,* 2nd ed. New York: Springer.

Campbell, K. W., R. Bret Durfee, and G. S. Hines. 1997. FedEx generates bid lines using simulated annealing. *Interfaces* 27 (2): 1–16.

Carter, M. W., and C. C. Price. 1988. *Local Improvement Heuristics.* Toronto, Canada: Department of Industrial Engineering, University of Toronto.

Coughlin, J. P., and R. H. Baran. 1995. *Neural Computation in Hopfield Networks and Boltzmann Machines.* Newark, NJ: University of Delaware Press.

Darwin, C. 1859. *On the Origin of Species by Means of Natural Selection, or Preservation of Favoured Races in the Struggle for Life*. London, UK: Murray.

Du, K.-L., and M. N. S. Swamy. 2010. *Search and Optimization by Metaheuristics: Techniques and Algorithms Inspired by Nature*. Cham, Switzerland: Springer International Publishing AG.

Ferreiro, A. M., J. A. García, J. G. López-Salas, and C. Vázquez. 2013. An efficient implementation of parallel simulated annealing algorithm in GPUs. *Journal of Global Optimization* 57 (3): 863–890.

Garcia, C., and G. Rabadi. 2011. A meta RaPS algorithm for spatial scheduling with release times. *International Journal of Planning and Scheduling* 1 (1–2): 19–31.

Gendreau, M., and J.-Y. Potvin. 2010. *Handbook of Metaheuristics*, Vol. 2. New York: Springer.

Glover, F., and M. Laguna. 1997. *Tabu Search*. Boston, MA: Kluwer Academic.

Glover, F. 1989. Tabu search part I. *ORSA Journal on Computing* 1 (3): 190–206.

Glover, F. W., and G. A. Kochenberger. 2006. *Handbook of Metaheuristics*, Vol. 57. New York: Springer Science & Business Media.

Goldberg, D. E. 1994. Genetic and evolutionary algorithms come of age. *Communications of the ACM* 37 (3): 113–120.

Goldberg, D. E, K. Deb, and J. H. Clark. 1992. Genetic algorithms, noise, and the sizing of populations. *Complex Systems* 6: 333–362.

Goldberg, D. E., and J. H. Holland. 1988. Genetic algorithms and machine learning. *Machine Learning* 3 (2): 95–99.

Goldberg, D. E. 1989. *Genetic Algorithms in Search, Optimization, and Machine Learning*. Reading, MA: Addison-Wesley.

Hentenryck, P. V., and L. Michel. 2009. *Constraint-based Local Search*. Cambridge, UK: The MIT press.

Holland, J. H. 1975. *Adaption in Natural and Artificial Systems*. Ann Arbor, MI: The University of Michigan Press.

Hughell, D. A., and J. P. Roise. 1995. Spatially explicit multi-objective analysis for timber and wildlife. *Quantitative Tools for Wildlife Analysis and Management Working Group at the SAF National Convention*, Portland, ME.

Kaplan, S., and G. Rabadi. 2013. A simulated annealing and meta-raps algorithms for the aerial refueling scheduling problem with due date-to-deadline windows and release time. *Engineering Optimization* 45 (1): 67–87.

Kirkpatrick, S. 1984. Optimization by simulated annealing: Quantitative studies. *Journal of Statistical Physics* 34 (5): 975–986.

Kirkpatrick, S., C. D. Gelatt, and M. P. Vecchi. 1983. Optimization by simulated annealing. *Science* 220 (4598): 671–680.

Lin, S., and B. W. Kernighan. 1973. An effective heuristic algorithm for the traveling-salesman problem. *Operations Research* 21 (2): 498–516.

Lucas, R. A., and C. C. Price. 1992. *Neural Computing Models and Parallel Simulated Annealing for Quadratic Assignment Problems*. Nacogdoches, TX: Stephen F. Austin State University.

Lundy, M., and A. Mees. 1986. Convergence of an annealing algorithm. *Mathematical Programming* 34 (1): 111–124.

Manikas, A. S., J. R. Kroes, and T. F. Gattiker. 2016. Metro meals on wheels Treasure Valley employs a low-cost routing tool to improve deliveries. *Interfaces* 46 (2): 154–167.

Metropolis, N., A. W. Rosenbluth, M. N. Rosenbluth, A. H. Teller, and E. Teller. 1953. Equation of state calculations by fast computing machines. *The Journal of Chemical Physics* 21 (6): 1087–1092.

Moraga, R. J. 2016. Metaheuristic for randomized priority search (Meta-RaPS): A tutorial. In *Heuristics, Metaheuristics and Approximate Methods in Planning and Scheduling*. Cham, Switzerland: Springer, pp. 95–108.

Price, C. C., and L. A. Wahsheh. 1999. Cascaded Boltzmann machines for combinatorial optimization. *Proceedings of the 4th Multiconference on Systemics*, Orlando, FL.

Rabadi, G., R. J. Moraga, and A. Al-Salem. 2006. Heuristics for the unrelated parallel machine scheduling problem with setup times. *Journal of Intelligent Manufacturing* 17 (1): 85–97.

Reeves, C. R. 1997. Genetic algorithms for the operations researcher. *INFORMS Journal on Computing* 9 (3): 231–250.

Resende, M. G. C., and C. C. Ribeiro. 2003. Greedy randomized adaptive search procedures. In F. Glover and G. Kochenberger (Eds.), *Handbook of Metaheuristics*. Dordrecht, the Netherlands: Kluwer Academic Publishers, pp. 219–249.

Santé, I., F. F. Rivera, R. Crecente, M. Boullón, M. Suárez, J. Porta, J. Parapar, and R. Doallo. 2016. A simulated annealing algorithm for zoning in planning using parallel computing. *Computers, Environment and Urban Systems* 59: 95–106.

Siarry, P. (Ed.). 2016. *Metaheuristics*, 1st ed. Cham, Switzerland: Springer.

Sivanandam, S. N., and S. N. Deepa. 2007. *Introduction to Genetic Algorithms*. Berlin, Germany: Springer Science & Business Media.

Taillard, E. 2016. Tabu search. In *Metaheuristics*. Cham, Switzerland: Springer, pp. 51–76.

Wang, C., D. Mu, F. Zhao, and J. W. Sutherland. 2015. A parallel simulated annealing method for the vehicle routing problem with simultaneous pickup–delivery and time windows. *Computers & Industrial Engineering* 83: 111–122.

Appendix: Review of Essential Mathematics—Notation, Definitions, and Matrix Algebra

A.1 Vectors

A **vector** is generally considered to be a quantity having both magnitude and direction. In some cases, it is convenient to think of a vector as a line segment beginning at the origin of an n-dimensional rectangular coordinate system and terminating at a point in the n-space. The components, or elements, of the vector are the projections of the vector onto each of the coordinate axes. These projections form an n-tuple and completely describe the vector.

More typically, a vector is described simply as a point X in n-space and is denoted as

$$X = (x_1, x_2 \ldots, x_n)$$

The set of all possible points, or n-tuples of real numbers, forms the real n-space, which is denoted by R^n.

If $X = (x_1, x_2, \ldots, x_n)$ and $Y = (y_1, y_2, \ldots, y_n)$ are vectors in R^n, then the sum $X + Y$ is an n-dimensional vector defined as

$$X + Y = (x_1 + y_1, x_2 + y_2, \ldots, x_n + y_n)$$

A vector X can be multiplied by a real number scalar a to obtain

$$\alpha X = (\alpha x_1, \alpha x_2, \ldots, \alpha x_n)$$

A **vector space** over the set of real numbers is a set of vectors for which addition and scalar multiplication are defined. Additionally, the operations in the vector space must satisfy a certain set of axioms, including commutative, associative, and distributive laws. The set of vectors must include an identity element, that is, the zero vector $(0, 0, \ldots, 0)$; and for every vector X, there must be an inverse $-X$, for which $X + (-X)$ is the zero vector.

A vector Y is a **linear combination** of vectors $X_1, X_2, \ldots, X_n$ if it can be expressed as

$$Y = \alpha_1 X_1 + \alpha_2 X_2 + \ldots + \alpha_n X_n$$

where the α_i are real numbers.

An n-dimensional vector space is said to be **spanned** by the set of vectors $\{X_1, X_2, \ldots, X_n\}$ if every vector in the space is some linear combination of $X_1, X_2, \ldots, X_n$. The set $\{X_1, X_2, \ldots, X_n\}$ is then called a spanning set for the vector space.

A set of vectors $\{X_1, X_2, \ldots, X_n\}$ is **linearly independent** if no one vector can be expressed as a linear combination of the other vectors in the set; that is, if the equation

$$\alpha_1 X_1 + \alpha_2 X_2 + \ldots + \alpha_n X_n = 0$$

can be satisfied *only* by setting all the α_i equal to zero. A set of vectors that is not **linearly independent** is **linearly dependent**. For example, two non-zero vectors X_1 and X_2 are linearly dependent if one of them is a non-zero scalar multiple of the other one; that is, if $\alpha_1 X_1 + \alpha_2 X_2 = 0$ for some scalars α_1 and α_2 not both zero.

A set of vectors $\{X_1, X_2, ..., X_n\}$ is a **basis** for an n-dimensional vector space if the set spans the space and is linearly independent. The **standard basis** of an n-dimensional space consists of a set of **unit vectors** that comprise a basis; that is, a set of vectors in which the i-th vector u_i has a 1 as the i-th element and zeros in all other positions. This standard basis is useful because of its simplicity and because of its obvious role as a basis for an n-dimensional vector space.

A.2 Matrices and Matrix Operations

A real matrix is a rectangular array of real numbers. Subscripts, such as i and j, can be used to index the rows and columns, respectively. A matrix A of m rows and n columns is called an m × n ("m by n") matrix and is written as:

$$A = (a_{ij}) = \begin{bmatrix} a_{11} & a_{12} & ... & a_{1n} \\ a_{21} & a_{22} & ... & a_{2n} \\ \vdots & \vdots & & \vdots \\ a_{m1} & a_{m2} & ... & a_{mn} \end{bmatrix}$$

where a_{ij} denotes the element in the i-th row and the j-th column. Any matrix A can be multiplied by a scalar α with the result that every element a_{ij} in A becomes the value αa_{ij} in the scalar product matrix.

Two matrices $A_{m \times n}$ and $B_{p \times q}$ can be added if m = p and n = q. The **sum** C = A + B is a matrix $C_{m \times n}$ in which the element c_{ij} is computed as $(a_{ij} + b_{ij})$. Two matrices $A_{m \times n}$ and $B_{p \times q}$ may be multiplied if n = p. The **product** C = AB is defined to be a matrix $C_{m \times q}$ in which the element c_{ij} is computed as:

$$c_{ij} = \sum_{k=1}^{n} a_{ik} b_{kj}$$

For the special case in which m = q = 1, we actually have just the product of two *vectors*. This product is called the inner product or **dot product**, and the dot product of two vectors $X = (x_1, x_2, ..., x_n)$ and $Y = (y_1, y_2, ..., y_n)$ is denoted and defined by:

$$X \bullet Y = (x_1 y_1 + x_2 y_2 + ... + x_n y_n)$$

which is consistent with the definition of general matrix multiplication.

Matrix addition and multiplication exhibit some of the properties of real arithmetic operations. For example, for matrices A, B, and C, the following properties hold:

$$A + B = B + A$$
$$(A + B) + C = A + (B + C)$$
$$A (B + C) = AB + AC$$
$$(AB)C = A(BC)$$

However, note that, in general, matrix multiplication is not commutative, so $AB \neq BA$ in general, and in fact these products may not even exist.

The **transpose** of an m × n matrix A is the n × m matrix A^T obtained by interchanging the roles of the rows and columns in A. Thus, if A is the matrix shown earlier, then A^T is the matrix whose elements have the same values as those in A, but arranged in the form

$$\begin{bmatrix} a_{11} & a_{21} & \cdots & a_{m1} \\ a_{12} & a_{22} & \cdots & a_{m2} \\ \vdots & \vdots & & \vdots \\ a_{1n} & a_{2n} & \cdots & a_{mn} \end{bmatrix}$$

Reversing the roles twice simply yields the original matrix, so for any matrix A, $(A^T)^T = A$.

For example, the following two matrices are the transpose of each other:

$$\begin{bmatrix} 3 & 2 & 4 \\ 7 & 1 & 5 \end{bmatrix} \begin{bmatrix} 3 & 7 \\ 2 & 1 \\ 4 & 5 \end{bmatrix}$$

A property of matrix multiplication and transposition is that $(AB)^T = B^T A^T$.

A **square matrix** is one for which m = n. The **main diagonal** of a square matrix A is the set of elements a_{ij} for which i = j, that is, $a_{11}, a_{22}, \ldots, a_{nn}$. An n × n matrix A is **symmetric** about the main diagonal if every element a_{ij} is equal to the element a_{ji}. A square matrix A is **triangular** (or *upper* triangular) if all the elements below the main diagonal have value zero; that is, $a_{ij} = 0$ for all i > j. For example, the following matrix is upper triangular:

$$\begin{bmatrix} 4 & 6 & 2 & 5 \\ 0 & 7 & 1 & 2 \\ 0 & 0 & 4 & 3 \\ 0 & 0 & 0 & 8 \end{bmatrix}$$

A matrix is *lower* triangular if all the elements above the main diagonal have value zero.

The **identity matrix** I is an n × n matrix whose columns are the standard basis, and in which the i-th column contains the i-th unit vector. The identity matrix contains ones along the main diagonal ($a_{ii} = 1$ for all i) and zeros elsewhere ($a_{ij} = 0$ for all $i \neq j$). This matrix has the property that $AI = IA = A$ for any n × n real matrix A.

The **rank** of a matrix A is the number of linearly independent rows (or columns) in A, and is denoted as **rank (A)**. A square matrix $A_{n \times n}$ having rank n (*full rank*) is called a non-singular matrix.

A square matrix A may have an **inverse** matrix A^{-1} such that $AA^{-1} = A^{-1}A = I$. Such an inverse exists if and only if rank (A) = n (or equivalently, if and only if A is non-singular), and in that case, the inverse A^{-1} is unique. The inverse of the inverse of a matrix A is the original matrix A; thus, $(A^{-1})^{-1} = A$. And if two matrices A and B have inverses A^{-1} and B^{-1}, respectively, then

$$(AB)^{-1} = B^{-1}A^{-1}$$

A.3 Linear Equations

A set of m linear equations in n variables is expressed as

$$a_{11}x_1 + a_{12}x_2 + \ldots + a_{1n}x_n = b_1$$

$$a_{21}x_1 + a_{22}x_2 + \ldots + a_{2n}x_n = b_2$$

$$\vdots \qquad \vdots \qquad \qquad \vdots$$

$$a_{m1}x_1 + a_{m2}x_2 + \ldots + a_{mn}x_n = b_m$$

The coefficients of the variables can be written as a matrix A, where

$$A = \begin{bmatrix} a_{11} & a_{12} & \ldots & a_{1n} \\ \vdots & \vdots & \vdots & \vdots \\ a_{m1} & a_{m2} & \ldots & a_{mn} \end{bmatrix}$$

The variables and right-hand sides of the equations can be written as column vectors, thus, $X = (x_1, x_2, \ldots, x_n)$ and $b = (b_1, b_2, \ldots, b_m)$. In this context, the matrix A can be viewed as an *operation* or *transformation* on the vector X, yielding the resulting vector b. This can be written as $AX = b$, and has the same meaning as the set of linear equations depicted earlier.

A solution to this set of linear equations is any vector X that satisfies the equations $AX = b$. A unique solution to a set of m independent linear equations in n variables exists if m = n and if the inverse of A exists. If m > n, there may be no solution. And if m < n, there are infinitely many solutions.

Techniques for solving a system of linear equations may involve the use of so-called **elementary row operations** on the equations. The application of any of the following row operations yields an equivalent system of equations and may simplify the solution process:

- Any two rows (equations) may be interchanged.
- Any row (equation) may be multiplied by a non-zero constant.
- Any row (equation) may be added to any other row (equation).

When m = n, if it is possible to transform the matrix of coefficients A into a triangular matrix by performing elementary row operations, then the system can be solved easily. For example if the system of equations appears as

$$\begin{bmatrix} a_{11} & a_{12} & \cdots & & a_{1n} \\ 0 & a_{22} & & & a_{2n} \\ 0 & 0 & a_{33} & \cdots & a_{3n} \\ \vdots & & & \ddots & \\ 0 & 0 & 0 & \cdots & a_{mn} \end{bmatrix} \begin{bmatrix} x_1 \\ x_2 \\ x_3 \\ \vdots \\ x_n \end{bmatrix} = \begin{bmatrix} b_1 \\ b_2 \\ b_3 \\ \vdots \\ b_m \end{bmatrix}$$

then we know that $a_{mn} \cdot x_n = b_m$, so we can easily find a value for x_n. Using this, we can then solve for x_{n-1} in the next to the last equation, and so on until finally we have a value for x_1.

An alternative approach to the solution process is to create an **augmented** matrix B consisting of the coefficient matrix A with one additional column containing the elements b_i. Then this new matrix B is an m row by (n + 1) column matrix, and each row of B represents one equation of the system of equations. Next apply the necessary elementary row operations to B that transform the original A portion of B into the identity matrix I. This will have the effect of causing the b portion of B to be transformed into a vector representing the solution to the system of equations.

A.4 Quadratic Forms

Let $A_{n \times n}$ be a symmetric matrix, and X be an n-element vector. The function f(X) defined as

$$f(X) = X^T A X$$

is called a quadratic form. Since X^T is of order 1 × n and A is n × n and X is n × 1, the product

$$\begin{bmatrix} x_1, x_2, \ldots, x_n \end{bmatrix} \begin{bmatrix} a_{11} & a_{12} & \cdots & a_{1n} \\ & & \vdots & \\ a_{n1} & a_{n2} & \cdots & a_{nn} \end{bmatrix} \begin{bmatrix} x_1 \\ \vdots \\ x_n \end{bmatrix}$$

exists and can be computed. Clearly, f(X) is a scalar value, and can be written as

$$f(X) = \sum_{i=1}^{n} \sum_{j=1}^{n} a_{ij} x_i x_j$$

Thus, f(X) is a sum of quadratic terms, and hence the name *quadratic form*.

In this context, the matrix A has one of the following characteristics:

Positive definite: If f(X) > 0 for all X ≠ 0

Positive semidefinite: If f(X) ≥ 0 for all X and there exists an X ≠ 0 for which f(X) = 0

Negative definite: If $f(X) < 0$ for all $X \neq 0$

Negative semidefinite: If $f(X) \leq 0$ for all X and there exists an $X \neq 0$ for which $f(X) = 0$

Indefinite: If none of the above

References and Suggested Readings

Cheney, W., and D. Kincaid. 2012. *Linear Algebra; Theory and Applications*, 2nd ed. Burlington, MA: Jones and Bartlett Learning.

Forbes, C., M. Evans, N. Hastings, and B. Peacock. 2010. *Statistical Distributions*, 4th ed. Hoboken, NJ: John Wiley & Sons.

Hadi, A. S. 1996. *Matrix Algebra as a Tool*. Belmont, CA: Duxbury Press.

Jennings, A. 1977. *Matrix Computation for Engineers and Scientists*. New York: John Wiley & Sons.

Kincaid, D., and W. Cheney. 2002. *Numerical Analysis: Mathematics of Scientific Computing*, 3rd ed. Providence, RI: American Mathematical Society.

Shiskowski, K. M., and Frinkle, K. 2011. *Principles of Linear Algebra with Mathematica*. New York: John Wiley & Sons.

Index

Note: Page numbers followed by f and t refer to figures and tables respectively.

Milton Keynes UK
Ingram Content Group UK Ltd.
UKHW051941071024
449327UK00026B/2122

9 781032 476063